Molecular Mechanisms of Cancer Development and Metastasis

Molecular Mechanisms of Cancer Development and Metastasis

Guest Editors

Babak Behnam
Hassan Fazilaty

MDPI

Basel • Beijing • Wuhan • Barcelona • Belgrade • Novi Sad • Cluj • Manchester

Guest Editors
Babak Behnam
Avicenna Biotech Research
Clarksburg, MD
USA

Hassan Fazilaty
Department of Molecular Life Sciences
University of Zurich
Zurich
Switzerland

Editorial Office
MDPI AG
Grosspeteranlage 5
4052 Basel, Switzerland

This is a reprint of the Special Issue, published open access by the journal *Cancers* (ISSN 2072-6694), freely accessible at: https://www.mdpi.com/journal/cancers/special_issues/BUS806XS17.

For citation purposes, cite each article independently as indicated on the article page online and as indicated below:

Lastname, A.A.; Lastname, B.B. Article Title. *Journal Name* **Year**, *Volume Number*, Page Range.

ISBN 978-3-7258-6730-1 (Hbk)
ISBN 978-3-7258-6731-8 (PDF)
https://doi.org/10.3390/books978-3-7258-6731-8

Contents

About the Editors

Babak Behnam

Babak Behnam's career is a journey across continents and disciplines, unified by a single goal: to translate complex genetic insights into tangible medical solutions. As Co-Founder and Chief Scientific Officer of Avicenna Biotech Research, he currently steers the company's scientific vision, IP strategy, and medical liaison efforts. He is also a Professor at the Floret Center for Advanced Genomics and Deputy Director of the JBI Nigerian Centre for Evidence-Based Healthcare, and has lent his expertise to American University as a Researcher-in-Residence. A medical doctor by training (IUMS, 1997), Dr. Behnam's quest to understand the fundamental mechanisms of disease led him to the UK for an M.Sc. in Molecular Medicine and a Ph.D. in Human Genetics from UCL (2005). Crossing the Atlantic, he secured an NIH Ruth L. Kirschstein fellowship to conduct postdoctoral research at the University of Michigan and UCF, where he delved into the genetics of cancer and mitochondrial function. His career reflects a rare ability to move between professional domains: from the academic rigor of a professorship at IUMS, where he founded a medical genetics lab and mentored countless students, to the clinical intensity of the NIH's Undiagnosed Disease Program, where he completed a fellowship in biochemical genetics. He even ventured into the regulatory landscape at NSF International, guiding critical FDA submissions. With over 80 publications, contributions to five major textbooks, and multiple patents to his name, Dr. Behnam's work, whether in cancer biology, addiction therapies, or global health metrics as a Senior Collaborator with the GBD, continues to push the boundaries of genomic medicine.

Hassan Fazilaty

Hassan Fazilaty is a molecular biologist and cancer researcher at the University of Zurich, Switzerland. He obtained his Ph.D. in Molecular Biosciences from the Universidad Autónoma de Madrid, following an M.Sc. in Human Genetics from Tehran University of Medical Sciences. He is currently a Group Leader at the Department of Gastroenterology and Hepatology, University Hospital Zurich, where he leads research on the reactivation of embryonic genetic programs in intestinal regeneration and cancer. His work, positioned at the intersection of developmental biology and oncology, explores how embryonic pathways are co-opted in tumor progression. His broader interests include translational oncology, developmental reprogramming, and molecular mechanisms of tissue repair.

Preface

The relentless progression and dissemination of cancer remain central challenges in oncology. This Special Issue of *Cancers* is dedicated to exploring the fundamental biological underpinnings of these processes, with a focus on the complex interplay between molecular alterations and the dynamic tumor microenvironment that collectively drives malignancy.

Our motivation stems from the critical need to integrate knowledge across scales—from genetic and epigenetic instability to tissue-level interactions—to develop more effective diagnostic and therapeutic strategies. This issue emphasizes the elucidation of core mechanisms such as dysregulated signaling pathways, epithelial-to-mesenchymal transition (EMT), and the supportive role of stromal components. Understanding how intrinsic cellular changes and extrinsic environmental factors converge to promote invasion, survival at distant sites, and therapy resistance is paramount.

Our goal was to provide a platform for disseminating high-quality research and comprehensive reviews that advance this understanding. By bridging disciplines including molecular biology, genetics, immunology, and clinical oncology, we aim to foster multidisciplinary perspectives. We hope that combining deep mechanistic insight with translational relevance will stimulate new directions in metastasis research and the development of targeted interventions, ultimately contributing to the ongoing battle against cancer's adaptive and aggressive nature.

Babak Behnam and Hassan Fazilaty
Guest Editors

cancers

MDPI

Editorial

Synthesizing Molecular Insights to Redefine the Battle Against Metastatic Cancer

Babak Behnam [1,2]

[1] Biology Department, College of Art and Sciences, American University, Washington, DC 20016, USA; bbehnam@american.edu or babak.behnam@gmail.com
[2] Floret Advanced Genomics & Bioinformatics Research Center, Lagos 100271, Nigeria

1. Introduction: The Frontier of Metastatic Biology

Metastasis represents the most formidable challenge in clinical oncology, accounting for over 90% of cancer-related deaths. This grim statistic underscores a persistent therapeutic gap: while we have made substantial progress in managing primary tumors, our ability to prevent or cure disseminated disease remains profoundly limited. This failure stems not from a lack of effort but from the extraordinary biological complexity of the metastatic cascade, a multi-step process involving dynamic cellular plasticity, adaptive survival mechanisms, and sophisticated hijacking of host systems. This Special Issue of *Cancers* directly addresses this complexity, assembling a series of investigations that dissect metastasis from the level of single molecules to systemic pathophysiology. The studies presented extend beyond the classical linear model of progression and toward an integrated understanding of metastasis as a systemic, evolutionarily adaptive disease. This Editorial synthesizes these insights, arguing that the path to transformative clinical breakthroughs lies in connecting these diverse molecular discoveries to create a new paradigm for interception and combination therapy.

This Special Issue transforms a segmented view of metastasis into an integrated model. The story is no longer solely about a cancer cell activating a single pathway to migrate. Instead, it is about a cell that dynamically rewires its signaling (AKT2, TCF12) (Contribution 4), remodels its own architecture and waste systems (vesicular trafficking) (Contribution 7), manipulates its energy sources and surroundings (TSGA10, FMD) (Contributions 6), and masterfully deceives the immune system (MACC1, T-cell duality) (Contribution 10) to complete its lethal journey.

The most promising future lies at the intersection of these layers. Could targeting endosomal trafficking impair the secretion of immune-suppressive signals like PD-L1? Might metabolic therapies that enhance TSGA10 function simultaneously starve a tumor and expose it to immune attack (Contribution 9)? Emerging paradigms, such as ghost mitochondria, serve as a crucial reminder that metastasis may be fueled by biological processes that we are only just beginning to describe.

2. An Integrated Overview: From Cellular Hijacking to Systemic Dysregulation

The contributions collectively reveal metastasis not as a single pathway but as a constellation of interconnected biological capabilities, each presenting unique vulnerabilities. The emerging portrait is of a cell that is metabolically agile, immunologically elusive, and capable of extraordinary phenotypic adaptation.

 https://doi.org/10.3390/cancers18020199

2.1. Mastery of Cellular Plasticity and Signaling Networks

At its core, metastasis is an exercise in cellular reprogramming. The epithelial-to-mesenchymal transition (EMT) remains a cornerstone of this plasticity, conferring invasive properties and therapy resistance. However, current research reveals EMT as a dynamic spectrum rather than a binary switch, regulated by intricate post-transcriptional and post-translational networks. The control of key EMT transcription factors like ZEB1 by ubiquitin ligases exemplifies how protein stability mechanisms are recruited to drive phenotypic fluidity (Contribution 2) [1]. Beyond EMT, new dimensions of plasticity have emerged, including the formation of polyploid giant cancer cells (PGCCs) (Contribution 8), stress-resistant entities that can spawn aggressive, genetically variant progeny through a primitive, depolyploidization cycle, contributing to tumor heterogeneity and relapse [2].

This plasticity is fueled by the precise rewiring of canonical signaling pathways. The PI3K-AKT-mTOR axis, a central regulator of growth and survival, demonstrates isoform-specific functionality in metastasis. For instance, AKT2, but not AKT1, has been shown to be specifically required for the metastatic dissemination of BRAF-mutant melanoma, where it regulates a distinct gene signature linked to glycolysis and invasion (Contributions 4 & 5) [3]. This underscores a critical theme: effective therapeutic targeting may require moving beyond pan-pathway inhibition toward disrupting specific molecular subroutines commandeered for dissemination. Furthermore, metastasis involves the creation of self-sustaining signaling loops. For example, the transcription factor TCF12 can directly activate the expression of TGF-β2, creating an autocrine/paracrine circuit that sustains both EMT and the immunosuppressive tumor microenvironment, thereby linking developmental pathways directly to metastatic progression (Contribution 5) [4].

2.2. Metabolic Reprogramming and Ecosystem Engineering

To survive the metabolically challenging journey from the primary site to a distant organ, disseminating tumor cells (DTCs) undergo profound metabolic reprogramming. This is not a passive response but an active, instructional program that enables every step of the metastatic cascade. A key orchestrator is the hypoxia-inducible factor (HIF) family, which drives a shift toward glycolysis (the Warburg effect) even in the presence of oxygen, while also promoting angiogenesis and invasion [5]. This metabolic shift generates biosynthetic precursors and manages oxidative stress, allowing cells to thrive in adverse conditions.

However, this metabolic addiction also presents a profound therapeutic vulnerability. Strategies like caloric restriction mimetics and fasting-mimicking diets (FMDs) exploit this by inducing a severe, selective metabolic stress on cancer cells while protecting normal tissues (Contribution 3). Crucially, as highlighted in this Special Issue, such interventions can synergize powerfully with chemotherapy by inhibiting pro-survival autophagy and, more importantly, by potentiating anti-tumor immunity through mechanisms involving T-cell-dependent clearance [6]. This demonstrates a fundamental principle: the intrinsic metabolism of the tumor cell, systemic host metabolism, and the immune response are inextricably linked nodes in the metastatic network.

A particularly fascinating frontier is the role of intercellular metabolic hijacking. Emerging evidence points to the direct transfer of organelles, most notably mitochondria, between cells in the tumor microenvironment. Cancer cells can use tunneling nanotubes (TNTs) to "steal" functional mitochondria from neighboring stromal cells, thereby replenishing their bioenergetic capacity and enhancing their metastatic fitness [7]. Conversely, they may donate dysfunctional mitochondria to immune cells, such as cytotoxic T lymphocytes, to impair their anti-tumor function (Contribution 11), a remarkable example of cellular piracy that blurs the line between metabolic and immune evasion strategies [8].

https://doi.org/10.3390/cancers18020199

This strategic co-option of host resources underscores a broader repertoire of emergent, non-genetic paradigms that enable metastasis. As summarized in Table 1, these advanced adaptations; ranging from organelle hijacking and the formation of polyploid giant cancer cells to the sculpting of alternative vascular networks (Contribution 8), represent fundamental shifts in our understanding of tumor biology. Critically, each paradigm reveals a distinct therapeutic vulnerability. Moving from mechanistic insight to clinical translation requires a focused effort to develop agents that can disrupt these specific processes, such as inhibiting tunneling nanotube formation to block mitochondrial trafficking or targeting the unique metabolic dependencies of polyploid cells. Thus, the table not only catalogs these emerging frontiers but also maps them directly to potential intervention strategies, highlighting a roadmap for the next generation of anti-metastatic therapeutics.

Table 1. Emerging metastatic paradigms and their therapeutic implications.

Paradigm	Core Mechanism	Potential Therapeutic Strategy
Mitochondrial Hijacking	Transfer of functional mitochondria to cancer cells via tunneling nanotubes (TNTs) to boost fitness.	Inhibiting TNT formation (e.g., via M-Sec inhibition); blocking mitochondrial uptake.
Immune Cell Sabotage	Donation of dysfunctional mitochondria from cancer cells to T cells to impair effector function.	Enhancing mitochondrial quality control in T cells; metabolic support for immunotherapy.
Polyploid Giant Cancer Cells (PGCCs)	Stress-induced polyploidization leading to depolyploidization and generation of aggressive, resistant progeny.	Targeting the depolyploidization process (e.g., Aurora kinases); exploiting vulnerabilities of PGCCs (e.g., autophagy dependence).
Vasculogenic Mimicry	Formation of perfusable, matrix-rich channels by tumor cells, independent of endothelial cells, to support perfusion.	Disrupting tumor cell plasticity programs or targeting key matrix modifiers (e.g., LOXL2).
Metabolic-Epigenetic Coupling	Oncometabolites (e.g., 2-HG, fumarate) directly inhibiting or activating epigenetic enzymes to lock in pro-metastatic gene expression.	Targeting mutant metabolic enzymes (IDH1/2 inhibitors); using epigenetic drugs to reverse dysregulated gene programs.

2.3. Subversion of the Immune System and Systemic Niches

Metastasis is unequivocally a systemic disease. The primary tumor does not merely shed passive cells into circulation; it actively manipulates the entire host to create a receptive environment for its progeny. A cornerstone of this manipulation is the formation of the pre-metastatic niche (PMN). Through secreted factors (like VEGF, TGF-β, and PTHrP) and extracellular vesicles (exosomes), the primary tumor "educates" distant organs, such as the lungs, liver, or bone marrow [9]. This education involves recruiting bone marrow-derived cells (e.g., myeloid-derived suppressor cells or MDSCs), remodeling the extracellular matrix, and increasing vascular permeability, all to create a supportive landing pad for circulating tumor cells.

Central to both PMN formation and the survival of established metastases is immunosuppression. Metastatic cells evade immune destruction through a multi-layered strategy: upregulating immune checkpoint ligands (e.g., PD-L1), secreting factors that recruit regulatory T cells (Tregs) and MDSCs, and expressing non-classical MHC molecules. Molecules like MACC1 have been identified as central hubs in this process, driving invasive growth while simultaneously upregulating PD-L1 and other immunosuppressive factors, thereby directly linking the mechanisms of motility and immune escape (Contribution 10) [10]. The immune landscape itself exhibits a critical duality, as seen in the context

 https://doi.org/10.3390/cancers18020199

of breast cancer, where T lymphocytes can be polarized toward pro-tumorigenic (Th2, Treg) or anti-tumorigenic (Th1, cytotoxic CD8+) phenotypes, with the balance dramatically influencing metastatic outcomes (Contribution 11) [11].

2.4. The Logistical Machinery of Spread

Beneath these high-order programs lies the critical, often overlooked, logistics of dissemination. The endosomal–lysosomal vesicular trafficking system emerges as a master regulator of the metastatic cascade (Contribution 7). Motor proteins like dynein and kinesins do not merely transport cargo; they direct the polarized delivery of integrins, matrix metalloproteinases (MMPs), and growth factor receptors to the leading edge of invading cells [12]. Furthermore, this machinery is essential for the secretion of exosomes, which act as long-range messengers to prepare pre-metastatic niches. By delivering oncogenic proteins, lipids, and nucleic acids (like miRNAs and circRNAs), tumor-derived exosomes can reprogram recipient cells in distant organs to support future colonization [13]. Targeting this logistical hub, for instance via inhibiting specific motor proteins or disrupting exosome biogenesis, represents a promising strategy to cripple both local invasion and systemic communication.

3. Conclusions: Charting a Path Toward Interceptive and Integrative Therapeutics

The collective narrative emerging from this Special Issue is one of convergence. Metastasis is no longer seen through the narrow lens of a single "driver" mutation or pathway but is understood as the integrated output of appointed plasticity, metabolism, and immune evasion programs, all orchestrated by a cell with remarkable adaptive intelligence. This refined understanding dismantles the old paradigm of treating metastasis as a late-stage complication and replaces it with an urgent imperative for early interception and systemic disruption.

The future of anti-metastatic therapy therefore lies in rational, mechanism-based combinations that attack the process on multiple fronts simultaneously (Contribution 1). The insights herein suggest several strategic pillars for this new therapeutic architecture:pt

(A) Therapies Targeting Adaptive States—Combating plasticity requires agents that can lock cells into a more vulnerable, differentiated state or eliminate plastic subpopulations (e.g., PGCCs, EMT hybrid cells). This may involve targeting epigenetic readers/writers, key transcription factor complexes, or stress-survival pathways like autophagy that permit transitions between states.

(B) Metabolic Warfare—Exploiting the metabolic dependencies of metastatic cells, especially during the vulnerable phases of dissemination and colonization, involves combining dietary interventions (e.g., FMD) with drugs that target oxidative phosphorylation, glutamine metabolism, or lipid synthesis. A key frontier disrupts the metabolic crosstalk between tumor cells and stromal cells in the niche.

(C) Disruption of Systemic Communication—Neutralizing the tools tumors use to engineer their own spread is a viable prophylactic strategy. This could involve inhibitors of exosome secretion, neutralization of key niche-educating factors (e.g., LOXL2, MFGE8), or blockers of tunneling nanotube formation to prevent mitochondrial hijacking.

(D) Immunotherapy 2.0 [14] via Re-educating the Ecosystem—Emphasizes combining modalities actively reprogram the metastatic microenvironment rather than relying on checkpoint blockade alone. This includes agents that deplete or reprogram MDSCs and TAMs, vaccines targeting metastasis-associated antigens, and adoptive cell therapies engineered to overcome metabolic suppression in the niche.

https://doi.org/10.3390/cancers18020199

(E) Advanced Biomarkers for Interception—To deploy these strategies effectively, we must detect the systemic process of metastasis earlier. The future lies in "liquid biopsy" 2.0 [15], extending beyond simple mutation detection in ctDNA to analyzing the phenotype of circulating tumor cells (CTCs), the cargo of exosomes, and immune cell profiles to assess the real-time activity of the metastatic cascade and the state of the pre-metastatic niche.

In conclusion, this Special Issue provides essential molecular and conceptual information for assembling a new, more effective war against metastatic cancer. By integrating the principles of cellular piracy, metabolic symbiosis, and ecological manipulation, we are building a unified theory of metastasis that is both more complex and more target-rich than ever before. The challenge is no longer purely biological; it is an engineering challenge of designing intelligent, timed, and layered therapeutic combinations. By embracing this integrative and interceptive paradigm, we can transform the clinical trajectory of metastatic disease from one of inevitable progression to one of durable control and prevention. This is the decisive frontier in oncology, and this Special Issue's research paves the way forward.

To translate this knowledge into clinical progress, the field must embrace combinatorial strategies. The future of anti-metastatic therapy likely rests on rational drug combinations that concurrently target a cancer cell's intrinsic drive to disseminate and its extrinsic ability to hide and thrive. The articles in this Special Issue provide a robust scientific foundation for building these strategies. By continuing to dissect the molecular mechanisms with the depth and breadth showcased here, we move closer to the ultimate goal: transforming metastasis from a terminal event into a controllable process.

Conflicts of Interest: The author declare no conflict of interest.

List of Contributions:

1. Encarnación, J.A.; Morillo Macías, V.; De la Fuente Muñoz, I.; Soria, V.D.; Fernández Fornos, L.; Antequera, M.A.; Rey, O.A.; García Martínez, V.; Alonso-Romero, J.L.; García Gómez, R. Apalutamide and Stereotactic Body Radiotherapy in Metastatic Hormone-Sensitive Prostate Cancer: Multicenter Real-World Study. *Cancers* **2025**, *17*, 2216.
2. Zhao, X.; Han, Z.; Liu, R.; Li, Z.; Mei, L.; Jin, Y. FBXO11 Mediates Ubiquitination of ZEB1 and Modulates Epithelial-to-Mesenchymal Transition in Lung Cancer Cells. *Cancers* **2024**, *16*, 3269.
3. Buono, R.; Tucci, J.; Cutri, R.; Guidi, N.; Mangul, S.; Raucci, F.; Pellegrini, M.; Mittelman, S.D.; Longo, V.D. Fasting-Mimicking Diet Inhibits Autophagy and Synergizes with Chemotherapy to Promote T-Cell-Dependent Leukemia-Free Survival. *Cancers* **2023**, *15*, 5870.
4. McRee, S.K.; Bayer, A.L.; Pietruska, J.; Tsichlis, P.N.; Hinds, P.W. AKT2 Loss Impairs BRAF-Mutant Melanoma Metastasis. *Cancers* **2023**, *15*, 4958.
5. Tian, Y.; Zhou, J.; Chai, X.; Ping, Z.; Zhao, Y.; Xu, X.; Luo, C.; Sheng, J. TCF12 Activates TGFB2 Expression to Promote the Malignant Progression of Melanoma. *Cancers* **2023**, *15*, 4505.
6. Amini, A.; Taghizadeh-Hesary, F.; Bracht, J.; Behnam, B. TSGA10 as a Model of a Thermal Metabolic Regulator: Implications for Cancer Biology. *Cancers* **2025**, *17*, 1756.
7. Nturubika, B.D.-D.; Logan, J.; Johnson, I.R.D.; Moore, C.; Li, K.L.; Tang, J.; Lam, G.; Parkinson-Lawrence, E.; Williams, D.B.; Chakiris, J.; et al. Components of the Endosome-Lysosome Vesicular Machinery as Drivers of the Metastatic Cascade in Prostate Cancer. *Cancers* **2025**, *17*, 43.
8. Krotofil, M.; Tota, M.; Siednienko, J.; Donizy, P. Emerging Paradigms in Cancer Metastasis: Ghost Mitochondria, Vasculogenic Mimicry, and Polyploid Giant Cancer Cells. *Cancers* **2024**, *16*, 3539.
9. Taghizadeh-Hesary, F.; Ghadyani, M.; Kashanchi, F.; Behnam, B. Exploring TSGA10 Function: A Crosstalk or Controlling Mechanism in the Signaling Pathway of Carcinogenesis? *Cancers* **2024**, *16*, 3044.
10. Torke, S.; Walther, W.; Stein, U. Immune Response and Metastasis—Links between the Metastasis Driver MACC1 and Cancer Immune Escape Strategies. *Cancers* **2024**, *16*, 1330.

 https://doi.org/10.3390/cancers18020199

11. Zareinejad, M.; Mehdipour, F.; Roshan-Zamir, M.; Faghih, Z.; Ghaderi, A. Dual Functions of T Lymphocytes in Breast Carcinoma: From Immune Protection to Orchestrating Tumor Progression and Metastasis. *Cancers* **2023**, *15*, 4771.

References

1. Zhang, P.; Wei, Y.; Wang, L.; Debeb, B.G.; Yuan, Y.; Zhang, J.; Yuan, J.; Wang, M.; Chen, D.; Sun, Y.; et al. ATM-mediated stabilization of ZEB1 promotes DNA damage response and radioresistance through CHK1. *Nat. Cell Biol.* **2014**, *16*, 864–875. [CrossRef] [PubMed]
2. White-Gilbertson, S.; Voelkel-Johnson, C. Giants and Monsters: Unexpected Characters in the Story of Cancer Recurrence. *Adv Cancer Res.* **2020**, *148*, 201–239. [PubMed]
3. Chin, Y.R.; Yoshida, T.; Marusyk, A.; Beck, A.H.; Polyak, K.; Toker, A. Targeting Akt3 signaling in triple-negative breast cancer. *Cancer Res.* **2014**, *74*, 964–973. [CrossRef] [PubMed]
4. Sánchez-Tilló, E.; Liu, Y.; de Barrios, O.; Siles, L.; Fanlo, L.; Cuatrecasas, M.; Darling, D.S.; Dean, D.C.; Castells, A.; Postigo, A. EMT-activating transcription factors in cancer: Beyond EMT and tumor invasiveness. *Cell. Mol. Life Sci.* **2012**, *69*, 3429–3456. [CrossRef] [PubMed]
5. Semenza, G.L. HIF-1: Upstream and downstream of cancer metabolism. *Curr. Opin. Genet. Dev.* **2010**, *20*, 51–56. [CrossRef] [PubMed]
6. Di Biase, S.; Lee, C.; Brandhorst, S.; Manes, B.; Buono, R.; Cheng, C.W.; Cacciottolo, M.; Martin-Montalvo, A.; de Cabo, R.; Wei, M.; et al. Fasting-Mimicking Diet Reduces HO-1 to Promote T Cell-Mediated Tumor Cytotoxicity. *Cancer Cell* **2016**, *30*, 136–146. [CrossRef] [PubMed]
7. Dong, L.F.; Kovarova, J.; Bajzikova, M.; Bezawork-Geleta, A.; Svec, D.; Endaya, B.; Sachaphibulkij, K.; Coelho, A.R.; Sebkova, N.; Ruzickova, A.; et al. Horizontal transfer of whole mitochondria restores tumorigenic potential in mitochondrial DNA-deficient cancer cells. *eLife* **2017**, *6*, e22187. [CrossRef]
8. Saha, T.; Dash, C.; Jayabalan, R.; Khiste, S.; Kulkarni, A.; Kurmi, K.; Mondal, J.; Majumder, P.K.; Bardia, A.; Jang, H.L.; et al. Intercellular nanotubes mediate mitochondrial trafficking between cancer and immune cells. *Nat. Nanotechnol.* **2022**, *17*, 98–106. [CrossRef]
9. Peinado, H.; Zhang, H.; Matei, I.R.; Costa-Silva, B.; Hoshino, A.; Rodrigues, G.; Psaila, B.; Kaplan, R.N.; Bromberg, J.F.; Kang, Y.; et al. Pre-metastatic niches: Organ-specific homes for metastases. *Nat. Rev. Cancer* **2017**, *17*, 302–317. [CrossRef]
10. Korshunov, A.; Sahm, F.; Zheludkova, O.; Golanov, A.; Stichel, D.; Schrimpf, D.; Ryzhova, M.; Potapov, A.; Habel, A.; Meyer, J.; et al. DNA methylation profiling is a method of choice for molecular verification of pediatric WNT-activated medulloblastomas. *Neuro Oncol.* **2019**, *21*, 214–221. [CrossRef] [PubMed]
11. DeNardo, D.G.; Brennan, D.J.; Rexhepaj, E.; Ruffell, B.; Shiao, S.L.; Madden, S.F.; Gallagher, W.M.; Wadhwani, N.; Keil, S.D.; Junaid, S.A.; et al. Leukocyte complexity predicts breast cancer survival and functionally regulates response to chemotherapy. *Cancer Discov.* **2011**, *1*, 54–67. [CrossRef] [PubMed]
12. Steffan, J.J.; Dykes, S.S.; Coleman, D.T.; Adams, L.K.; Rogers, D.; Carroll, J.L.; Williams, B.J.; Cardelli, J.A. Supporting a role for the GTPase Rab7 in prostate cancer progression. *PLoS ONE* **2014**, *9*, e87882. [CrossRef] [PubMed]
13. Hoshino, A.; Costa-Silva, B.; Shen, T.L.; Rodrigues, G.; Hashimoto, A.; Tesic Mark, M.; Molina, H.; Kohsaka, S.; Di Giannatale, A.; Ceder, S.; et al. Tumour exosome integrins determine organotropic metastasis. *Nature* **2015**, *527*, 329–335. [CrossRef] [PubMed]
14. Morad, G.; Helmink, B.A.; Sharma, P.; Wargo, J.A. Hallmarks of response, resistance, and toxicity to immune checkpoint blockade. *Cell* **2021**, *184*, 5309–5337. [CrossRef] [PubMed]
15. Heitzer, E.; Haque, I.S.; Roberts, C.E.S.; Speicher, M.R. Current and future perspectives of liquid biopsies in genomics-driven oncology. *Nat. Rev. Genet.* **2019**, *20*, 71–88. [CrossRef] [PubMed]

 https://doi.org/10.3390/cancers18020199

cancers

MDPI

Article

TCF12 Activates TGFB2 Expression to Promote the Malignant Progression of Melanoma

Youjia Tian [1,2], Jiang Zhou [3], Xinxin Chai [1,2], Zejun Ping [1,2], Yurong Zhao [1,2], Xin Xu [1,2], Chi Luo [4,*] and Jinghao Sheng [1,2,3,4,*]

1 Affiliated Hangzhou First People's Hospital, Zhejiang University School of Medicine, Hangzhou 310006, China; tianyj@zju.edu.cn (Y.T.); 12018090@zju.edu.cn (X.C.); 22118833@zju.edu.cn (Z.P.); 0620803@zju.edu.cn (Y.Z.); xxuxin@zju.edu.cn (X.X.)
2 Liangzhu Laboratory, Zhejiang University, Hangzhou 310012, China
3 Cancer Center, Zhejiang University, Hangzhou 310058, China; 22018055@zju.edu.cn
4 Zhejiang Provincial Key Laboratory of Bioelectromagnetics, Zhejiang University School of Medicine, Hangzhou 310058, China
* Correspondence: chi.luo@alumni.tufts.edu (C.L.); jhsheng@zju.edu.cn (J.S.)

Simple Summary: Melanoma is the deadliest form of skin cancer, with the BRAF(V600E) mutation being the most prevalent driver mutation. Despite targeted therapies against BRAF(V600E) mutation and immune checkpoint-blocking antibodies providing treatment options for patients, the heterogeneous nature of melanoma significantly limits treatment efficacy. Understanding diverse regulatory mechanisms in melanoma will shed light on improving the current treatment modalities. In this study, we explored the function of a novel transcriptional activator, TCF12, in melanoma progression. We found that the expression level of TCF12 is elevated as melanoma progresses, and high expression is strongly associated with poor survival outcomes in melanoma patients. Functionally, TCF12 enhances melanoma proliferation and metastasis, as well as the sensitivity to BRAF(V600E)-targeted therapy. Mechanistically, TGFB2 is the direct transcriptional target of TCF12, mediating its pro-tumorigenic effects. Collectively, our study supported the oncogenic functions of TCF12 in melanoma, revealing it as a potential target to improve the efficacy of BRAF(V600E)-targeted therapy.

Abstract: As one of the most common malignant tumors, melanoma is a serious threat to human health. More than half of melanoma patients have a BRAF mutation, and 90% of them have a BRAF(V600E) mutation. There is a targeted therapy for patients using a BRAF(V600E) inhibitor. However, no response to treatment is generally inevitable due to the heterogeneity of melanoma. Coupled with its high metastatic character, melanoma ultimately leads to poor overall survival. This study aimed to explore the possible mechanisms of melanoma metastasis and identify a more effective method for the treatment of melanoma. In this paper, we report that TCF12 expression is higher in melanoma, especially in metastatic tumors, through analyzing data from TCGA. Then, cell proliferation, colony formation, and transwell assays show that the upregulated expression of TCF12 can promote proliferation and metastasis of melanoma cells in vitro. The same result is confirmed in the subcutaneous tumor formation assay. Moreover, TGFB2 is identified as a direct downstream target of TCF12 by RNA-seq, qPCR, immunoblotting, ChIP, and a dual luciferase reporting assay. Interestingly, depletion of TCF12 can sensitize melanoma to BRAF inhibition both in vitro and in vivo. Overall, our results demonstrate that TCF12 promotes melanoma progression and can be a potential tumor therapeutic target.

Keywords: melanoma; TCF12; RNA-seq; TGFB2; BRAF(V600E)

1. Introduction

Melanoma is a malignant tumor that originates from normal epidermal melanocytes or pre-existing nevus cells [1]. The global incidence of melanoma has been steadily increasing in recent years, posing a significant health challenge [2]. Unfortunately, a major concern is the late-stage diagnosis of most patients, when invasion and metastasis have already occurred, leading to higher mortality rates [3]. The landscape of malignant melanoma treatment has witnessed a transformative shift with the emergence of targeted therapies directed against BRAF(V600E) mutation and immune checkpoint-blocking antibodies [4–6]. These groundbreaking advancements have substantially improved survival rates among patients with advanced-stage disease. Nevertheless, treatment responses remain heterogeneous [7,8]. A comprehensive understanding of the underlying mechanisms contributing to the heightened risk of recurrence holds tremendous potential for enhancing clinical management and optimizing patient outcomes through tailored surveillance and adjuvant treatment strategies.

BRAF(V600E) is the most prevalent mutation in melanoma, regulating tumor growth, invasion, and metastasis [9]. Among the diverse downstream pathways of BRAF(V600E), transcriptional regulation presents an essential one [10]. For example, BRAF(V600E) is known to directly enhance the stability and activity of microphthalmia-associated transcription factor (MITF), the master regulator of melanocyte development and differentiation, leading to increased cell proliferation, survival, and resistance to therapies [11,12]. Additionally, we have previously reported that BRAF(V600E) negatively regulates the expression and activity of the transcriptional coactivator PGC1α (peroxisome proliferator-activated receptor gamma coactivator 1-alpha), fine-tuning melanoma metabolism and transcriptional programs to balance tumor growth and metastatic spreading [13,14]. However, it is still not fully understood whether other novel transcriptional factors and programs act downstream of BRAF(V600E) in the regulation of melanoma progression.

Transcription factor 12 (TCF12), also known as HTF4 or HEB, is a member of the helix-loop-helix (HLH) protein family. It plays a crucial role in cell development and differentiation across various tissues, including skeletal muscle, neurons, mesenchymal tissues, and lymphocytes [15–18]. TCF12 can form homodimers or heterodimers with other members of the HLH family to activate gene expression [19]. Recent studies have highlighted the contributions of TCF12 to the progression of different tumor types, including colorectal, pancreatic, liver, and ovarian cancers [20–23]. Additionally, TCF12 expression has also been found to be associated with advanced tumor stages and poor prognosis in breast and lung cancers [24,25]. Interestingly, our early studies suggested that TCF12 may be a functional partner of PGC1α, likely playing a role in melanoma metastasis [13,14]. However, whether TCF12 regulates melanoma progression and, if yes, by what mechanism, remain elusive to date.

In this study, we explored the functional involvement and molecular mechanisms of TCF12 in melanoma. We found that the expression level of TCF12 is elevated as melanoma progresses, and high expression is strongly associated with poor survival outcomes in melanoma patients. Functionally, we found that TCF12 can enhance melanoma proliferation and metastasis, as well as sensitivity to BRAF(V600E)-targeted therapy. Mechanistically, we identified TGFB2 as a direct transcriptional target of TCF12, mediating its pro-tumorigenic effects. Collectively, our study supported the oncogenic functions of TCF12 in melanoma, revealing it as a potential target to improve the efficacy of BRAF(V600E)-targeted therapy.

2. Materials and Methods

2.1. Cell Lines and Cell Culture

The human melanoma cell line A375 and murine melanoma cell line YUMM1.7 utilized in this study were purchased from the American Type Culture Collection (ATCC). The human kidney cell line HEK293T was purchased from the Cell Bank of the Chinese Academy of Sciences. The cells were regularly tested for mycoplasma contamination using the MycoAlert Mycoplasma Detection Kit (Lonza, Basel, Switzerland) and verified by morphological observation. Cells were cultured in Dulbecco's Modified Eagle's Medium (DMEM, 11965084, Gibco, Billings, MT, USA) supplemented with 10% fetal bovine serum (FBS, 10099141, Gibco, Billings, MT, USA), 100 U/mL penicillin, and 100 mg/mL streptomycin (15140122, Gibco, Billings, MT, USA). They were kept at 37 °C in a humidified incubator containing 5% CO_2. The medium was replaced every 2–3 days, and upon attaining 80–90% confluency, the cells underwent subculturing.

2.2. Reagents and Antibodies

The following primary antibodies were used for immunoblotting: TCF12/HEB (D2C10) rabbit mAb (Cell Signaling Technology, Danvers, MA, USA, dilution: 1:1000), TGFB2 rabbit pAb (A3640) (ABclonal, Woburn, MA, USA, dilution: 1:500), alpha tubulin monoclonal antibody (1E4C11)-HRP (Proteintech, Rosemont, IL, USA, dilution: 1:3000), and beta actin monoclonal antibody (7D2C10)-HRP (Proteintech, dilution: 1:3000). The following primary antibodies were used for immunohistochemical (IHC) staining: TCF12/HEB antibody (14419-1-AP) (Proteintech, dilution: 1:100), TGFB2 rabbit pAb (A3640) (ABclonal, dilution: 1:50), and anti-Ki-67 rabbit pAb (ab15580) (Abcam, dilution: 1:400). The following primary antibodies were used for immunoprecipitation: TCF12/HEB (D2C10), rabbit mAb (Cell Signaling Technology, 2 μL), and rabbit control IgG (AC005) (ABclonal, 2 μL).

The BRAF(V600E) inhibitor PLX4032 (#S1267) was purchased from Selleck Chemicals (Houston, TX, USA). The MEK inhibitor trametinib (#SD5973) was purchased from Beyotime (Shanghai, China). MG132 (ab141003) was purchased from Abcam (Cambridge, United Kingdom). Melanoma tissue microarray (DC-Mel21020) was purchased from Shaanxi Avila Biotechnology (Shaanxi, China).

2.3. RNA Interference

For TCF12 knockdown experiments, short hairpin RNA (shRNA) oligonucleotides were cloned into the lentiviral vector pLKO.1. Lentiviruses were produced in HEK293T cells by co-transfection with packaging vectors pMD2G and psPAX2 using Lipofectamine 3000 (L30000015, Invitrogen, Waltham, MA, USA) according to the manufacturer's instructions. Lentivirus-containing supernatants were collected 48 h after transfection, filtered through a 0.45 μm filter, and used to infect A375 and YUMM1.7 melanoma cells. Polybrene (8 μg/mL) was added to enhance infection efficiency. Infected cells were selected with 2 μg/mL of puromycin for at least 4 days before subsequent experiments. The shRNA clones targeting TCF12 were as follows: shTCF12-1: 5′-CCATCCCATAATGCACCAATT-3′; shTCF12-2: 5′-GCTGTGATTATGGTGAACATA-3′; shTcf12-1: 5′-TGTATGTCACTGTGGCTAGT-3′; shTcf12-2: 5′-CAGTCTTGATTTCTGTTGGAAC-3′. siTgfb2-1: 5′-GACCCUACUUCAGAAU CGUTT-3′; siTgfb2-2: 5′-GAGGGAUCUUGGAUGGAAATT-3′.

2.4. Cell Proliferation and Colony Formation Assays

To evaluate cell proliferation, A375 and YUMM1.7 cells were seeded in 6-well plates at a density of 5×10^3 cells per well in triplicate. The cell number was counted at the indicated time points using a hemocytometer. For drug treatment experiments, cells were exposed to either DMSO (vehicle control) or PLX4032 the day after seeding, followed by cell counting at the indicated time points.

To assess the clonogenic potential of cells, A375 and YUMM1.7 cells were seeded in 6-well plates at a density of 1×10^2 cells per well in triplicate. After two weeks of incubation, colonies were fixed with 100% ethanol for 10 min and stained with a 0.5% crystal violet solution in 25% methanol for 20 min. Excess stain was washed with water, and the plates were air-dried before counting the colonies.

2.5. Migration and Invasion Assays

Cell migration was assessed using transwell chambers with an 8 μm pore size (Corning Life Science). YUMM1.7 EV/Tcf12 (1×10^4), YUMM1.7 shScr/shTcf12 (2×10^4), A375 EV/TCF12 (2×10^4), or A375 shScr/shTCF12 (3×10^4) cells were suspended in 0.1 mL of FBS-free medium and seeded into the upper chamber. The lower chamber was filled with a medium containing 10% FBS as a chemoattractant. Cells were incubated at 37 °C in a humidified incubator containing 5% CO_2. After incubation, non-migrated cells in the upper chamber were gently removed with a cotton swab. Migrated cells attached to the lower surface of the membrane were fixed with 4% paraformaldehyde for 20 min and stained with a 0.5% crystal violet solution in 25% methanol for 20 min. The membrane was then rinsed with water, air-dried, and mounted onto a glass slide. Cells from three random fields were imaged under 10× or 20× magnification using a Nikon Ti-s inverted microscope, and the total number of migrated cells was quantified.

For the invasion assay, the upper chamber was coated with Matrigel (Corning) before seeding the cells. The subsequent steps were the same as the migration assay. Cells that had invaded through the Matrigel-coated membrane were fixed, stained, and quantified as described above.

2.6. Immunoblotting

A375 and YUMM1.7 cells underwent lysis in RIPA buffer (50 mM Tris-HCl pH 7.4, 150 mM NaCl, 1% NP-40, 0.5% sodium deoxycholate, 0.1% SDS, protease and phosphatase inhibitors). Whole-cell extracts were harvested, and protein concentration was determined using the BCA Protein Assay Kit (P0010, Beyotime).

Subsequently, equal amounts of protein were separated through sodium dodecyl sulfate-polyacrylamide gel electrophoresis (SDS-PAGE) and electro-transferred nitrocellulose or polyvinylidene fluoride (PVDF) membranes. To prevent nonspecific binding, the membranes were incubated in a blocking solution of 5% non-fat dry milk or bovine serum albumin (BSA) in Tris-buffered saline with 0.1% Tween 20 (TBST) for 1 h at room temperature. Following this, the membranes were incubated in a blocking solution containing the appropriate primary antibodies overnight at 4 °C.

After washing with TBST three times, the membranes were incubated in a blocking solution containing secondary antibodies conjugated by horseradish peroxidase (HRP) for 1 h at room temperature. The protein levels were visualized using an enhanced chemiluminescence (ECL) kit (Bio-Rad, Hercules, CA, USA), and images were captured using a chemiluminescence imaging system.

2.7. Quantitative Real-Time PCR

Total RNA was isolated from cells using the TRIzol reagent (Life Technologies, Carlsbad, CA, USA) according to the manufacturer's instructions. A total of 2 μg of RNA was reverse transcribed into complementary DNA (cDNA) using the M-MLV reverse transcriptase (Takara, Kusatsu, Japan) and random primers. The cDNA was used as a template for real-time PCR analysis, which was carried out using SYBR Green PCR Master Mix (Takara) on a real-time PCR system. Experimental Ct values were normalized to the housekeeping genes, and relative mRNA expression was calculated using the $2^{-\Delta\Delta CT}$ method. Each sample was analyzed in triplicate, and the mean Ct value was used for calculations. The primer sequences used for RT-qPCR are shown in Table S1.

2.8. Immunohistochemical (IHC) Staining

The tissue samples were prepared by fixing them in 10% buffered formalin for an overnight period, followed by preservation in 70% ethanol prior to being embedded in paraffin. They were then sectioned and stained with hematoxylin and eosin (H&E). The resulting paraffin sections (4 μm in thickness) were subjected to deparaffinization in xylene and sequentially rehydrated using a descending concentration of ethanol. For optimal antigen exposure, these sections were immersed in 0.1 M citrate buffer (pH 6.0) and brought to a boil either via microwave or water bath. Once cooling to room temperature, the sections underwent washing with phosphate-buffered saline (PBS). Endogenous peroxidase activity was neutralized using 3% hydrogen peroxide. To block non-specific interactions, sections were treated with 10% normal goat serum for 30 min at room temperature. The sections were then incubated with indicated antibodies at 4 °C overnight. After PBS washing, the sections were incubated with biotinylated secondary antibodies. This was followed by an incubation with streptavidin-horseradish peroxidase (HRP) conjugate using the mouse/rabbit two-step assay kit (Mouse/Rabbit Polymer Assay Detection System, ZSGB-Bio, Beijing, China). The immunoreactivity was visualized using 3,3′-diaminobenzidine (DAB) as the chromogen, and the sections were counterstained with hematoxylin. All tissue slides were photographed using a Leica DM2000 upright microscope. The immunostaining was scored based on the positive percentage and staining intensity of positively staining cells by a pathologist blinded to the experimental conditions.

2.9. Animal Experiments

All animal experiments were designed and conducted following the protocol approved by the Medical Experimental Animal Care Commission of Zhejiang University (#ZJU20220217). Six- to eight-week-old male C57BL/6 mice (purchased from Shanghai Slack Laboratory Animal Co., Ltd., Shanghai, China) were used for this study. Mice were housed in a controlled environment under a 12 h dark/12 h light cycle, with food and water provided ad libitum.

Before the injection, YUMM1.7 cells were detached using 0.5 mM EDTA in PBS and washed with 1× PBS. A total of 1×10^5 YUMM1.7 melanoma cells with vector control or Tcf12 overexpression were injected subcutaneously into the flanks of mice in 0.1 mL of PBS, and the tumor development was monitored every other day. Tumor volume was calculated based on the equation V = (width (in mm)^2 × length (in mm))/2. For melanoma cells with Tcf12 knockdown, 2×10^5 cells were used for subcutaneous injection.

For the drug administration experiment, tumor-bearing mice were given 20 mg/kg PLX4032 (in 10% NMP and 90% PEG) or vehicle via oral administration every other day after subcutaneous tumors could be detected.

To establish a tumor lung metastasis model, a total of 1 or 2×10^5 YUMM1.7 melanoma cells with Tcf12 overexpression or knockdown were injected into the tail veins of C57BL/6 mice in 0.2 mL of PBS. Lung tissues were harvested 3–4 weeks after injection. The survival time of the mice was recorded to generate the survival curve.

For histological analysis, subcutaneous tumors and lungs from mice were collected, fixed, paraffin-embedded, and sectioned for hematoxylin and eosin (H&E) or immunohistochemical (IHC) staining.

2.10. RNA-Sequencing Analysis

RNA was harvested from YUMM1.7 cells of both the stable negative control (shScr) and Tcf12 knockdown (shTcf12-1) utilizing the TRIzol reagent (Life Technologies). The total RNA concentration and purity were validated by a NanoDrop™ One Microvolume UV–vis spectrophotometer (Thermo Scientific, Waltham, MA, USA).

Novogene Co., Ltd. (Beijing, China) conducted the RNA-seq analysis. Briefly, the RNA quality was verified by an Agilent 2100 bioanalyzer (Agilent Technologies, Santa Clara, CA, USA). Oligo-dT magnetic beads facilitated the purification of mRNA with poly-A tails, which were subsequently fragmented into small pieces through divalent cations at a heightened temperature. These fragments then underwent first-strand cDNA synthesis with the aid of random hexamer primers and reverse transcriptase. This was followed by a second-strand cDNA synthesis using DNA polymerase I and RNase H. The double-stranded cDNA fragments were subjected to end repair, A-tailing, adapter ligation, and PCR amplification. The final library quantification was carried out with the Qubit 2.0 fluorometer (Life Technologies), and an Agilent 2100 bioanalyzer (Agilent Technologies) verified the insert size and calculated the library concentration.

The resulting cDNA library was sequenced by an Illumina NovaSeq 6000 platform, producing 150 bp paired-end reads. Raw read processing eliminated low-quality reads and adapter sequence contaminants. The refined, high-quality clean reads were then mapped to the reference genome via HISAT2. Transcript abundance estimation and differential expression evaluations were conducted utilizing StringTie and DESeq2, respectively. Differentially expressed genes (DEGs) were identified based on criteria: a false discovery rate (FDR) < 0.05 and an absolute fold change ≥ 2.

The DEGs underwent a functional enrichment analysis utilizing the Gene Ontology (GO) and Kyoto Encyclopedia of Genes and Genomes (KEGG) pathway databases. The significance of enrichment was determined using a corrected p-value < 0.05.

2.11. Chromatin Immunoprecipitation (ChIP)

YUMM1.7 melanoma cells with Tcf12 overexpression or Tcf12 knockdown were crosslinked by 1% paraformaldehyde, and then 10% glycine was added to quench untreated paraformaldehyde. Cells from each dish were scraped into PBS and centrifuged at 800 rpm for 5 min. The pellets were resuspended in swelling buffer for 10 min of rotation at 4 °C, centrifuged at 1000 rpm for 15 min at 4 °C, and resuspended in SDS lysis buffer. Following sonication, equal amounts of lysates were incubated with IgG or TCF12 antibodies overnight at 4 °C.

The lysates were then incubated with protein A/G-magnetic beads (Thermo Scientific) for 2 h to precipitate the antibody-bound chromatin. The beads were washed sequentially with low-salt buffer, high-salt buffer, LiCl buffer, and TE buffer to remove any nonspecific interactions. The immunoprecipitated chromatin was then eluted from the beads and reverse crosslinked at 65 °C overnight to obtain the DNA.

After treatment with RNase A and Proteinase K to degrade RNA and protein contaminants, DNA was purified using a PCR purification kit (Qiagen, Venlo, The Netherlands) according to the manufacturer's instructions. The purified DNA was then subjected to quantitative real-time PCR (qPCR) using SYBR Green PCR Master Mix (Takara) to analyze the enrichment of specific genomic regions. The primer sequences used for qPCR are shown in Table S2.

2.12. Luciferase Reporter Assay

The fragment of the Tgfb2 promoter was inserted into the pGL3 luciferase reporter vector, which was synthesized by Miaoling Biotechnology, Heze, China. The fragments containing mutated binding sites were constructed using the Fast Mutagenesis System Kit (TransGene Biotech, Beijing, China). YUMM1.7 cells were seeded in 24-well plates and co-transfected with pGL3 vectors containing either the wild-type or mutated Tgfb2 promoter, pRL-TK Renilla luciferase vector (for normalization), and either the Tcf12 overexpression or Tcf12 knockdown plasmid using Lipofectamine 3000 (Invitrogen) according to the manufacturer's instructions. Forty-eight hours after transfection, cells were harvested and lysed using a passive lysis buffer. The luciferase activity of both firefly and Renilla luciferases was analyzed using a dual-luciferase reporter assay system (Promega, Madison, WI, USA) according to the manufacturer's instructions. The total light intensity was measured using a VarioskanTM LUX microplate reader (Thermo Scientific). The firefly luciferase activity was normalized to Renilla luciferase activity to account for any differences in transfection efficiency. The relative luciferase activity was calculated by comparing the normalized luciferase activity of cells transfected with the Tcf12 overexpression plasmid to that of cells transfected with the control vector.

2.13. Statistical Analysis

No statistical methods were used to predetermine sample size for in vivo and in vitro experiments, but at least three biologically independent samples were used per experimental group and condition. The data were presented as mean ± standard deviation (SD). Statistical analysis was conducted employing the GraphPad Prism 9.5 software. Comparisons involving two sets were made through a two-tailed unpaired Student's t-test. In assessing survival, we utilized the Kaplan–Meier approach to generate survival curves, and differences were calculated by a p-value below 0.05, which was deemed to indicate statistical significance.

3. Results

3.1. TCF12 Expression Is Positively Correlated with Poor Prognosis in Melanoma Patients

To examine the expression of TCF12 in melanoma patients, we utilized the GEPIA (Gene Expression Profiling Interactive Analysis) online platform to profile TCF12 mRNA levels in melanoma patients and normal tissues [26]. The results revealed that melanoma tissues had a notably elevated expression of TCF12 compared to adjacent normal skins (Figure 1a). Moreover, patients with an increased TCF12 expression level showed reduced overall survival (Figure 1b), suggesting a role for TCF12 in melanoma progression. Indeed, we confirmed that the expression of TCF12 increases as disease progresses, as evidenced by higher levels in metastatic tissues than in primary melanomas (Figure 1c,d). Interestingly, we also found that even within the primary melanomas, the vertical growth phase (VGP) tumors, which have penetrated deeper into the skin layers, showed a higher level of TCF12 compared to tumors in the radial growth phase (RGP), which represent the very early stage of melanoma (Figure 1e). To further validate the findings from the public dataset, we performed TCF12 immunohistochemistry (IHC) on a melanoma tissue array. In line with the bioinformatic analyses, our results demonstrated that the TCF12 protein was upregulated in melanoma and elevated progressively as the tumor advanced (Figure 1f,g). These findings suggest a potential functional role for TCF12 in melanoma progression and metastasis.

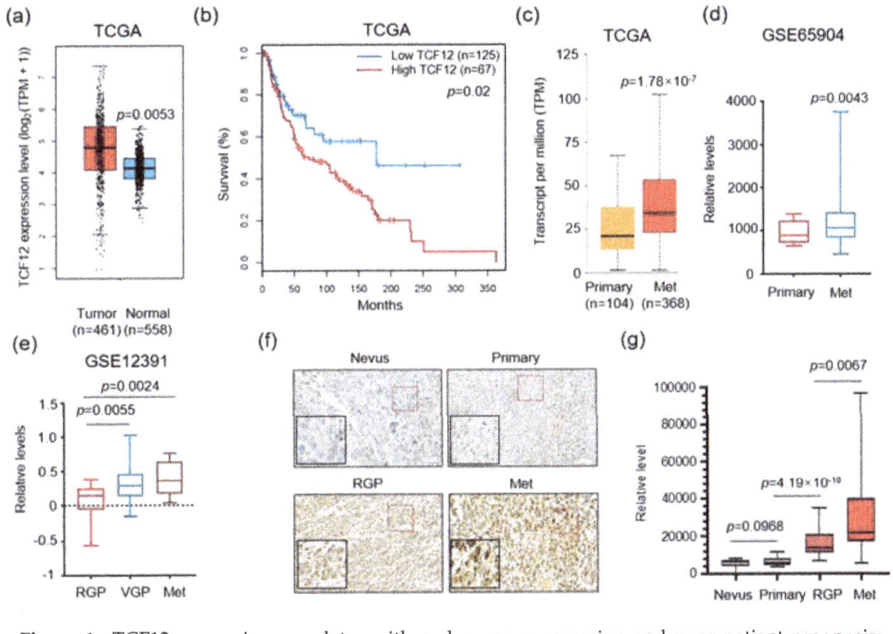

Figure 1. TCF12 expression correlates with melanoma progression and poor patient prognosis: (**a**) Boxplot illustrating the expression of TCF12 in melanoma tissues compared with normal skin samples, data derived from The Cancer Genome Atlas (TCGA) database; (**b**) Kaplan–Meier survival analysis comparing overall survival in melanoma patients grouped by low and high TCF12 expression levels; (**c**,**d**) analysis of relative TCF12 expression in primary melanomas compared to metastatic tissues (Met) using TCGA database (**c**) and the GSE65904 dataset (**d**); (**e**) comparative analysis of TCF12 expression levels across different stages of melanoma: radial growth phase (RGP), vertical growth phase (VGP), and metastatic phase (Met). Statistical significance is based on comparison with the RGP group; (**f**,**g**) representative images of immunohistochemistry (IHC) staining for TCF12 in melanoma tissue samples (red square, 200×; black square, 400×) (**f**), and corresponding histological analysis quantifying TCF12 protein levels (**g**). Scale bar: 100 μm. Statistical significance is based on comparison indicated in the graphs.

3.2. TCF12 Enhances Melanoma Cell Proliferation In Vitro and Tumorigenicity In Vivo

To explore the biological function of TCF12 in melanoma, we first generated stable TCF12 knockdown in both human and mouse melanoma cells (A375 and YUMM1.7) (Figures 2a,b and S1a,b). Proliferation and colony formation assays revealed that TCF12 knockdown led to the inhibition of cell growth and reduced colony formation capacity (Figures 2c,d and S1c,d). Subsequently, we carried out ectopic TCF12 expression experiments (Figures S2a,b and S1e,f) and found that TCF12 overexpression promoted melanoma cell proliferation and colony formation (Figures S2c,d and S1g,h). Next, we tested whether TCF12 affected the tumorigenesis of melanoma in vivo in an immunocompetent background. Knockdown of TCF12 significantly impaired the growth of YUMM1.7 tumors, as evidenced by slower tumor growth kinetics (Figure 2e) and smaller tumor size at the endpoint (Figure 2f,g), in line with a lower proliferation index by Ki67 IHC (Figure 2h,i). On the other hand, overexpression of TCF12 greatly accelerated YUMM1.7 tumor progression (Figure S2e–g), consistent with more Ki67+ cells in the tumor tissues (Figure 2h,i). Collectively, these results indicate that TCF12 has an oncogenic function and promotes melanoma tumorigenesis both in vitro and in vivo.

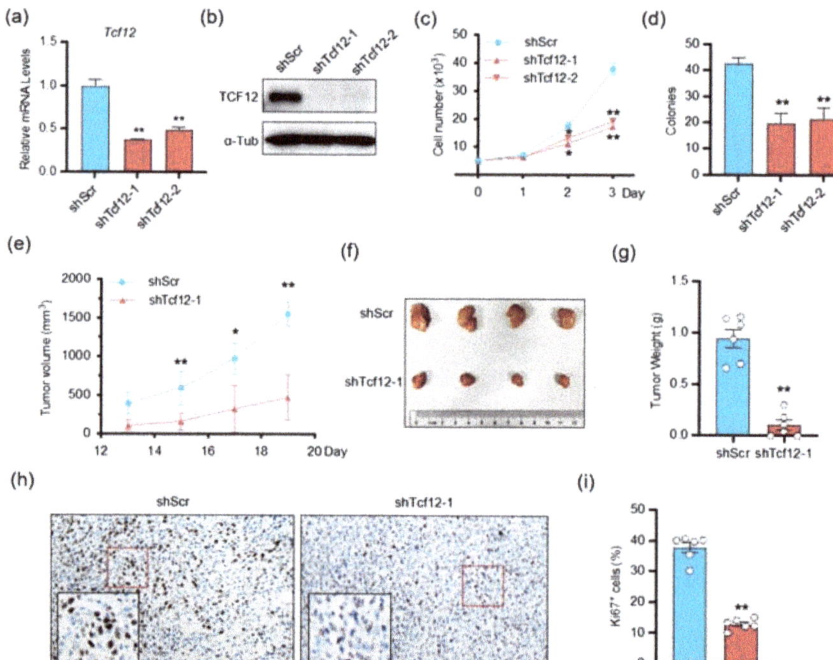

Figure 2. TCF12 enhances melanoma cell proliferation in vitro and tumorigenicity in vivo: (**a**,**b**) qPCR (**a**) and immunoblot (**b**) analysis of TCF12 level in YUMM1.7 cell lines after TCF12 knockdown. shScr: control shRNA, shTcf12-1/2: mouse TCF12-specific shRNA; (**c**,**d**) cell proliferation (**c**) and colony formation capability (**d**) in the TCF12 knockdown cells as compared to control cells; (**e**) tumor growth curves in mice injected with YUMM1.7 cells with TCF12 knockdown, n = 6 mice per group; (**f**) representative tumor image from control (shScr) and TCF12-knockdown (shTcf12-1) mice; (**g**) tumor weights comparison between control and TCF12 knockdown groups; (**h**,**i**) Ki67 immunohistochemistry staining representative images (red square, 200×; black square, 400×) (**h**) and subsequent analysis (**i**) in tumor tissues. Scale bar: 50 μm. Statistical significance is based on comparison with shScr group. * $p < 0.05$, ** $p < 0.01$.

3.3. TCF12 Promotes Melanoma Cell Migration, Invasion In Vitro and Metastasis In Vivo

We continued to investigate whether TCF12 in melanoma affected cell migration and metastasis. In vitro transwell migration and Matrigel invasion assays showed that TCF12 knockdown reduced the migration and invasion of melanoma cells (Figures 3a,b and S3a,b), whereas TCF12 overexpression greatly enhanced their migratory and invasive capacities (Figures 3c,d and S3c,d). Similarly, knockdown of TCF12 significantly compromised the metastatic outgrowth of melanoma cells in mice after tail-vein injection, as revealed by improved animal survival (Figure 3e) and fewer macroscopic and microscopic tumor nodules in the lung (Figure 3f–h). In contrast, overexpression of TCF12 in melanoma enabled the cells to establish larger and larger tumor nodules in the lung upon tail-vein injection (Figure 3i–k). Taken together, these findings indicate that TCF12 promotes melanoma metastasis in vitro and in vivo.

3.4. TGFB2 Is a Direct Downstream Target Gene of TCF12

To elucidate the key molecules involved in TCF12-induced proliferation and metastasis in melanoma, we performed RNA-seq analysis on YUMM1.7 cells with scramble control or TCF12 knockdown. The volcano plot showed that 372 and 109 genes were significantly downregulated and upregulated upon TCF12 depletion in melanoma cells, respectively (Figure 4a). Gene ontology enrichment analysis found that the upregulated genes were enriched in muscle developmental and functional programs, while the downregulated genes were enriched in cell mobility, extracellular matrix organization, and proliferation (Figure 4b). Consistently, KEGG pathway analysis also revealed similar enrichment (Figure 4c). Particularly, the detection of ECM–receptor interaction, focal adhesion, cGMP, cAMP, and PI3K-AKT signaling pathways among the top lists for downregulated genes was consistent with our observation that knockdown of TCF12 impaired melanoma proliferation and metastasis. Subsequently, we screened a set of melanoma-related target genes with TCF12 binding sites by RT-qPCR (Figure S4 and Table S3) and verified that TCF12 knockdown significantly decreased *Tgfb2* transcription (Figure 4d), leading to reduced protein expression (Figure 4e). The uncropped Western blots are shown in Figure S5. We further employed IHC on the matched mouse melanoma tissues to validate that TCF12 depletion decreased TGFB2 protein while TCF12 overexpression boosted TGFB2 levels (Figure 4f,g).

To determine whether TCF12 can directly modulate the transcription of TGFB2, chromatin immunoprecipitation (ChIP) coupled with qPCR was performed in melanoma cells. The results showed that TCF12 can occupy the *Tgfb2* promoter, which was significantly decreased upon TCF12 depletion (Figure 5a) but largely enhanced by TCF12 overexpression (Figure 5b). To further test whether the TCF12 binding is active, we constructed *Tgfb2* promoter-driven luciferase reporters with either wild-type or E-box-deficient sequences (Figure 5c). Luciferase assays showed that knockdown of TCF12 significantly impaired the reporter activity, while overexpression of TCF12 boosted the signal; however, depletion of the E-box, which mediates TCF12 binding, completely abolished the reporter activity (Figure 5c). In summary, we identified that TCF12 can directly bind to the *Tgfb2* promoter and activate its expression in melanoma.

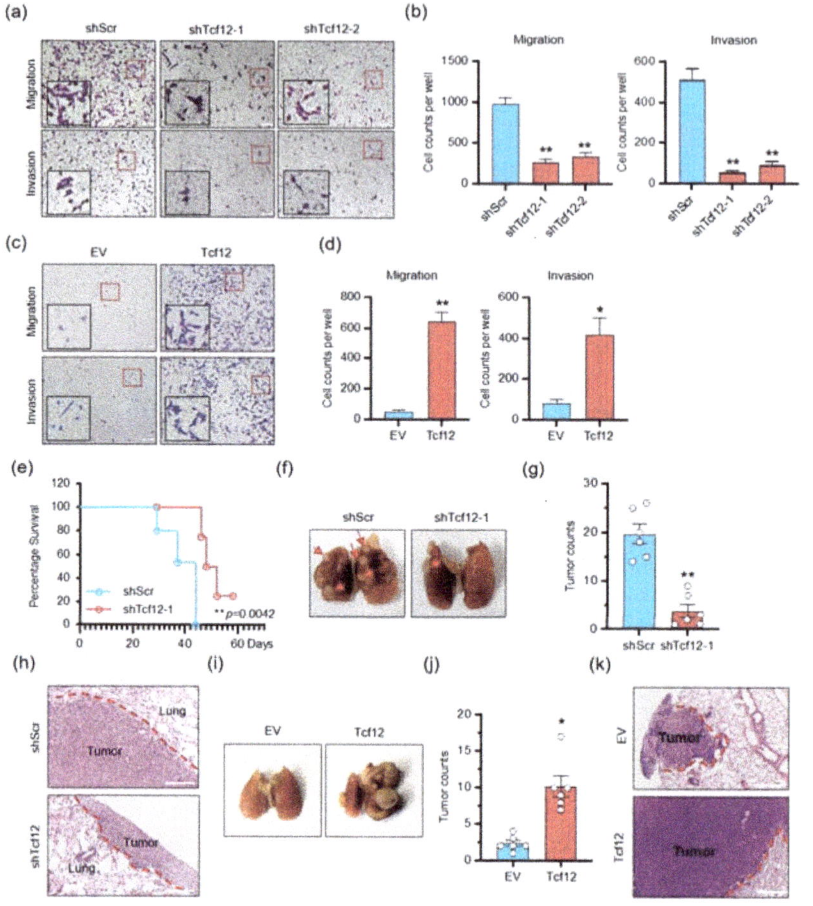

Figure 3. TCF12 promotes melanoma cell migration, invasion in vitro and metastasis in vivo: (**a**,**b**) The representative images (red square, 200×; black square, 400×) (**a**) and cell count analysis (**b**) of transwell migration and Matrigel invasion assays of YUMM1.7 cells after TCF12 knockdown. shScr: control shRNA, shTcf12-1/2: mouse TCF12-specific shRNA; (**c**,**d**) the representative images (red square, 200×; black square, 400×) (**c**) and cell count analysis (**d**) of transwell migration and Matrigel invasion assays of YUMM1.7 cells following TCF12 overexpression. EV: empty vector expression; Tcf12: mouse TCF12 overexpression; (**e**) Kaplan–Meier survival curves of mice after tail-vein injection of TCF12 knockdown YUMM1.7 cells, n = 6 per group; (**f**–**h**) representative images of lung metastasis (**f**), hematoxylin and eosin (H&E) staining of lung sections (**h**), and the number of tumor nodules (**g**) in mice injected via tail vein with control (shScr) and TCF12 knockdown (shTcf12) YUMM1.7 cells. Scale bar: 100 µm. Representative images of lung metastasis in mice tail-vein injected with control (shScr) and TCF12 knockdown (shTcf12) YUMM1.7 cells; (**i**–**k**) representative images of lung metastasis (**i**), H&E staining of lung sections (**k**), and the number of tumor nodules (**j**) in mice tail-vein injected with control (EV) and TCF12 overexpression (Tcf12) YUMM1.7 cells, n = 6 mice per group. Scale bar: 100 µm in H&E staining. Statistical significance is based on comparison with shScr group or EV group. * $p < 0.05$, ** $p < 0.01$.

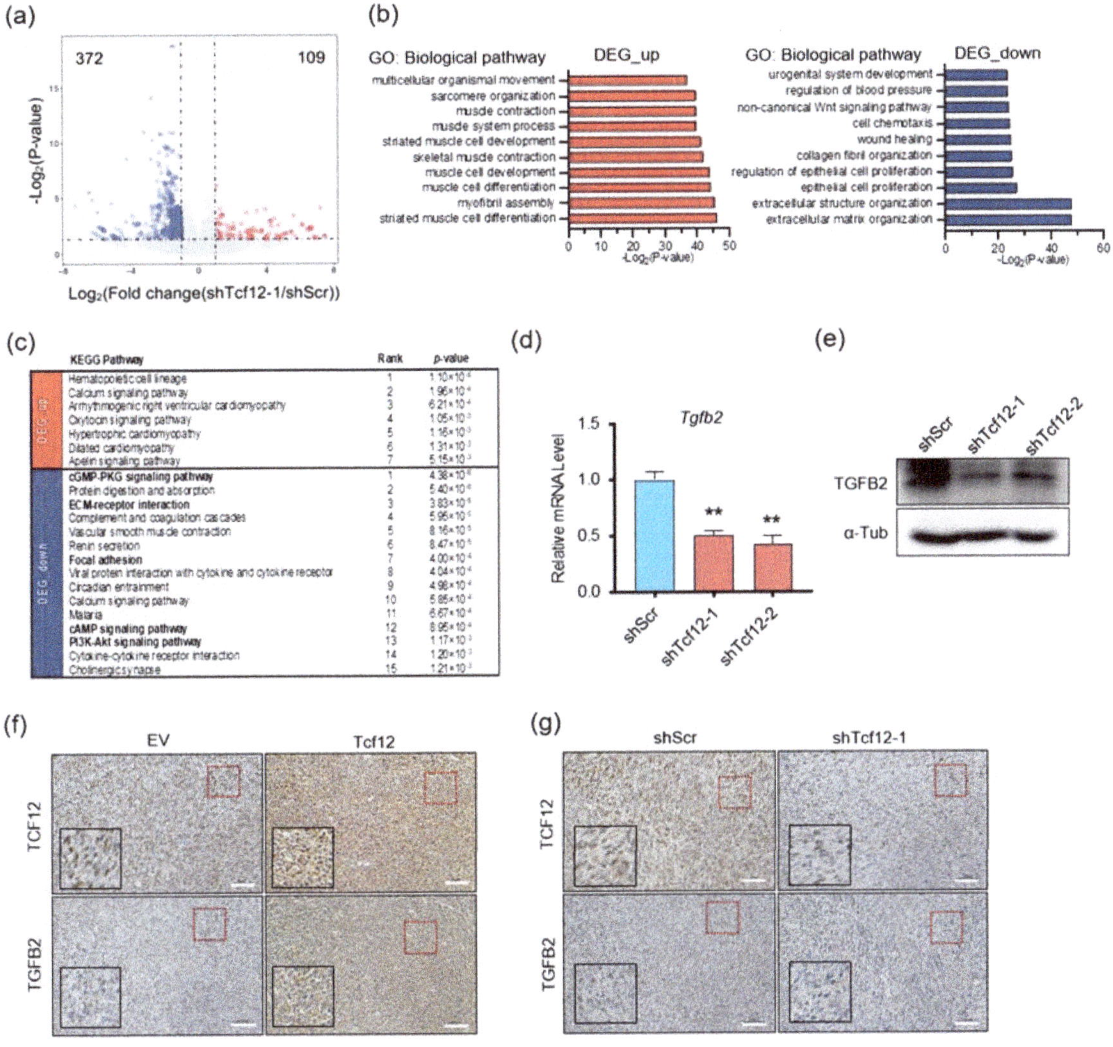

Figure 4. TGFB2 is a downstream target of TCF12: (**a**) Volcano plot of differentially expressed genes in YUMM1.7 cells with TCF12 knockdown (shTcf12-1) compared to control (shScr), analyzed by RNA-seq (blue, genes down-regulated; red, genes up-regulated); (**b**,**c**) gene ontology (GO) enrichment (**b**) and KEGG pathway (**c**) analysis for biological processes of the differentially expressed genes upon TCF12 knockdown; (**d**,**e**) qPCR (**d**) and immunoblot (**e**) analysis of TGFB2 expression in YUMM1.7 cells upon TCF12 knockdown; (**f**,**g**) representative images of TGFB2 IHC staining in mouse melanoma tissues with TCF12 knockdown (**g**) or TCF12 overexpression (**f**) (red square, 200×; black square, 400×). Scale bar: 50 μm. Statistical significance is based on comparison with shScr group. ** $p < 0.01$.

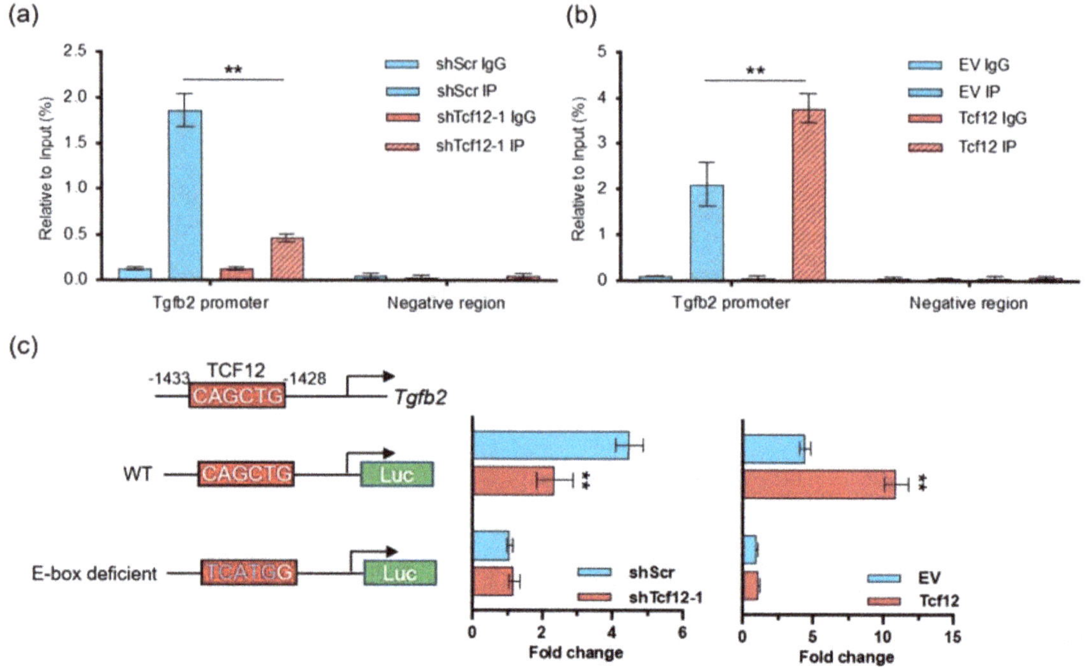

Figure 5. TCF12 directly binds to and activates the *Tgfb2* promoter: (**a**) Chromatin immunoprecipitation (ChIP) coupled with qPCR analysis showing the binding of TCF12 to the *Tgfb2* promoter in YUMM1.7 cells, which is decreased upon TCF12 knockdown; (**b**) ChIP−qPCR analysis showing the increased binding of TCF12 to the *Tgfb2* promoter in YUMM1.7 cells upon TCF12 overexpression; (**c**) the schematic diagrams show the *Tgfb2* promoter−driven luciferase reporters with either wild-type (WT) or E-box deficient sequences. Luciferase activity was measured and displayed as fold change. Statistical significance is based on comparison with the control group (shScr or EV). ** $p < 0.01$.

3.5. TGFB2 Is Essential for TCF12-Induced Cell Proliferation, Migration and Invasion In Vitro

Given the critical functions of the TGF-β pathway in cancer biology [27–30], we speculated whether TGFB2 mediates the tumorigenic activity of TCF12 in melanoma. We found that, despite the fact that depletion of TGFB2 on its own did not alter melanoma proliferation, it was able to block the growth advantage conferred by TCF12 overexpression (Figure 6a,b). Similarly, TGFB2 knockdown significantly suppressed the migratory and invasive capabilities induced by TCF12 overexpression (Figure 6c,d). Together, we confirmed that TGFB2 was essential for TCF12 to induce melanoma proliferation, migration, and invasion.

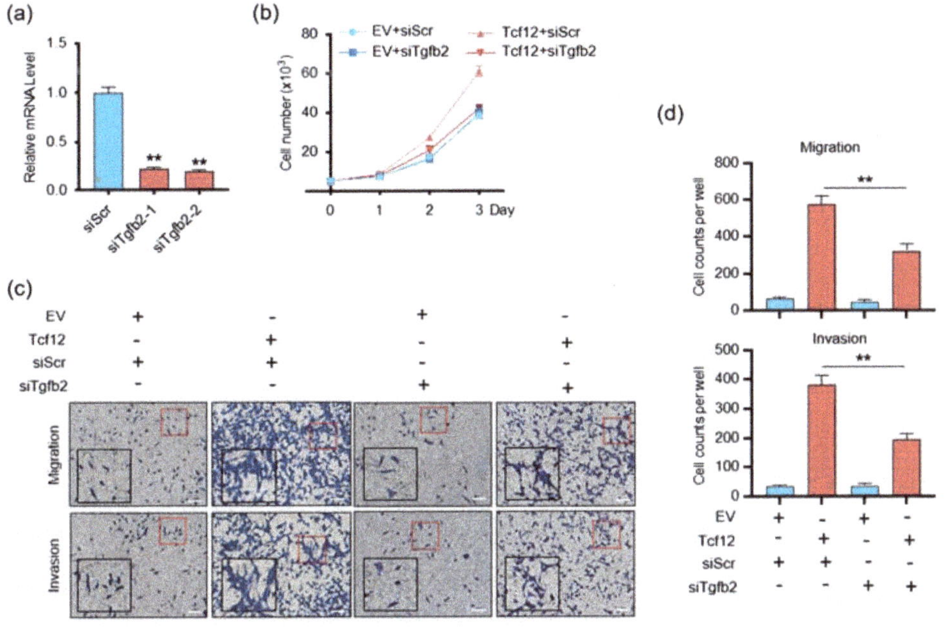

Figure 6. TGFB2 is essential for TCF12−induced cell proliferation, migration, and invasion in vitro: (**a**) qPCR analysis of TGFB2 level in YUMM1.7 cell lines after TGFB2 knockdown. siScr: control siRNA, siTgfb2-1/2: TGFB2 knockdown siRNA; (**b**) cell proliferation analysis of YUMM1.7 cells with TGFB2 knockdown, TCF12 overexpression, or their combination; (**c**,**d**) representative images (red square, 200×; black square, 400×) (**c**) and cell count analysis (**d**) of transwell migration and Matrigel invasion assays in YUMM1.7 cells with TGFB2 knockdown, TCF12 overexpression, or their combination. EV: empty vector expression, Tcf12: TCF12 overexpression, siScr: control siRNA, siTgfb2: TGFB2−specific siRNA. Scale bar: 100 μm. Statistical significance is based on comparison with control (Tcf12 + siScr) group. ** $p < 0.01$.

3.6. Depletion of TCF12 Sensitizes Melanoma to BRAF Inhibition

BRAF mutations represent the most prevalent oncogenic event in melanoma, resulting in constitutive activation of the BRAF-MEK-ERK MAPK pathway [31–33]. To test whether there is any correlation between BRAF signaling and TCF12, we treated BRAF-mutant A375 cells with the BRAF(V600E) inhibitor PLX4032 or the MEK inhibitor trametinib. We found that inhibition of the MAPK pathway did not alter the transcript of TCF12 (Figure 7a), but reduced the expression level of TCF12 protein (Figure 7b), suggesting a post-transcriptional regulation. We then measured the TCF12 protein stability in melanoma cells with or without BRAF inhibition and found that suppression of the BRAF pathway greatly reduced TCF12 protein stability (Figure 7c,d). We further found that the BRAF inhibition-facilitated TCF12 degradation was mediated by the proteosome, as the concurrent addition of the proteosome inhibitor MG132 markedly prevented the reduction of TCF12 protein by PLX4032 (Figure 7e).

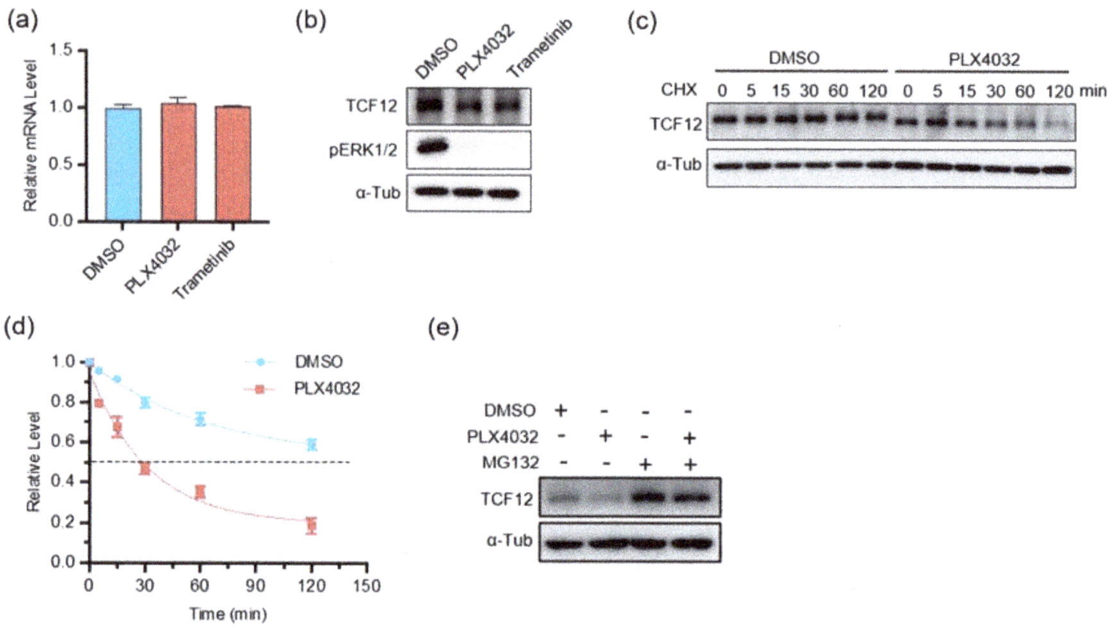

Figure 7. MAPK pathway inhibition reduces TCF12 protein stability: (**a**) qPCR analysis of TCF12 transcript in A375 cells treated with PLX4032 or trametinib for 24 h; (**b**) immunoblot analysis of TCF12 protein in A375 cells treated with PLX4032 or trametinib for 24 h. α-Tub: α-tubulin as internal control; (**c**,**d**) TCF12 protein stability analysis in A375 cells treated with PLX4032 for different time points. CHX: cycloheximide; (**e**) immunoblot analysis of TCF12 protein in A375 cells treated with PLX4032 alone or in combination with proteasome inhibitor MG132.

Given that the BRAF pathway regulates TCF12 stability, we speculated whether TCF12 is involved in the sensitivity to BRAF-targeted therapy. We found that depletion of TCF12 substantially reduced melanoma proliferation in response to PLX4032, as measured by the cell growth curve (Figure 8a) and colony formation (Figure 8b). Consistently, the tumors with TCF12 deletion treated with PLX4032 showed the slowest growth kinetics (Figure 8c–e) and contained the smallest percentage of Ki67+ proliferating cells (Figure 8f,g). These data support the hypothesis that depletion of TCF12 sensitizes mutant melanoma to BRAF inhibitor therapy.

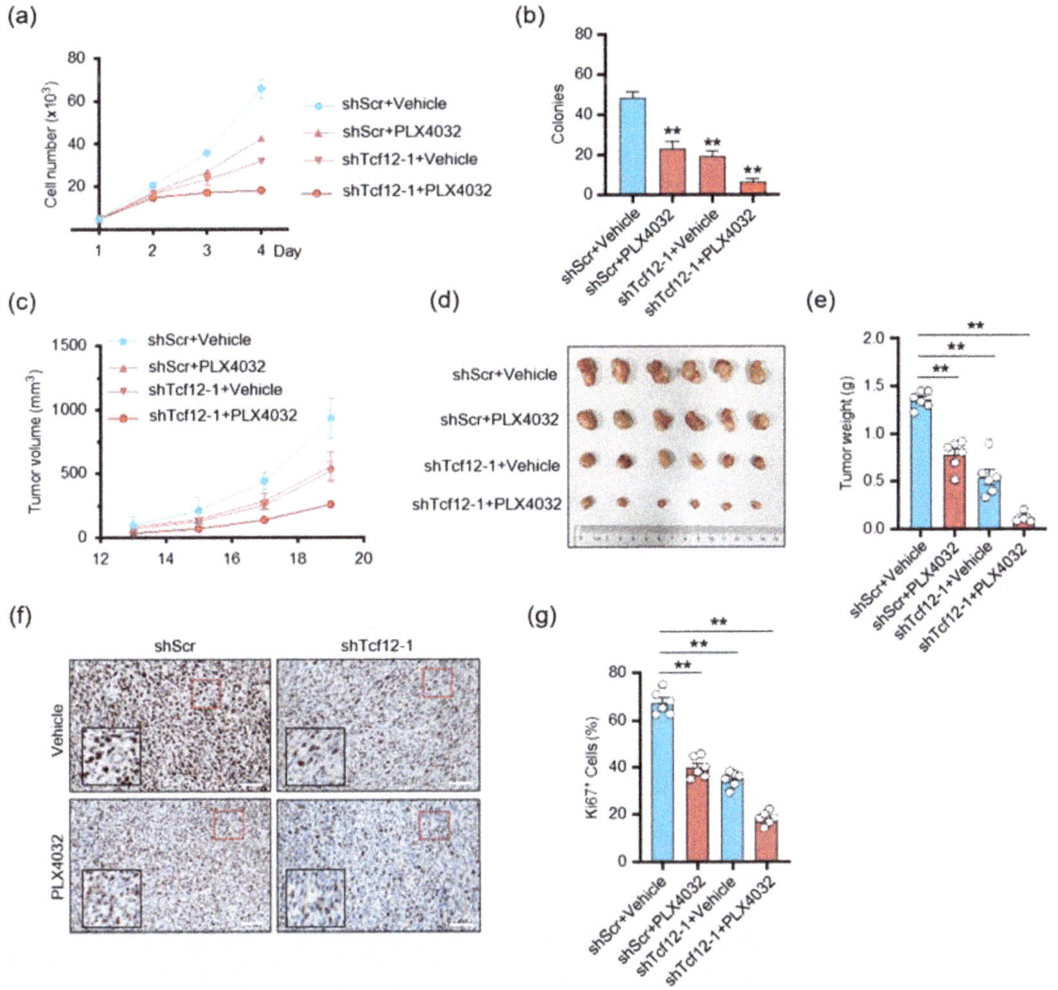

Figure 8. Depletion of TCF12 sensitizes melanoma to BRAF inhibition: (**a**,**b**) Cell proliferation (**a**) and colony formation (**b**) analysis of cells with TCF12 knockdown, PLX4032 treatment, or their combination; (**c**) tumor growth curves in mice injected cells with TCF12 knockdown and treated with PLX4032, n = 6 mice per group; (**c**,**g**) tumor growth curves (**c**), representative tumor image (**d**), tumor weights (**e**), representative images of Ki67 IHC (red square, 200×; black square, 400×) (**f**) and subsequent analysis (**g**) from control (shScr + vehicle), PLX4032 treatment (shScr + PLX4032), TCF12 knockdown (shTcf12-1 + vehicle), and their combination (shTcf12-1 + PLX4032) groups. Scale bar: 50 μm. Statistical significance is based on comparison with control group. ** $p < 0.01$.

4. Discussion

Our findings shed light on a novel aspect of melanoma pathogenesis, uncovering the oncogenic functions of TCF12 and its regulatory mechanism within the disease (Figure 9). We found that TCF12 is upregulated in melanoma, and high expression is correlated with disease progression and a poorer prognosis. TCF12 enhances melanoma cell proliferation, metastasis, and sensitivity to BRAF(V600E)-targeted therapy. Importantly, we discovered that TCF12 exerts its oncogenic effects partly through transcriptional activation of TGFB2.

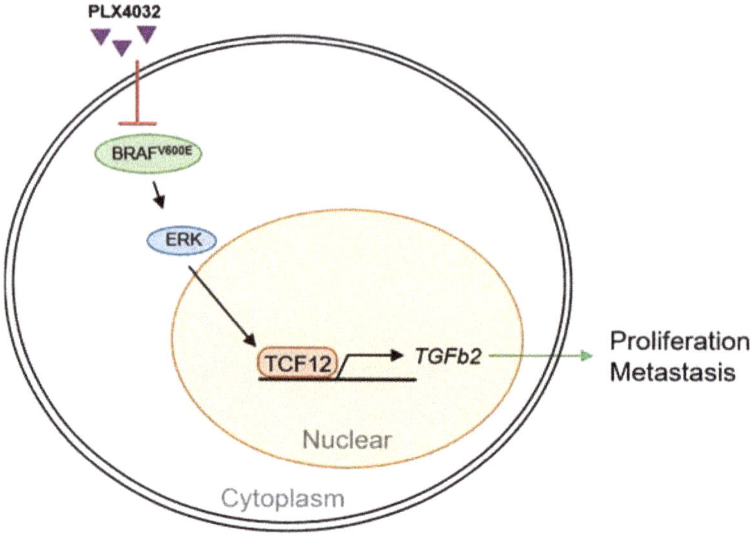

Figure 9. Schematic summary of TCF12 functions in melanoma pathogenesis.

Melanoma exhibits a significant burden of genetic alterations that can regulate a myriad of biological processes, including cell proliferation, survival, differentiation, migration, and metastasis [34,35]. Mutations in BRAF, specifically the V600E variant, are the most common genetic alterations in melanoma, occurring in approximately 50% of cases [36]. Our study suggests that TCF12 is a new player in this intricate network of genetic interactions that contribute to melanoma progression. Moreover, our results indicate that the interaction between TCF12 and the BRAF pathway is crucial in mediating the tumor's response to BRAF-targeted therapy.

The TGF-β signaling pathway has well-documented roles in tumor progression, promoting epithelial-to-mesenchymal transition, invasion, and metastasis in several cancers [37,38]. In melanoma, previous studies have reported a paradoxical role of TGF-β signaling [39]. On one hand, TGF-β has been shown to inhibit melanoma initiation by suppressing cell proliferation [40]. On the other hand, it promotes later stages of tumor progression by enhancing invasion and metastasis [41]. TGF-β protein is predominantly found in active melanocytes, whereas quiescent melanocytes exhibit minimal to non-existent TGF-β levels [42]. Typical melanocytes predominantly express TGFB1 and TGFB3. However, TGFB2 expression incrementally rises from initial to metastatic melanoma phases, suggesting its potential role in melanoma's malignant evolution [43]. Notably, only a fraction of melanoma patients displays signs of TGFB2 reduction, hinting that its presence is not a primary event in melanoma development but rather associated with tumor progression [44]. Our finding that TGFB2 is a direct transcriptional target of TCF12 suggests a mechanism by which TCF12 contributes to melanoma progression. However, further studies are needed to fully elucidate the functional roles of TGFB2 in this context and whether its effects are context dependent.

Interestingly, we observed that TCF12 protein expression is regulated post-transcriptionally by the BRAF/MEK/ERK pathway. This observation implies a potential feedback loop wherein BRAF mutations upregulate TCF12, which in turn promotes tumor progression. Furthermore, we found that TCF12 depletion sensitizes melanoma cells to BRAF inhibition, suggesting that TCF12 may represent a potential therapeutic target for enhancing the efficacy of current BRAF-targeted therapies. This is of great clinical significance given the emergence of drug resistance as a significant problem in the treatment of BRAF-mutated melanomas.

We believe our findings offer promising avenues for future research and add to the growing body of knowledge that will hopefully lead to improved melanoma patient outcomes. However, the translation of these findings into clinical applications will require further investigations, including preclinical studies and potentially clinical trials, to validate the efficacy and safety of targeting TCF12 in the treatment of melanoma.

5. Conclusions

Our study provides new insights into the molecular mechanisms underlying melanoma progression and reveals a potential therapeutic target for melanoma treatment.

Supplementary Materials: The following supporting information can be downloaded at: https://www.mdpi.com/article/10.3390/cancers15184505/s1, Figure S1: TCF12 enhances melanoma A375 cell proliferation in vitro; Figure S2: Overexpression of TCF12 enhances melanoma cell proliferation in vitro and tumorigenicity in vivo; Figure S3: Overexpression of TCF12 promotes melanoma cell migration and invasion in vitro; Figure S4: Analysis of potential melanoma-related target genes of TCF12; Figure S5: The original western blot figures; Table S1: The primer sequences used for RT-qPCR; Table S2: The primer sequence used for ChIP-qPCR; Table S3: Differentially expressed gene after TCF12 knockdown.

Author Contributions: Conceptualization, J.S. and C.L.; data curation, Y.T., X.C., Z.P., X.X. and J.S.; formal analysis, Y.T., J.Z. and X.X.; funding acquisition, J.S. and C.L.; investigation, Y.T.; methodology, Y.T., J.S. and C.L.; project administration, J.S.; resources, Y.T.; supervision, J.S. and C.L.; validation, Y.T., X.C., Y.Z. and J.S.; visualization, Y.T.; writing—original draft, Y.T., J.S. and C.L.; writing—review and editing, J.S. and C.L. All authors have read and agreed to the published version of the manuscript.

Funding: This work was supported by grants from National Natural Science Foundation of China (32071289, 32200960 and 91957105); Zhejiang Provincial Natural Science Foundation (LR20H160004); Fundamental Research Funds for the Central Universities (No. 2021QNA7009); Leading Innovative and Entrepreneur Team Introduction Program of Zhejiang (No. 2021R01012); Leading Innovation and Entrepreneur Team of Hangzhou (No. TD2020006).

Institutional Review Board Statement: The study was conducted in accordance with the Declaration of Helsinki, and approved by the Medical Experimental Animal Care Commission of Zhejiang University (#ZJU20220217).

Informed Consent Statement: Not applicable.

Data Availability Statement: The data presented in this study are available in the paper and/or the supplementary materials. Additional data related to this paper are available on request from the corresponding author.

Conflicts of Interest: The authors declare that the research was conducted in the absence of any commercial or financial relationships that could be construed as a potential conflict of interest.

References

1. Centeno, P.P.; Pavet, V.; Marais, R. The journey from melanocytes to melanoma. *Nat. Rev. Cancer* **2023**, *23*, 372–390. [CrossRef]
2. Siegel, R.L.; Miller, K.D.; Fuchs, H.E.; Jemal, A. Cancer statistics. *CA A Cancer J. Clin.* **2022**, *72*, 7–33. [CrossRef] [PubMed]
3. Nascentes Melo, L.M.; Kumar, S.; Riess, V.; Szylo, K.J.; Eisenburger, R.; Schadendorf, D.; Ubellacker, J.M.; Tasdogan, A. Advancements in melanoma cancer metastasis models. *Pigment Cell Melanoma Res.* **2023**, *36*, 206–223. [CrossRef] [PubMed]
4. Poulikakos, P.I.; Sullivan, R.J.; Yaeger, R. Molecular Pathways and Mechanisms of BRAF in Cancer Therapy. *Clin. Cancer Res. Off. J. Am. Assoc. Cancer Res.* **2022**, *28*, 4618–4628. [CrossRef]
5. Ziogas, D.C.; Theocharopoulos, C.; Koutouratsas, T.; Haanen, J.; Gogas, H. Mechanisms of resistance to immune checkpoint inhibitors in melanoma: What we have to overcome? *Cancer Treat. Rev.* **2023**, *113*, 102499. [CrossRef] [PubMed]
6. Merlino, G.; Herlyn, M.; Fisher, D.E.; Bastian, B.C.; Flaherty, K.T.; Davies, M.A.; Wargo, J.A.; Curiel-Lewandrowski, C.; Weber, M.J.; Leachman, S.A.; et al. The state of melanoma: Challenges and opportunities. *Pigment Cell Melanoma Res.* **2016**, *29*, 404–416. [CrossRef] [PubMed]
7. Bashash, D.; Zandi, Z.; Kashani, B.; Pourbagheri-Sigaroodi, A.; Salari, S.; Ghaffari, S.H. Resistance to immunotherapy in human malignancies: Mechanisms, research progresses, challenges, and opportunities. *J. Cell. Physiol.* **2022**, *237*, 346–372. [CrossRef]
8. Ganesh, K.; Massagué, J. Targeting metastatic cancer. *Nat. Med.* **2021**, *27*, 34–44. [CrossRef]
9. Ascierto, P.A.; Kirkwood, J.M.; Grob, J.J.; Simeone, E.; Grimaldi, A.M.; Maio, M.; Palmieri, G.; Testori, A.; Marincola, F.M.; Mozzillo, N.; et al. The role of BRAF V600 mutation in melanoma. *J. Transl. Med.* **2012**, *10*, 85. [CrossRef]
10. Śmiech, M.; Leszczyński, P.; Kono, H.; Wardell, C.; Taniguchi, H. Emerging BRAF Mutations in Cancer Progression and Their Possible Effects on Transcriptional Networks. *Genes* **2020**, *11*, 1342. [CrossRef]

11. Ngeow, K.C.; Friedrichsen, H.J.; Li, L.; Zeng, Z.; Andrews, S.; Volpon, L.; Brunsdon, H.; Berridge, G.; Picaud, S.; Fischer, R.; et al. BRAF/MAPK and GSK3 signaling converges to control MITF nuclear export. *Proc. Natl. Acad. Sci. USA* **2018**, *115*, E8668–E8677. [CrossRef] [PubMed]
12. Vachtenheim, J.; Ondrušová, L. Microphthalmia-associated transcription factor expression levels in melanoma cells contribute to cell invasion and proliferation. *Exp. Dermatol.* **2015**, *24*, 481–484. [CrossRef] [PubMed]
13. Luo, C.; Balsa, E.; Perry, E.A.; Liang, J.; Tavares, C.D.; Vazquez, F.; Widlund, H.R.; Puigserver, P. H3K27me3-mediated PGC1α gene silencing promotes melanoma invasion through WNT5A and YAP. *J. Clin. Investig.* **2020**, *130*, 853–862. [CrossRef] [PubMed]
14. Luo, C.; Lim, J.H.; Lee, Y.; Granter, S.R.; Thomas, A.; Vazquez, F.; Widlund, H.R.; Puigserver, P. A PGC1α-mediated transcriptional axis suppresses melanoma metastasis. *Nature* **2016**, *537*, 422–426. [CrossRef]
15. Di Rocco, G.; Pennuto, M.; Illi, B.; Canu, N.; Filocamo, G.; Trani, E.; Rinaldi, A.M.; Possenti, R.; Mandolesi, G.; Sirinian, M.I.; et al. Interplay of the E box, the cyclic AMP response element, and HTF4/HEB in transcriptional regulation of the neurospecific, neurotrophin-inducible vgf gene. *Mol. Cell Biol.* **1997**, *17*, 1244–1253. [CrossRef]
16. Li, Y.; Brauer, P.M.; Singh, J.; Xhiku, S.; Yoganathan, K.; Zúñiga-Pflücker, J.C.; Anderson, M.K. Targeted Disruption of TCF12 Reveals HEB as Essential in Human Mesodermal Specification and Hematopoiesis. *Stem Cell Rep.* **2017**, *9*, 779–795. [CrossRef]
17. Parker, M.H.; Perry RL, S.; Fauteux, M.C.; Berkes, C.A.; Rudnicki, M.A. MyoD synergizes with the E-protein HEB beta to induce myogenic differentiation. *Mol. Cell. Biol.* **2006**, *26*, 5771–5783. [CrossRef]
18. Yi, S.; Yu, M.; Yang, S.; Miron, R.J.; Zhang, Y. Tcf12, A Member of Basic Helix-Loop-Helix Transcription Factors, Mediates Bone Marrow Mesenchymal Stem Cell Osteogenic Differentiation In Vitro and In Vivo. *Stem Cells* **2017**, *35*, 386–397. [CrossRef]
19. Yoon, S.J.; Foley, J.W.; Baker, J.C. HEB associates with PRC2 and SMAD2/3 to regulate developmental fates. *Nat. Commun.* **2015**, *6*, 6546. [CrossRef]
20. Lee, C.-C.; Chen, W.-S.; Chen, C.-C.; Chen, L.-L.; Lin, Y.-S.; Fan, C.-S.; Huang, T.-S. TCF12 protein functions as transcriptional repressor of E-cadherin, and its overexpression is correlated with metastasis of colorectal cancer. *J. Biol. Chem.* **2012**, *287*, 2798–2809. [CrossRef]
21. Wang, L.; Tang, Y.; Wu, H.; Shan, G. TCF12 activates MAGT1 expression to regulate the malignant progression of pancreatic carcinoma cells. *Oncol. Lett.* **2022**, *23*, 62. [CrossRef] [PubMed]
22. Yang, J.; Zhang, L.; Jiang, Z.; Ge, C.; Zhao, F.; Jiang, J.; Tian, H.; Chen, T.; Xie, H.; Cui, Y.; et al. TCF12 promotes the tumorigenesis and metastasis of hepatocellular carcinoma via upregulation of CXCR4 expression. *Theranostics* **2019**, *9*, 5810–5827. [CrossRef] [PubMed]
23. Gao, S.; Bian, T.; Zhang, Y.; Su, M.; Liu, Y. TCF12 overexpression as a poor prognostic factor in ovarian cancer. *Pathol. Res. Pract.* **2019**, *215*, 152527. [CrossRef] [PubMed]
24. Cui, Y.; Zhao, M.; Yang, Y.; Xu, R.; Tong, L.; Liang, J.; Zhang, X.; Sun, Y.; Fan, Y. Reversal of epithelial-mesenchymal transition and inhibition of tumor stemness of breast cancer cells through advanced combined chemotherapy. *Acta Biomater.* **2022**, *152*, 380–392. [CrossRef]
25. Tai, G.; Fu, H.; Bai, H.; Liu, H.; Li, L.; Song, T. Long non-coding RNA GLIDR accelerates the tumorigenesis of lung adenocarcinoma by miR-1270/TCF12 axis. *Cell Cycle (Georget. Tex.)* **2021**, *20*, 1653–1662. [CrossRef]
26. Tang, Z.; Li, C.; Kang, B.; Gao, G.; Li, C.; Zhang, Z. GEPIA: A web server for cancer and normal gene expression profiling and interactive analyses. *Nucleic Acids Res.* **2017**, *45*, W98–W102. [CrossRef]
27. Nguyen, D.X.; Bos, P.D.; Massagué, J. Metastasis: From dissemination to organ-specific colonization. *Nat. Rev. Cancer* **2009**, *9*, 274–284. [CrossRef]
28. Plantefaber, L.C.; Hynes, R.O. Changes in integrin receptors on oncogenically transformed cells. *Cell* **1989**, *56*, 281–290. [CrossRef]
29. Hamidi, H.; Ivaska, J. Every step of the way: Integrins in cancer progression and metastasis. *Nat. Rev. Cancer* **2018**, *18*, 533–548. [CrossRef]
30. Steeg, P.S. Targeting metastasis. *Nat. Rev. Cancer* **2016**, *16*, 201–218. [CrossRef]
31. Fagin, J.A.; Krishnamoorthy, G.P.; Landa, I. Pathogenesis of cancers derived from thyroid follicular cells. *Nat. Rev. Cancer* **2023**. *ahead of print*. [CrossRef] [PubMed]
32. Delyon, J.; Vallet, A.; Bernard-Cacciarella, M.; Kuzniak, I.; Reger de Moura, C.; Louveau, B.; Jouenne, F.; Mourah, S.; Lebbé, C.; Dumaz, N. TERT Expression Induces Resistance to BRAF and MEK Inhibitors in BRAF-Mutated Melanoma In Vitro. *Cancers* **2023**, *15*, 2888. [CrossRef] [PubMed]
33. Marranci, A.; Prantera, A.; Masotti, S.; De Paolo, R.; Baldanzi, C.; Podda, M.S.; Mero, S.; Vitiello, M.; Franchin, C.; Laezza, M.; et al. PARP1 negatively regulates MAPK signaling by impairing BRAF-X1 translation. *J. Hematol. Oncol.* **2023**, *16*, 33. [CrossRef] [PubMed]
34. Mahumud, R.A.; Shahjalal, M. The Emerging Burden of Genetic Instability and Mutation in Melanoma: Role of Molecular Mechanisms. *Cancers* **2022**, *14*, 6202. [CrossRef] [PubMed]
35. Zhang, T.; Dutton-Regester, K.; Brown, K.M.; Hayward, N.K. The genomic landscape of cutaneous melanoma. *Pigment Cell Melanoma Res.* **2016**, *29*, 266–283. [CrossRef] [PubMed]
36. Davies, H.; Bignell, G.R.; Cox, C.; Stephens, P.; Edkins, S.; Clegg, S.; Teague, J.; Woffendin, H.; Garnett, M.J.; Bottomley, W.; et al. Mutations of the BRAF gene in human cancer. *Nature* **2002**, *417*, 949–954. [CrossRef] [PubMed]
37. Derynck, R.; Turley, S.J.; Akhurst, R.J. TGFβ biology in cancer progression and immunotherapy. *Nat. Rev. Clin. Oncol.* **2021**, *18*, 9–34. [CrossRef]

38. Seoane, J.; Gomis, R.R. TGF-β Family Signaling in Tumor Suppression and Cancer Progression. *Cold Spring Harb. Perspect. Biol.* **2017**, *9*, a022277. [CrossRef]
39. Baba, A.B.; Rah, B.; Bhat, G.R.; Mushtaq, I.; Parveen, S.; Hassan, R.; Hameed Zargar, M.; Afroze, D. Transforming Growth Factor-Beta (TGF-β) Signaling in Cancer-A Betrayal Within. *Front. Pharmacol.* **2022**, *13*, 791272. [CrossRef]
40. Krasagakis, K.; Krüger-Krasagakes, S.; Fimmel, S.; Eberle, J.; Thölke, D.; von der Ohe, M.; Mansmann, U.; Orfanos, C.E. Desensitization of melanoma cells to autocrine TGF-beta isoforms. *J. Cell Physiol.* **1999**, *178*, 179–187. [CrossRef]
41. Javelaud, D.; Mohammad, K.S.; McKenna, C.R.; Fournier, P.; Luciani, F.; Niewolna, M.; André, J.; Delmas, V.; Larue, L.; Guise, T.A.; et al. Stable overexpression of Smad7 in human melanoma cells impairs bone metastasis. *Cancer Res.* **2007**, *67*, 2317–2324. [CrossRef] [PubMed]
42. Javelaud, D.; Alexaki, V.I.; Mauviel, A. Transforming growth factor-beta in cutaneous melanoma. *Pigment Cell Melanoma Res.* **2008**, *21*, 123–132. [CrossRef] [PubMed]
43. Berking, C.; Takemoto, R.; Schaider, H.; Showe, L.; Satyamoorthy, K.; Robbins, P.; Herlyn, M. Transforming growth factor-beta1 increases survival of human melanoma through stroma remodeling. *Cancer Res.* **2001**, *61*, 8306–8316. [PubMed]
44. Cosgarea, I.; McConnell, A.T.; Ewen, T.; Tang, D.; Hill, D.S.; Anagnostou, M.; Elias, M.; Ellis, R.A.; Murray, A.; Spender, L.C.; et al. Melanoma secretion of transforming growth factor-β2 leads to loss of epidermal AMBRA1 threatening epidermal integrity and facilitating tumour ulceration. *Br. J. Dermatol.* **2022**, *186*, 694–704. [CrossRef] [PubMed]

cancers MDPI

Review

Dual Functions of T Lymphocytes in Breast Carcinoma: From Immune Protection to Orchestrating Tumor Progression and Metastasis

Mohammadrasul Zareinejad , Fereshteh Mehdipour, Mina Roshan-Zamir, Zahra Faghih * and Abbas Ghaderi *

Shiraz Institute for Cancer Research, School of Medicine, Shiraz University of Medical Sciences, Shiraz 71348-45505, Iran; m.rasul.z.nejad@gmail.com (M.Z.); mehdipourf@sums.ac.ir (F.M.); minarzs95@gmail.com (M.R.-Z.)

* Correspondence: faghihz@sums.ac.ir (Z.F.); ghaderia@sums.ac.ir (A.G.); Tel.: +98-71-32303687 (Z.F. & A.G.); Fax: +98-71-32304952 (Z.F. & A.G.)

Simple Summary: New insights into the foundation of cellular and molecular cancer immunology have revealed that immune cells play crucial roles in the development and growth of breast cancer (BC). T-cells are one of the most important cells in the tumor microenvironment and are divided into several subtypes including helper, cytotoxic, and regulatory subsets according to their transcription factors, markers, and functions. This article provides a comprehensive review of contradictory functions of various T-cell subsets in the prognosis and treatment of patients with BC, and crosstalk between tumor cells and T-cells. The literature shows that the role of T-cells in BC immunity depends on a variety of factors, including the tumor type or subtype, the stage of the disease, the localization of the cells in the tumor tissue and the presence of different cells or cytokines.

Abstract: Breast cancer (BC) is the most common cancer type in women and the second leading cause of death. Despite recent advances, the mortality rate of BC is still high, highlighting a need to develop new treatment strategies including the modulation of the immune system and immunotherapies. In this regard, understanding the complex function of the involved immune cells and their crosstalk with tumor cells is of great importance. T-cells are recognized as the most important cells in the tumor microenvironment and are divided into several subtypes including helper, cytotoxic, and regulatory T-cells according to their transcription factors, markers, and functions. This article attempts to provide a comprehensive review of the role of T-cell subsets in the prognosis and treatment of patients with BC, and crosstalk between tumor cells and T-cells. The literature overwhelmingly contains controversial findings mainly due to the plasticity of T-cell subsets within the inflammatory conditions and the use of different panels for their phenotyping. However, investigating the role of T-cells in BC immunity depends on a variety of factors including tumor types or subtypes, the stage of the disease, the localization of the cells in the tumor tissue and the presence of different cells or cytokines.

Keywords: breast cancer; immune system; T lymphocyte; helper subset; cytotoxic subset; regulatory subset

1. Introduction

Breast cancer (BC) is a clinically and histologically heterogeneous disease, consisting of different subtypes with various prognoses [1]. Its incidence, mortality, and survival rates vary among different ethnicities and populations and depend on genetic and environmental factors, lifestyle, and population structure [2]. Despite recent advances in BC treatment, its mortality rates continue to rise especially in developing countries, highlighting the need for developing new treatment strategies including targeting driver mutations and modulation of the immune system [3].

New insights into the cellular and molecular mechanisms involved in cancer development have revealed that immune cells play crucial roles in the development and growth of BC as well. With their dual functions, immune cells can produce a pro-tumorigenic inflammatory environment on one hand, and cause tumor rejection on the other. Since the effect of an immune response is largely determined by the type of stimulated immune response, understanding the crosstalk between tumor cells and the cells of the immune system is of great importance. In this regard, analyses of the total BC tumor tissue have detected a broad spectrum of genes related to immune responses, reflecting both innate and adaptive immunity including T-cell metagene (a set of related or co-regulated genes with similar expression patterns, functions, and regulatory elements) [4], B-cell metagene [5,6], or related signaling pathways [7]. Consistent with genetic studies, phenotyping investigations have also indicated an increase in leukocyte infiltration along with the tumor growth [8]. The composition of infiltrating lymphocytes is very heterogeneous and includes T-cells, both CD4^{+} and CD8^{+} subtypes, and to a lower extent, B-cells, macrophages, and NK cells [8,9]. Although there is no agreement on the specific immune cell subset, generally, an immune-enriched signature is considered an indicator of an active anti-tumor immune response and better clinical outcomes [10]. However, some studies reported the association of higher numbers of tumor-infiltrating lymphocytes (TILs) subpopulations with aggressive phenotypes of the disease [11,12]. However, this could be a primary consequence of the immune response to the tumor growth; it also indicates that the role of immune cells in the BC prognosis, progress, and/or response to treatment is still controversial. Besides different molecular subtypes of BC which have different behaviors [13], one of the most important reasons for the discrepancy and lack of coordination among the studies is that only one or a few major cell types (i.e., CD3^{+}) have been investigated regardless of their functional (stimulatory or inhibitory) subgroups.

The specific influences of individual immune subsets in the breast tumor microenvironment have been addressed in several studies. Their results generally indicate a mixture of activation and suppression in TILs and suggest that their position and prevalence within the tumor dictate responsiveness. In many of these studies, CD3^{+} T-cells, as critical regulators of adaptive immune responses, were reported as the major leukocyte population detected in the breast tumor tissue. Among various T-cell subsets, CD4^{+} T-cells (in some cases CD8^{+} lymphocytes) are more prevalent in the tumor microenvironment and peripheral blood [14–16]. In most cases, T-cells (both CD4^{+} and CD8^{+}) in the tumor microenvironment display an activated phenotype represented by an increased expression of activation markers (i.e., CD69, CD25, CD95, CD44, and HLA-DR) and concomitant decrease in naive T-cells markers (CD45RA and CCR7). The activation markers, however, do not necessarily mean that intratumoral T-cells are fully functional [17,18]. In line with this, recent molecular profiling studies demonstrate that infiltrating CD4^{+} T-cells are effector memory cells, containing all helper subpopulations including Th1, Th2, Th17, Tfh, and regulatory T-cells (Tregs) [19]. However, they represent a restricted repertoire of receptors [20], helper cytokines and chemokines, implying suboptimal activation levels [19]. The complexity of CD4^{+} T lymphocytes (presumably Th2 or Tregs or both) in conjunction with CD8^{+} T lymphocytes and CD68^{+} tumor-associated macrophages (TAMs) were also reported that could be predictive of overall survival (OS) and relapse-free survival (RFS) in node-positive human BC [19,21]. These results, in addition to our observations, indicated that although there are no significant differences in the major populations of lymphocytes (CD3^{+} T-cell, and its main subgroups, CD4^{+} helper and CD8^{+} cytotoxic, or B-cells), with the progress of the disease, functional subgroups of CD4^{+} and CD8^{+} T lymphocytes show significant variations [22,23]. This confirms that the investigation of general markers, like CD4 and CD8, alone does not fully represent their functional status. Therefore, a comprehensive effort is needed to investigate the main components of the adaptive immune system involved in tumor growth, both effectors and regulators, in BC patients with different pathological properties. In what follows, we reviewed the available literature and highlighted significant recent discoveries demonstrating the contradictory functions of various T-cell subsets as key regulators of immune responses during BC growth and progression (Figures 1 and 2).

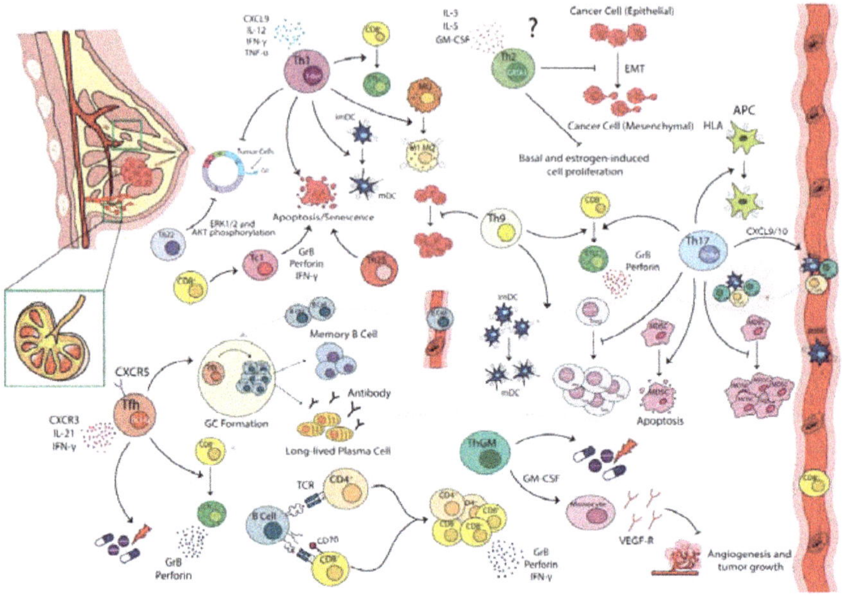

Figure 1. Anti-tumor functions of T-cell subsets in breast cancer. Through the secretion of inflammatory mediators, i.e., IFNγ and TNFα, **Th1** arrests the cell cycle in G1/G0 and induces apoptosis and senescence in the breast tumor cells. Besides direct effects, Th1 increases antigen presentation, proliferation and cytolytic function of CTLs, activates M1-macrophages and leads to maturation of DCs. Intriguingly, there are some reports that TSLP-stimulated **Th2** could reprogram tumor cells to terminal differentiation, and directly block carcinogenesis and EMT through secreting cytokines, i.e., IL-3, IL-5, and GM-CSF. It could also inhibit basal and estrogen-induced cell proliferation. **Th17** produces anti-tumor chemokines, attracts and stimulates NK and DCs, induces MHC-I and II expression, decreases proliferation and induces apoptosis in MDSCs. It also enhances CTL activity and crosstalk with Th1 but reciprocally regulates Tregs. **Th22** reduces tumor growth and promotes cell cycle arrest via the reduction of ERK1/2 and AKT phosphorylation. **Th25** induces apoptosis in tumor cells. Besides the direct inhibitory effect on tumor cells, **IL-9** and IL-9-producing T-cells promote tumor-specific T-cell responses, particularly CTLs, through the activation, survival and secretion of granzyme B, perforin and IFN-γ from T-cells. In addition, they recruit leukocytes including DCs into tumor tissues, increase their survival, and enhance antigen-presentation in the tumor draining lymph nodes. **ThGM** inhibits tumor progression and metastasis through its effects on monocytes and secretion of soluble VEGFR-1, which in turn inactivates the VEGF and blocks angiogenesis. In addition, GM-CSF overexpressing cells exert more sensitivity to anti-cancer drugs. **Tfh** is the main source of CXCL13 in breast tumors, directing B-cells and promoting lymphoid structure and GCs formation. It constitutes one of the important components of GCs in both draining lymph nodes and tertiary lymphoid structures (TLS) in tumor bed, where it provides help for production of immunoglobulins, and induces effector memory B-cells and plasma cells. It also regulates the activation and recruitment of CTLs. On the other side, B-cells could present or cross-present antigens to $CD4^+$ and $CD8^+$ T-cells. Engagement of CTLs with bystander B-cells through CD27/CD70 contact also leads to the proliferation and survival of $CD8^+$ T-cells. **Abbreviations:** APC: antigen-presenting cell, CTL: cytotoxic T-cells, DC: dendritic cells, EMT: epithelial–mesenchymal transition, GC: germinal center, NK: natural killer, MDSC: myeloid-derived suppressor cells, Th: T helper, Tfh: T follicular helper, Treg: regulatory T-cells, TSLP: thymic stromal lymphopoietin, VEGF-R: vascular endothelial growth factor receptor.

2. Th1/Th2 Paradigm in Breast Carcinoma

In tumor immunology, it is generally believed that Th1 cells are often critical components in anti-tumor immune responses due to their ability to produce IFN-γ, activate macrophages, and boost the killer CD8$^+$ T-cells, whereas type 2 helper responses can promote cancer development or metastasis [24]. Clinical evaluation of human BCs also showed the presence of Th2 lymphocytes accompanied by an increase in the frequency of Tregs during cancer development. In this regard, the ratio of Th2/Th1 cells in primary tumors as well as sentinel lymph nodes (SLNs), where Th2 cells are more frequent than Th1 cells, has been shown to be positively correlated with tumor stage, metastatic lymph nodes, larger tumor size, and reduced OS [17,25–27]. In addition, it was shown that the cellular immune responses, from dendritic cell (DC) maturation to Th1 responses, were also less active in SLNs than in non-SLNs in these patients before metastasis. It has been suggested that as the tumor grows, immunosuppressive secretions from the tumor are drained into the SLNs and change the immune responses in both the tumor site and draining LNs in favor of a reduced Th1/Th2 ratio. These results support the idea that changes in the immune profile of SLNs can provide a niche for tumor LN metastases, which in turn accelerates further tumor growth and spreading. After metastasis, DC maturation was found to be triggered and followed by the upregulation of Th1 responses, which could be a reflection of antigen-specific priming in SLNs; however, Th2 and Tregs responses upregulated in parallel [28]. Immunophenotyping analysis of the intracellular IFN-γ and IL-4 also indicated a shift toward the Th2 phenotype in whole blood. The secretion of IFN-γ was strongly impaired in BC, whereas the levels of TNF-α and IL-1β were comparable with those obtained from normal subjects [29]. In line with these findings, our results confirmed an increased accumulation of Th2 and Tregs in the breast tumor-draining lymph nodes (TDLNs) of patients with at least one involved lymph node (LN$^+$ patients). Conversely, a decrease in IFN-γ production was observed with tumor progression from stage I to stage III. The expression intensity of IFN-γ also showed a strong correlation with the frequency of IFN-γ-producing Th1 cells, indicating that the activity of the Th1 cells decreases along with the decline in their numbers [22]. Overall, these findings indicate that metastasis might be accelerated by inflammatory Th2 responses along with the reduction in Th1 responses. Similarly, Fracol et al. showed a reduction in anti-HER-3 IFN-γ immunity during breast tumorigenesis, highlighting the important role of the Th1 cells in surveilling tissues against overexpressed antigens. Tumor recurrence and incomplete response to neoadjuvant therapy also correlated with the suppression of Th1 immune responses, and accordingly, Th1 status could be introduced as a prognostic factor in patients with invasive BC [30].

Gene expression analysis further confirmed the results of the clinical studies. While a decrease in Th1-related genes in tumor tissues compared with the controls was observed in BC [31], unsupervised profiling of BC stroma demonstrated that a gene signature functionally enriched in Th1-type immune response elements is associated with favorable prognosis (>98%, 5-year survival) [32]. Consistently, an IFN-based gene signature has also been detected in triple-negative patients who are more likely to remain metastasis-free and independently predicted improved RFS and OS [33,34]. Similarly, the gene expression profiling of purified CD4$^+$ T-cells from primary tumors, axillary LNs, and peripheral blood of BC patients suggested that a Th1 signature including 12 genes, e.g., chemokine (C-X-C motif) ligand 9 (CXCL9) and IFN-γ, predicts better survival in the human epidermal growth factor receptor 2 (HER2$^+$) BC subtype [19]. These observations imply that a tumor environment enriched in type 1 cytokines would result in an increased antigen presentation, proliferation, and cytolytic function of cytotoxic T-cells (CTLs) on one hand, and PD-L1 expression, growth arrest in G_1/G_0, HER2 oncogene inactivation, on the other hand, and consequently the apoptosis and senescence of the breast tumor [19,35–38]. Consistently, the response to HER2-targeted therapies has been correlated with the presence of Th1 immunity, which increases MHC-I expression and promotes tumor cell recognition by CTLs and cytolysis [36,39]. Additionally, it has been shown that the presence of Th1-mediated immunity increases the efficiency of BC treatments [36]. Accordingly, the majority of immunotherapies aim to restore Th1 immunity and shift Th2 toward Th1 cell response [25,40].

In line with these findings from human tumors, studies on transgenic mouse models of aggressive mammary adenocarcinoma demonstrated that the elimination of endogenous T-cells significantly reduced pulmonary metastases with no effects on developing primary tumors. Further studies demonstrated that metastasis specifically depended on $CD4^+$ T-cells, as their absence reduced the metastasis rate. They found that $CD4^+$ T-cells expressing high levels of IL-10, IL-4, and IL-13 compared with those expressing IFNγ or IL-17 enhanced pulmonary metastasis indirectly through the induction of M2-macrophages. M2 macrophages increase the invasive behavior of malignant mammary epithelial cells by promoting intracellular signaling cascades such as the epidermal growth factor (EGF) pathway. This phenomenon was independent of cytotoxic T-cells, indicating that pro-tumor activity of $CD4^+$ T-cells did not involve the suppression of CTLs [41]. Spontaneous breast carcinomas also developed more quickly in HER2/neu transgenic mice when T-cells were depleted. However, it could be considered as evidence for slowing tumor growth through T-cell-mediated immunosurveillance; yet, the blockade of IL-13 further enhanced this effect, which confirmed a role for the type 2 immune response in promoting tumor growth [42]. Assessing the anti-tumor and biological activities of endogenous type 1 and type 2 effector T-cell subpopulations at the primary site of the mammary tumor also revealed differential infiltration kinetics of effector T-cell subpopulations at the tumor site along with a general delay in $CD3^+$ T-cells infiltration. Cytokine profiling showed elevated IL-4-producing Th2 cells earlier than IFN-γ-producing Th1 cells, which remained noticeably higher and was linked with tumor growth and metastases. It suggested that this initial increase in IL-4-producing Th2 cells might antagonize and/or postpone the emergence of a more favorable Th1 immune response in untreated animals. As a result, infiltrating lymphocytes would experience immunological ignorance and/or anergy, which would ultimately encourage the development and proliferation of the tumor [18]. As revealed by in vitro studies, tumor growth could be directly enhanced by IL-4 and IL-13 through the activation of their receptors on epithelial cells [43]. Consistently, intense IL-13 staining in breast TILs was observed in tumor cells which, along with the expression of phosphorylated signal transducer and activator of transcription-6 (pSTAT6), suggested that IL-13, in fact, delivers growth signals to cancer cells. It seems that BC educates DC cells in a manner to induce IL-13 secretion by Th2 lymphocytes, and facilitates their development [44]. IL-4 also promoted tumor cell survival by making them resistant to apoptosis [45]. It was also shown that BC cell-derived thymic stromal lymphopoietin (TSLP) fosters an inflammatory Th2 microenvironment by prompting OX40L expression on DCs. Antibodies neutralizing TSLP or OX40L inhibited IL-13 production and tumor growth in a xenograft model [46]. However, Boieri et al. proposed an opposite role for Th2 cells in BC, as they showed that Th2 cells stimulated by TSLP could reprogram tumor cells and induce their terminal differentiation. They could also directly block carcinogenesis and Epithelial-to-Mesenchymal Transition (EMT) through the secretion of cytokines such as IL-3, IL-5, and GM-CSF [47]. There are also reports on human breast carcinoma cell lines, particularly estrogen receptor (ER) α-expressing lines, showing that IL-4 and IL-13 could inhibit basal and estrogen-induced cell proliferation in vitro, and in xenograft models [48,49]. In addition, it was also shown that baseline IL-4/13 signaling is implicated in normal mammary gland development [50]. Epidemiological studies also implied less susceptibility to BC in patients with allergic diseases, a phenomenon believed to be mediated by inflammatory Th2 cells [51,52]. These findings suggest a more complicated role for Th2 cells in BC influenced by the threshold of activation, along with the complexity, plasticity, or the involvement of new subsets sharing effector molecules, i.e., ThGM or Th25 (discussed later), as well.

3. Th17 in Breast Carcinoma

Th17 cells constitute the third subset of effector T helper cells with a potent inflammatory nature, characterized by their distinctive cytokine, IL-17A [53]. Although the contribution of Th17 and its related molecules in infection and autoimmunity is well-documented, its role in tumor immunity remains elusive. The tumor-infiltrating Th17 cells are reported in various cancers such as BC with both pro- and anti-tumor properties. In most of these studies, IL17-producing T-cells have been associated with disease progression, worse prognosis, triple-negative molecular subtypes, shorter disease-free survival (DFS), and genes related to the tumors' proliferation and survival [54–56]. The pro-tumor properties of Th17 cells in BC (Figure 2) are attributed to multiple mechanisms, including the regulation of angiogenesis, induction of pro-invasive factors (i.e., IL-17, IL-22, and IL-23), metalloproteinases (MMPs) that promote proliferation, survival, and the invasion of malignant cells, interfering with $CD8^+$ T-cell migration through the activation of STAT3 signaling and subsequent reduction of CXCR3 expression [57–60]. The role of IL-17 in angiogenesis is enhanced through the induction of angiogenic factors such as MMP-2, MMP-9, vascular endothelial growth factor (VEGF), and CXCL8 [61]. IL-17 also supports inflammation in the tumor microenvironment indirectly by inducing tumor progression locus 2 (TPL2) [62].

The data also indicate that tumor cells, tumor-associated fibroblasts, and myeloid-derived suppressor cells (MDSCs) not only produce the chemokines mediating Th17 recruitment but also create a pro-inflammatory milieu and a cell–cell contact framework that enhance Th17 cell differentiation and expansion in the breast tumor microenvironment. Accordingly, its neutralization could lead to the reduction of tumor growth and migration of tumor cells to the secondary tumor site [54,61,63–65]. Direct interaction of CD40 on MDA-MB231 cells with CD40L on T-cells was shown to upregulate Transforming Growth Factor-beta (TGF-β), induce Th17 differentiation, and increase the proliferation of tumor cells via STAT-3 signaling [66]. It was also shown that tumor cells through secretion of some mediators, i.e., Prostaglandin E2 (PGE2) and CXCL1/5, led to Th17 expansion and CXCL1 production, and subsequently BC growth and development [63,64]. On the other hand, the IL-17 response was found to promote recruitment, suppressive functions, and the release of anti-apoptotic factors from MDSCs [61,67]. In contrast, an anti-tumor role for Th17 cells and their related molecules has been suggested by independent groups, including us, through the observation of negative correlations of the frequency of Th17 cells with the size and stage of BC [22,68] or longer survival [69]. In this regard, we observed that the mean expression of IL-17 in Th17 cells declined significantly in node-positive patients, and in those in higher stages [22]. Yang et al. also demonstrated an association between the frequency of Th17 lymphocytes and a more favorable prognosis in BC. Nearly all cancerous tissue specimens showed an increased infiltration of Th17 cells compared with the normal breast tissues. A negative association was also observed between the prevalence of Th17 cells and TNM-stage, the invasion of the blood vessel, and higher numbers of metastatic LNs [68]. A bioinformatics analysis also indicated an association between the Th17 metagene and good prognosis and longer survival in Triple-negative BC (TNBC) patients [69]. In another study, it was shown that the accumulation of both Th17 and Tregs in the breast tumor microenvironment occurred at the early stage of the disease; however, with tumor progression, Th17 infiltration gradually decreased, while Treg infiltration increased and resulted in Treg dominance in advanced stages of BC [70]. In addition, the frequency of Th17 in peripheral blood was lower in $HER2^+$ patients compared with the healthy ones, while it increased following a treatment with Trastuzumab. It also had an inverse correlation with Tregs in metastatic patients [71]. The anti-tumor activities of Th17 could be attributed to multiple mechanisms (Figure 1), indirectly through fighting bacterial and viral infections, known to play a significant role in the pathogenesis of many cancers, or directly through the regulation of Tregs in a reciprocal manner, the induction of MHC-I and II expression, the enhancement of CTL activity, and crosstalk with Th1 [57,72]. IL-17 produced by Th17 cells was shown to synergize with IFN-γ in the induction of anti-tumor chemokines CXCL9 and CXCL10 that attract and stimulate NK, DC, and T-cell responses [73]. It was also revealed that a low level of IL-17 could lead to MDSC differentiation in vitro, yet, IL-17 could decrease cell proliferation and induce apoptosis in MDSCs [74]. In summary, Th17 cells play a complex role in BC as well, with both pro-tumor and anti-tumor effects. While their pro-tumor properties are primarily associated with the upregulation of angiogenic factors or the secretion of mediators by the other cells leading to Th17 expansion, the anti-tumor properties could result from Th17 interactions with effector immune cells such as Th1 and CTLs.

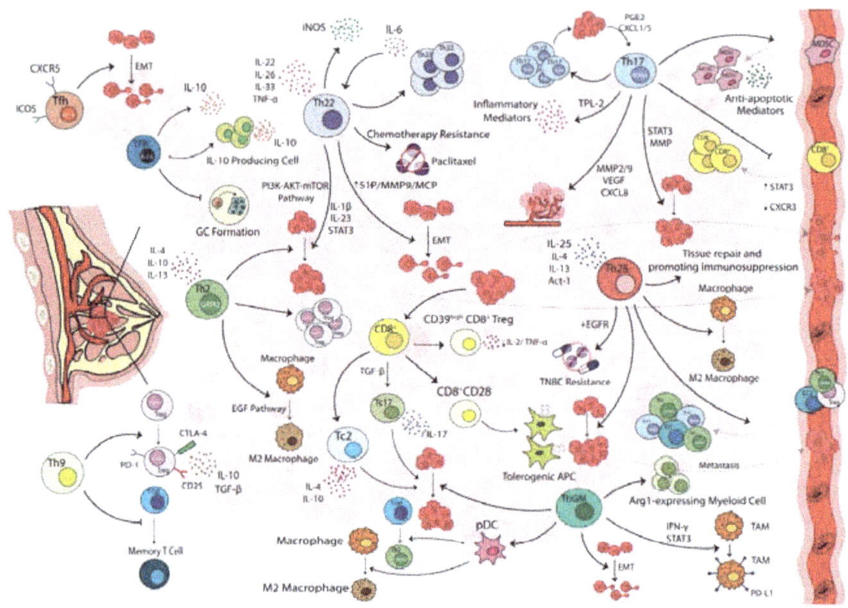

Figure 2. Pro-tumor functions of T-cell subsets in breast cancer. Th2 increases the frequency of Tregs and induces M2-macrophage differentiation and subsequently promotes tumor survival, growth, and metastases. **Th17** regulates angiogenesis, induces pro-invasive factors (i.e., IL-17, IL-22, and IL-23, MMPs) and promotes the proliferation, survival, and invasion of malignant cells. It could also interfere with CTLs migration through the activation of STAT3 signaling, and reduction of CXCR3 expression. IL-17 supports inflammation in the tumor microenvironment indirectly by inducing TPL2, and promotes the recruitment, suppressive functions, and release of anti-apoptotic factors from MDSCs. **Th22** enhances tumor cells migration by activating the STAT3/MAPKs/AKT pathway. It also stimulates S1P production in MSCs and promotes the chemotactic migration of breast tumor cells and their metastasis to bone marrow. IL-22 accelerates this process by increasing MCP-1 and MMP-9 in MSCs. In addition, following the upregulation of IL-22, the PI3K-AKT-mTOR pathway is activated and increases the growth, migration, and invasion of the tumor cells. It also renders them resistant to anti-mitotic and anti-EGFR therapeutic agents. **Th25** promotes immunosuppression through the recruitment and activation of type 2 immune cells (Th2 and M2-macrophages). **ThGM** increases the expression of PD-L1 on TAMs via an IFN-γ/STAT3-dependent pathway and induces an EMT/stemness-like invasive phenotype or accumulates Arginase 1 expressing myeloid cells. It also induces pDC, which in turn deviates naive $CD4^+$ T-cells and macrophages into a pro-tumorgenic Th2 and pro-invasive TAM-like phenotype, respectively. **Th9** enhances the immunosuppressive function of Tregs, prevents immunological memory formation, and promotes the development of hematological malignancies. **Tfh** plays a role in the regulation of EMT during lymph node metastasis. **TFR** controls Tfh and GC responses and prevents antibody production. It also increases IL-10 directly or indirectly via the differentiation of IL-10-producing B-cells. **$CD8^+$ T-cells** might differentiate into pro-tumor and regulatory cells ($CD39^{high}CD8^+$ Tregs, $CD8^+CD28^-$ T-cells, Tc2, and Tc17 cells) and promote tumor growth. **Abbreviations:** Arg1: arginase 1, APC: antigen-presenting cell, CTL: cytotoxic T-cells, DC: dendritic cells, EGF: epidermal growth factor, EGFR: epidermal growth factor

receptor EMT: epithelial–mesenchymal transition, iNOS: inducible nitric oxide synthase, GC: germinal center, NK: natural killer, MDSC: myeloid-derived suppressor cells, MMP: matrix metalloproteinases, MCP: monocyte chemoattractant protein, pDC: plasmacytoid dendritic cells, PGE2: prostaglandin E2, S1P: sphingosine-1-phosphate, Th: T helper, Tfh: T follicular helper, TFR: T follicular regulatory, Treg: regulatory T-cells, TPL2: tumor progression locus 2, TSLP: thymic stromal lymphopoietin, TAMs: tumor-associated macrophages, TNBC: triple-negative breast cancer, VEGF-R: vascular endothelial growth factor receptor.

4. IL-25-Producing T-Cells

IL-25-producing cells were recently proposed as a potentiate subset of helper T-cells, Th25, identified by the production of Act1, IL-4, IL-13, and IL-25 (also known as IL-17E). This subset and its main cytokine, IL-25, is commonly involved in response to extracellular pathogens and plays a major role in autoimmune diseases and allergic inflammation [75–77]. It is also suggested that Th25 is involved in the processes promoting immunosuppression or tissue repair along with Th2 and M2-macrophages [78]. Nevertheless, defining the role of Th25 cells in cancer is still in the early stages, but some studies have suggested that IL-25, its key cytokine, plays a role in modulating the immune responses to tumors, including BC [79]. IL-25 is a member of the IL-17 family with a unique structure and function. In this connection, it was demonstrated that IL-25, similar to IL-17A, could accelerate the proliferation and survival of tumor cells through the same oncogenic signaling pathways [80]. IL-25 was also shown to have a shared pathway with the epidermal growth factor receptor (EGFR) and rendered TNBC resistant to anti-mitotic and anti-EGFR therapeutic agents [69,80,81]. Bioinformatics analyses also showed similar results in the luminal subtype where high expression of ER was associated with higher expression of IL-25, which in turn suppressed IL-17 signaling and recruited Th17 cells [82]. Similarly, in a transgenic mouse model of spontaneous BC, it was shown that IL-25 promoted the recruitment of type 2 immune cells and facilitated lung metastasis, whereas blocking IL-25 remarkably decreased the type 2 response (Th2, M2-macrophages, and IL-10) in the tumor microenvironment but significantly increased the expression of IL-12 and the activity of CTLs to kill tumor cells [83]. However, it should be noted that in this study and other similar investigations, the function of Th25 cells might be mistakenly misinterpreted as that of Th2 cells due their shared transcription factors, cytokine requirements for development, and similar functions. Reduced tumorigenicity was also observed when the IL-25 receptor, and IL-25RB, were deleted in the drug-resistant BCs [84]. Concordantly, the expression of IL-25 and its receptor was significantly higher in breast tumors compared with the normal samples while it was undetectable in most normal breast tissues [80,83]. However, some studies reported this expression just on tumor-infiltrating CD4$^+$ T-cells, and infrequently on tumor cells [83], or could not find IL-25 in the T-cells isolated from the tumoral and adjacent tissues of the patients with different molecular subtypes [69]. Differential expression of IL-25R was also observed in peripheral blood where it was highly expressed in malignant patients and showed direct associations with poor prognoses, i.e., higher grade/stage tumors, and decreased survival [84,85]; however, the serum level as well as mRNA expression of IL-25 in peripheral blood mononuclear cells was remarkably higher in the healthy controls than in the malignant cases [85].

On the other hand, there are some studies introducing IL-25 as one of the endogenous factors secreted by non-malignant mammary epithelial cells or tumor-associated fibroblasts conferring high cytotoxic activity on BC cells without affecting non-malignant mammary cells [86–88]. It was proposed that IL-25 differentially induces caspase-mediated apoptosis through the differential expression of its receptor, IL-25R, on poor prognoses tumors rather than on non-malignant cells [84,86–88]. In line with these findings, a combination of rIL-25 and IL-17B silencing (siIL-17B) or a single chain against IL-25R could provide a strong inhibition in cancer progression along with decreased VEGF expression, reduced cell viability, and stimulated apoptosis in BC cells [88,89]. Collectively, Th25 seems to be mostly involved in tumor progression through the secretion of IL-25, which leads to the promotion of type 2 immune responses. However, limited studies also suggest an anti-turmeric role for IL-25 by induction of apoptosis in tumors which requires further investigations.

5. IL-22-Producing T-Cells

A subset of T-cells, namely Th22, was introduced which mainly produces IL-22 but not any of the other subsets' related-cytokines, i.e., IFN-γ (Th1), IL-4 (Th2), and IL-17 (Th17) cytokines. These cells produce multiple cytokines including IL-22, IL-26, IL-33, and TNF-α; however, they exert their function mainly through the secretion of IL-22. IL-22 is a member of the IL-10 family, and its expression is stimulated by inflammatory cytokines, i.e., IL-1β, IL-6, IL-21, and IL-23. It binds to IL-22R (a heterodimer of IL-22R1 and IL-10R2) which is selectively expressed by non-immune cells. Besides the production of anti-microbial peptides (β-defensin-2 and β-defensin-3, and peptides of the S100 family), this cytokine induces the proliferation and differentiation of epithelial cells, which contributes to the pathogenesis of some autoimmune diseases and cancers [90–92]. Th22 and IL-22 were studied more in gastro-intestinal cancers in which, besides higher frequency of Th22 and IL-22 expression, IL-22 directly promotes the proliferation of colonic epithelial cells and increases their production of nitric oxide synthase and subsequently increases colonic inflammation and carcinogenesis [93]. In BC, however, there are limited studies; Wang et al. demonstrated that the frequency of Th22 increases in tumor tissues compared with para-tumoral and normal breast tissues. Furthermore, in vitro analysis on TNBC cell lines (MDA-MB-231 and MDA-MB-468) also showed that through the activation of STAT3/MAPKs/AKT pathway, IL-22 enhances tumor cell migration and their resistance to paclitaxel chemotherapy in a dose-dependent manner [94]. Our observation also revealed, on average, more than 2% of CD4$^+$ lymphocytes in TDLNs of patients with BC produced IL-22; however, the majority of them simultaneously expressed IFN-γ and to a lesser extent, IL-17. The percentage of the Th22 subset (CD4$^+$IL-22$^+$IFN-γ^-IL-17$^-$) was extremely low (about 0.6%). We observed that the mean percentage of IL-22-producing CD4$^+$ cells significantly increased in patients with higher stages and in those with higher numbers of involved nodes. A non-significant increase in the frequency of Th22 lymphocytes was also found in patients with late stages and higher involved nodes [95]. An increased level of IL-22 was also observed in the sera of patients with BC [96]. In tumor tissue, it was observed that IL-22 and IL-22R1 were mostly expressed in tumor cells and to a lesser extent in stromal cells. The IL-22 expression in tumors was a poor prognostic factor for OS, and along with tumor-IL-22R1, was positively associated with the infiltration of CD68-positive TAM, which together displayed the worst prognosis outcomes regarding both OS and RFS [97]. These findings collectively suggest a pro-tumorigenic function for IL-22-producing cells in patients with BC, a concept further supported by in vitro and animal model studies.

Similarly, in the 4T1 mouse model (a model similar to metastatic triple-negative BCs), the level of IL-22 mRNA showed an increase in tumor tissues compared with normal mammary tissues. In addition, exogenous IL-22 increased the proliferation of BC cells in a STAT3/IL-1β/IL-23-dependent manner. Thus, blocking IL-22 activity may lessen the progression of tumors induced by IL-1 and IL-23 [98]. Consistently, Rasé et al. demonstrated that Th22 along with Th17 were detectable very early in the mass and remained until the final day. In the peripherals, when the tumors were well-established, the frequency of Th22 significantly increased over time. Addressing the mechanism underlying Th22 cell recruitment, using 4T1-IL-6-KO mouse mammary carcinoma, they observed that in the absence of IL-6, total CD4$^+$ helper cells including Th17 significantly expanded in the tumors, whereas Th22 and MDSC frequencies reduced in all tissues. Accordingly, they concluded that IL-6 might facilitate tumor growth and confer immunotherapy resistance through tumor cell polarization and expansion of the Th22 cell population [99]. Another investigation revealed an increased IL-22 level in the higher stage of breast carcinoma, and its depletion limited tumor invasion and progression and decreased tumor burden. IL-22 deletion was associated with the reduced expression of the transcription factors involved in the EMT [100]. IL-22 also stimulates sphingosine-1-phosphate (S1P) production in mesenchymal stem cells (MSCs) and promotes the chemotactic migration of breast tumor cells toward MSCs, thereby enhancing BC metastasis to bone marrow. Furthermore, IL-22 accelerates this process by increasing the expression of monocyte chemoattractant protein 1 (MCP-1) and MMP-9 activity in MSCs [101]. In addition, it was observed that following the upregulation of IL-22 and HOXB-AS5 (a long non-coding RNA located in *HOX* gene clusters), the PI3K-AKT-mTOR pathway was activated and increased the growth, migration, and invasion of the MDA-MB-231 BC cell-line [102]. Voigt et al. proposed another mechanism in which breast tumor cells induce IL-22 production by memory $CD4^+$ T-cells in an IL-1-dependent manner. The IL-1 (β in humans, α in mice) activated the NLRP3 Inflammasome in the tumor microenvironment and induced IL-22 production in various $CD4^+$ T-cells including Th22, Th17, and Th1 cells [103]. However, Weber et al. reported a protective role for IL-22 in the BC mouse model as they observed that the treatment of EMT6 cells with IL-22 not only did not induce angiogenesis and apoptosis but also reduced tumor growth and promoted cell cycle arrest through the reduction of ERK1/2 and AKT phosphorylation [104]. Despite this, most studies, including ours, suggest a pro-tumorigenic role for Th22/IL-22 in BC (Figure 2).

6. IL-9-Producing T-Cells

IL-9 and IL-9-producing T-cells (termed Th/Tc9), are mainly introduced as pro-inflammatory mediators involved in the pathogenesis of a variety of autoimmune diseases and allergic inflammations. However, it has also been reported that IL-9 can provide a tolerogenic environment, highlighting IL-9 as a pleiotropic cytokine with both positive and negative effects on immune responses [105]. Nevertheless, how IL-9 and Th/Tc9 cells contribute to the immune responses to cancer remains to be elucidated.

Controversial data were reported about the role of Th9/Tc9 in various types of cancer. On the one hand, IL-9 enhances the immunosuppressive function of Tregs, prevents immunological memory formation, and promotes the development of hematological malignancies via its effect as a growth factor [105–107]. On the other hand, some studies support the protective role of Th9/Tc9, even superior to Th1/Tc1 subsets, especially in solid tumors like BC [105]. Besides the direct inhibitory effect of IL-9 on tumor cells [108,109], it was shown that IL-9 and IL-9-producing T-cells provide a unique inflammatory environment in tumor tissues, which promotes tumor-specific T-cell responses, particularly CTLs. In addition, it recruits leukocytes, i.e., DCs, into tumor tissues, increases their survival, and enhances antigen-presentation in draining lymph nodes [105,110].

Limited controversial data are available in BC, as well. Through a longitudinal investigation of soluble factors in the sera of BC patients, Carlsson et al. found that patients with metastatic lesions had higher amounts of IL-9 in their serum over time. This finding suggested a relationship between IL-9 and tumor progression, or tumor load [111]. A higher IL-9 level was also observed in both patients' sera and in the circulating $CD4^+$ T-cells than in those of the healthy ones. IL-9-expressing Th9 cells were more abundant in the $CCR4^-CCR6^-CXCR3^-$ subset and produced an elevated level of IL-10 and IL-21 following activation. These Th9 cells could mediate higher cytotoxicity in $CD8^+$ T-cells via IL-9 and IL-21 expression, and IL-9 blocking led to a significant decrease in $CD8^+$ T-cells' cytotoxicity. Accordingly, it could be concluded that Th9 exerts anti-tumor activity in BC at least indirectly by promoting $CD8^+$ T-cell inflammation [112]. In another study, IL-9-producing $CD8^+$ T-cells were present in some resected BC tumors while IL-9R was almost present on $CD8^+$ T-cells in all patients, yet in different degrees. Although there was no association between the expression of IL-9 and its receptor, IL-9R, and the clinical characteristics, IL-9 production in $CD8^+$ T-cells from patients with BC was higher than in the healthy controls with more expression in IL-$9R^{Hi}CD8^+$ T-cells. The IL-$9R^{Hi}CD8^+$ T-cell subset also exhibited lower expression of inhibitory molecules, i.e., KLRG-1, PD-1, and Tim-3, on one hand, but higher inflammatory mediators, i.e., IL-2 and IL-17, and lower IFN-γ following the activation compared with the IL-$9R^{Low}$ subset. Together, these data indicated an inflammatory response for IL-9-producing CD8$^+$ T-cells in BC, and the low expression of PD-1 on IL-9R^{Hi} CD8 $^+$ T-cells may indicate that these cells are more resistant to inhibitory mechanisms in the tumor microenvironment, enabling them to perform effector functions more effectively [113]. Evaluating the role of IL-9 on cancer development in a mouse model of BC (HER2$^+$ and TN models) also demonstrated that IL-9 deficiency through neutralizing antibodies or deleting endogenous IL-9 led to priming host tumor-specific T-cells, acquiring immunologic memory, and the early rejection of BC [114]. These results represent IL-9 as an inhibitor of adaptive responses, so blocking IL-9 could be proposed as a therapy that limits tumor growth; however, the exact role of these cells in immune responses to breast tumors and their trafficking pattern need further in vivo and in vitro studies (Figure 2).

7. GM-CSF-Producing T-Cells

Another distinct subset of T-cells, named ThGM, has been introduced, which mainly produces GM-CSF but not IFNγ and IL-4 [115]. However, not limited to helper cells, this subset consists of up to 2% of all helper T-cells and is characterized by the expression of various chemokines and chemokine receptors such as CCR10$^+$, CCR4$^+$, CCR6$^+$, and CXCR3 [116]. It has been shown that GM-CSF-producing T-cells play a role in the pathogenesis of various inflammatory and autoimmune diseases [116,117], yet the frequency and function of GM-CSF-producing lymphocytes have not been widely studied in cancers [118]. Nevertheless, GM-CSF is well-known for its immune-modulatory functions as it can exert both suppressive and stimulatory effects on tumor cells [119]. Limited studies on BC also show both pro- and anti-tumorigenic roles for this cytokine or its corresponding T-cell subsets (Figures 1 and 2) [118,120,121]. Accordingly, we observed lower GM-CSF plasma levels in the patients with higher stages of BC than in those with a lower stage of the disease, which could be attributed to GM-CSF suppression following tumor growth and progression [122]. Concordantly, it was shown that in the murine model of BC, GM-CSF inhibited tumor progression and metastasis through their effects on monocytes and the secretion of soluble VEGFR-1, which in turn inactivates the VEGF and blocks angiogenesis [121]. In addition, GM-CSF overexpressing MCF-7 cells exerted more sensitivity to anti-cancer drugs in Chaubey's study [120].

On the other hand, it is also proposed that higher GM-CSF in the serum is related to the BC metastasis and increased production of GM-CSF in cancer patients (probably due to inflammatory milieu, i.e., in response to TNF-α and LPS), supports angiogenesis, and promotes tumor growth and progression [123,124]. GM-CSF from activated lymphocytes was also observed to increase the expression of PD-L1 on TAMs via an IFN-γ/STAT3 dependent pathway, and provided an immunosuppression [125]. Multiple reports also showed that primary breast tumor cells or mesenchymal-like BCs could aberrantly produce GM-CSF [126,127]. This BC-derived GM-CSF seems to play a pro-tumorigenic role and as an endogenous signal, develops an immunosuppressive microenvironment by inducing plasmacytoid DCs (pDCs)'s activation, which in turn deviates naive CD4$^+$ T-cells and macrophages into a regulatory Th2 response and pro-invasive TAM-like phenotype, respectively [128], inducing an EMT/stemness-like invasive phenotype [129] or accumulating Arginase 1 expressing myeloid cells [127]. These conditions were also associated with the more aggressive BC subtypes, metastasis, and reduced survival [126,127,129,130]. In line with these, GM-CSF neutralization significantly reduced tumor growth and metastases probably in part due to the modulation of the tumor microenvironment, the reduction of angiogenesis, and the immunosuppressive cells within the tumor [119]. These discrepancies propose that GM-CSF effects might be source-, dose-, and context-dependent, and further studies are required to clarify the mechanisms by which GM-CSF affects breast tumors. Despite these studies, most observations, in addition to the immunostimulatory effects of GM-CSF in immunotherapies alone or in combination with chemotherapy [131,132], suggest a protective role for this cytokine in the BC tumor setting or its therapeutic potential in the treatment of BC (Figure 1).

8. Helper and Regulatory Follicular T-Cells in Breast Carcinoma

Follicular helper T-cells (Tfh) are one of the most prevalent and significant populations of effector T-cells with specialized functions in lymphoid tissues. The expression of CXCR5 enables them to localize within B-cell follicles, where they provide help to B-cells. They also produce IL-21, a potent stimulator of the B-cell differentiation into antibody-forming cells. Tfh cells also express several markers, including the CXCL13 chemokine, the PD-1 and inducible co-stimulator (ICOS) co-stimulatory/inhibitory molecules, and the Bcl6 transcription factor [133,134]. The dysregulated function of Tfh or its related molecules, i.e., ICOS or IL-21, has been reported in certain autoimmune diseases or immunodeficiencies [133]. Few studies have considered the role of these cells in cancer as the origin of follicular lymphoma or other hematologic cancers, as well [135]. However, except for a few recent investigations studying Tfh cells in breast carcinoma, the role of these cells in the immunity against solid tumors has been poorly demonstrated.

In a comprehensive study on infiltrating $CD4^+$ T-cells in untreated invasive primary breast tumors, the Gallo group reported, for the first time, the presence of Tfh cells among infiltrating lymphocytes in BC. Most of these CXCL13-producing Tfh cells were in the germinal centers (GCs) of tertiary lymphoid structures (TLS), predominantly located adjacent to the tumor bed, and constituted one of the important components of these structures in the breast tumors [19]. However, through ex vivo functional studies, they later showed signals initiating from ICOS/ICOSL interactions led to Tfh differentiation and migration to the B-cell follicle. It was observed that just a subset of Th1-oriented Tfh with the activated phenotype (PD-1^{hi}ICOS^{int}) was functional, could produce IFN-γ, provided help for the production of immunoglobulin, induced effector memory B-cells, and was correlated with better disease outcomes [136]. A similar subset was also observed in TDLNs of patients with BC in our previous study with the mean frequency of 2.19 ± 1.59, which showed a positive relationship with both B-cells and Th2 cells, highlighting the collaboration of Tfh cells in configuring antibody responses [22].

Further analysis revealed the association of a Tfh-related gene signature composed of eight genes including *CXCL13* with a greater chance of responding to preoperative chemotherapy and improved survival in patients who have not received treatment [19]. Immunohistochemistry staining also confirmed a correlation between CXCL13 expression and the degree of immune infiltration [19]. The activation of the CXCL13–CXCR5 chemokine axis was previously shown in BC [137]. Although CXCL13 overexpression in BC tissues and increased serum level of this chemokine in patients with metastatic disease suggested a role for CXCL13 in the BC progression [138], the activation of this axis was negatively associated with determinants of a poor prognosis, including axillary node involvement and high histological grade and improved outcome in HER2 overexpressing BC [137]. In addition, the positive correlation of CXCL13 with mesenchymal markers (Vimentin, N-cadherin, Snail, Slug, and MMP9) and negative correlation with E-cadherin implied that the CXCL13–CXCR5 axis could have a role in the regulation of EMT during lymph node metastasis [138,139].

In a complementary study, it has been further demonstrated that the main source of CXCL13 in human BC was $CD4^+$ TILs (and to a lesser degree, $CD8^+$ TILs), but not follicular DCs. It was proposed that after consuming IL-2 by Tregs and their proliferation at the tumor site, the IL-2 level for effector TILs decreased. Under this condition, some activated $CD4^+$ TILs upregulated CXCL13 and were differentiated to CXCL13-producing $CXCR5^-$ Tfh (TfhX13). It has been shown that suppressive cytokines (e.g., TGF-β) of Tregs had no effects on the accumulation of these cells. Thus, the frequency of TfhX13 gradually increased at the tumor site, which in turn, directed B-TILs' migration and subsequently promoted lymphoid structure and GCs' formation. These cells also regulated the activation and recruitment of $CD8^+$ lymphocytes. Since long-lived B-cells and plasma cells are generated in the GCs and play a crucial role in the elimination of tumor residues, the balance between TfhX13 and Treg-TILs seems to be an important issue. In fact, TfhX13 differentiation might be a feedback response that participates in the second round of humoral and cell-mediated immune responses to overcome Treg-mediated immune suppression. However, TfhX13 cells intriguingly expressed considerable levels of PD-1, ICOS, and an intermediate level of intracellular cytotoxic T-lymphocyte-associated protein 4 (iCTLA-4), and accordingly can be considered as an important target for the modulation of the immune system by checkpoint inhibition in BC [136,140]. The higher expression of PD-1 and TIM-3 on Tfh cells was also observed in the peripheral blood of patients with BC. It has been shown that TIM-3$^+$ Tfh cells expressed higher levels of PD-1 and were considered to possess an exhausted phenotype as they had lower proliferation, and CXCL13 and IL-21 production. Moreover, B-cells cocultured with these TIM-3$^+$ Tfh cells produced less IgM, IgG, and IgA [141].

Collectively, it can be concluded that a combined activated/suppressed TILs profile with higher TLS and GC along with increased Tfh cells in extensively infiltrated BCs suggests that patients with an organized immune response to their tumors, specifically those with detectable Tfh signature, are more likely to respond better to preoperative chemotherapy or have improved postoperative DFS [19].

Besides the specialized T subset with helper function in the follicles, a novel subgroup of Foxp3$^+$ regulatory T-cells has been reported recently, which similar to Tfh cells, expresses Bcl6 and enters the follicles by upregulating CXCR5. These cells, designated as follicular regulatory T-cells or TFRs, along with other regulatory cells, seem to play a crucial role in regulating Tfh cells and GC responses and preventing auto-antibody production [142]. The presence and role of this subset in tumor immunity, especially humoral immune responses, have been rarely studied [143,144]. We previously reported a subgroup of CD4$^+$Bcl6$^+$CXCR5$^{int/hi}$, which were positive for the Foxp3 marker, in TDLNs of patients with BC. This subgroup comprises about one percent of helper cells; however, no significant changes were observed in the frequency of these cells following tumor cells' infiltration to lymph nodes or progression of the disease [22]. No study has been done on the origin and differentiation of TFR cells in BC, though in a study on human follicular lymphoma, it has been suggested that mesenchymal stromal cells support the viability of TFRs and also mediate differentiation of Tfh to TFR cells through the upregulation of Foxp3 in Tfh cells [145]. These cells with a demethylated *FOXP3* gene could suppress Tfh functionality in a glycoprotein A repetitions predominant (GARP)-associated TGF-β-dependent manner [136]. These cells also showed increased frequency in the peripheral blood of BC patients and functionally increased IL-10 by directly producing IL-10 and indirectly by stimulating the differentiation of IL-10-producing B-cells [146]. A study on Bcl6FC mice, which has a specified deletion in the *BCL6* gene in Foxp3$^+$ T-cells and is commonly used as a mouse model for studying TFR, concordantly revealed that TFR cells promote B-cell growth and entry into the dark zone of the GCs through IL-10 production [147]. These findings, collectively, suggest an essential role for TFR cells and their relative balance with Tfh in regulating GC-dependent antibody responses.

9. Cytotoxic T-Cells and Their Effector Subsets

Classically, CTLs have been considered the key component of effective anti-tumor immunity [148]. In BC, according to the reports, CD8$^+$ lymphocytes are among the most frequent infiltrating subpopulations in the tumor, and those tumors with high infiltration of CD8$^+$ T-cells had a better prognosis and survival [148,149]. The gene expression profile of BC stroma also showed that the gene signatures related to CTLs and NK cell activities are predictive of a good outcome with a 5-year survival rate of more than 98% [32]. Concordantly, the frequency of tumor-infiltrating CD8$^+$ T-cells showed a negative association with advanced stages and metastasis but a positive correlation with RFS and OS [149,150].

It was also shown that the majority of infiltrating CTLs in tumor beds exhibited effector (CD8$^+$CD28$^+$), memory (CD45RO$^+$), or activated phenotype (CD25$^+$, CD69$^+$, HLA-DR$^+$) and the naive subset (expressing CD45RA and CCR7) compromised the least proportion [148,151,152]. The frequency of CD8$^+$CD28$^+$CD25$^+$ effector cells was found to be negatively correlated with the prevalence of distant metastasis in infiltrative ductal carcinoma of the breast [148]. However, there are also some reports showing the association of CD8$^+$ T-cells with survival rate just in some tumor subtypes [153,154], or indicating no obvious association [155] or even an association with poor outcome [26,156].

The discrepancies in the results may be due to the activation status as well as the heterogeneity in the expression profiles of these cells in differential states of anti-tumor immune responses, since similar to CD4$^+$ helper lymphocytes, CTLs are also classified into different subsets based on their transcription factors, cytokine profiles, and effector functions, which could dramatically affect the outcome of the host–tumor interaction [157]. In this regard, we observed that the mean expression of IFN-γ in Tc1 lymphocytes in TDLNs was lower in patients with positive nodes and late stages, whereas the frequencies of Tc2 and Tc17 were higher in advanced stages of BC. These differences were more pronounced in patients whose tumor type was infiltrative ductal carcinoma. Higher frequencies of Tc2 and Tc17 were also observed in patients with more involved nodes (N3) compared with the node-negative ones [23]. In this regard, it was shown that in the breast tumor microenvironment, CD8$^+$ T-cells are induced to produce IL-17 (Tc17) by TGF-β, which in turn enhances tumor growth directly in tumor-bearing mice and suppresses apoptosis. Consistently, knocking down the receptor of IL-17 increased apoptosis and reduced tumor growth [158]. Albeit later, Tc2 effector subpopulation was also demonstrated to be localized in the tumor site along with Th2 with higher IL-4 and IL-10 production which seemed to be correlated to tumor growth and metastasis in untreated mice [18].

Although there are many reports that CTLs in the tumor microenvironment had an activated phenotype, there is no guarantee that intratumoral T-cells are fully functional [17]. In this connection, the proliferation capacity and IFN-γ production of intratumoral CD8$^+$ T-cells were observed to be reduced in response to T-cell receptor stimulation in higher stages [159]. The anti-tumor activity of CTLs was also found to be impaired because of the significant loss of the CD3ζ chain and the CD3-complex dysfunction in the early stage of BC [160]. In addition, T-cell exhaustion could occur due to repetitive and chronic exposure of CTLs to the antigens in the tumor microenvironment, which represents the loss of cytokine production, proliferation, and/or cytolytic activity [161]. In this regard, we observed the higher frequency of TIM-3$^+$CD8$^+$ cells in TDLNs of patients with high-grade BC. The expression level of TIM-3 in CD8$^+$ and CD4$^+$ T-cells was also higher in patients with more involved lymph nodes [162]. Similarly, the co-expression of PD-1 and LAG-3 was demonstrated to be associated with exhaustion in CD8$^+$ T-cells in patients with TNBC [163]. However, cytokine production, degranulation, and cytotoxic capacity of PD1-expressing CD8$^+$ T-cells were reported to be maintained in BC, whereas the same population showed reduced function and exhaustion in melanoma [151,164]. In addition, it was revealed that the tumor microenvironment changes the anti-tumor function of CD8$^+$ T-cells through the induction of the CD39 expression (an immunosuppressive ATP ectonucleotidase) and the reduction of effector markers. These CD39^{high}CD8$^+$ Tregs were only detected in tumor tissues and metastatic lymph nodes but not in the peripheral blood. Upregulation of CD39 was accompanied by the expression of inhibitory receptors and reduced IL-2 and TNF secretion, and tumor growth, indicating CTLs' exhaustion [165]. Controversially, Tallón de Lara et al. introduced CD39$^+$PD-1$^+$CD8$^+$ T-cells, but not all CD8$^+$ T-cells, as a protective population that correlated with DFS [166]. In addition, a population of CD8$^+$ T-cells with the suppressive phenotype (CD8$^+$CD28$^-$) was also detected in the tumor microenvironment [148]. These senescent cells may exert an immunosuppressive function partly through the induction of tolerogenic antigen-presenting cells (APC) [167]. These findings further confirm the considerable heterogeneity of CD8$^+$ T-cells in the breast tumor microenvironment and emphasize that several parameters should be considered in the study of CD8$^+$ T-cells, including the type and even subtype of cancers, and the expression of immune checkpoint inhibitors or activation markers.

10. Memory T-Cells

Memory T-cells are crucial players in the immunity against tumors due to their potential for providing a quick and sustained immune response to the tumors [168]. Hence, several studies showed the importance of CD45RO$^+$ memory T-cells in BC prognosis as well. Immunohistochemical evaluations revealed that CD45RO$^+$ lymphocytes were the most frequent immune subset in both the periphery and invasive margin of breast tumors indicating previous antigen priming in tumor tissues, while their infiltrations in normal-like tissues adjacent to the tumors were remarkably low [169–171]. The frequency of these cells in invasive margins was observed to be strongly higher than in the center of tumors [11]. However, Schnellhardt et al. reported relatively low density for CD45RO$^+$ lymphocytes in BC tissues along with poor epithelial infiltration [172]. Nevertheless, the findings collectively showed that higher infiltrations of CD45RO$^+$ lymphocytes were significantly associated with smaller tumor size, fewer metastatic lymph nodes and fewer peri-tumoral lymphatic invasions, lower histological grade, and TNM-stage [173,174], although we conversely observed this association with node metastasis, higher histological grade and stages [11]. The older patients (over 50) also had a higher frequency of CD45RO$^+$ lymphocytes, consistent with the research on immune system development suggesting an increase in the CD45RO$^+$ memory population with age [175]. Immunophenotyping the CD45RO-expressing lymphocytes infiltrating breast tumor tissues revealed that the majority of CD45RO$^+$ lymphocytes displayed the CD3 pan T-cells marker (more than 90%), followed by CD8$^+$, CD4$^+$, and CD16$^+$/CD56$^+$ NK cells. We also observed that the majority of the NK cells in BC had NKT phenotype expressing CD3 (more than 90%). CD45RO was also rarely expressed on CD14$^+$ monocytes and CD11c$^+$ DCs (less than 10%), with no expression on CD19$^+$ B-cells [unpublished data].

Studies mostly look into the patients generally without considering differences caused by the patients' molecular subtypes or heterogenicity that exists among memory T-cells. Accordingly, based on the expression of ER, PR, and HER2, we classified the patients into different subgroups: luminal and non-luminal (including HER2-enriched and TNBC) subtypes. Our findings indicated remarkably more infiltrations of CD45RO$^+$ lymphocytes in the non-luminal subtypes, with the highest frequency in HER2-enriched tumors, than the luminal type [11]. However, limited studies have investigated CD45RO$^+$ lymphocytes in BC with different molecular subtypes [11,176,177]; these observations are in accordance with studies reporting more infiltration of immune cells in non-luminal tumors [178,179]. One of the possible causes of the difference in lymphocytic infiltration among different molecular subtypes was proposed to be the modulatory effect of hormone receptors on immune responses to the tumor and lymphocytic composition in the TME [180–182].

In addition to tumor heterogeneity, it is now well-documented that there is a broad diversity among memory T-cells with unique properties within key functional features such as trafficking, localization, effector functions, and durability [183]. In this regard, CD103$^+$CD8$^+$ T-cells, recognized as tissue-resident memory T-cells (TRM), were observed to be a common effector subset in breast tumors with high expression of effector markers and immunological checkpoint molecules and were linked to improved survival [152]. Immunohistochemical analysis showed a correlation between CD103 positivity and tumor grade, tumor size, and ER/PR status. Additionally, co-expression of CD103$^+$/CD8$^+$ demonstrated a stronger predictive value and was associated with higher RFS and OS. Therefore, it was suggested that CD103-CD8 co-expression would be a more valuable prognostic marker than CD8 alone [184]. Another independent investigation demonstrated TRM enrichment (defined by CD39, CD69, and CD103 expression) in TNBC samples and its relationship to survival. However, in this study, TRM exhibited an exhausted phenotype, which could be reinvigorated to active form with the addition of immune checkpoint inhibitors [185]. Accordingly, the TRM cell is introduced as a prognostic factor that plays an essential role in BC surveillance and is a key mediator in immunotherapy success. We also assessed memory subsets in draining lymph nodes of patients with BC. We observed that most lymphocytes (both CD4$^+$ and CD8$^+$ subsets) in draining lymph nodes from patients with BC exhibited a memory phenotype, with the highest frequency for central memory T-cells (TCM) and the lowest for T memory stem cells (TSCM) subsets. Statistical analysis indicated that tumor-positive lymph nodes had higher frequencies of CD4$^+$ TSCM and TCM than lymph nodes without tumors. In addition, the frequency of TCM with low expression of CD45RO increased in the advanced stages, while TEM cells decreased [186]. Thus, we hypothesized that in an attempt to provide a pool of memory and effector T-cells, TSCM cells proliferate following long-term exposure to tumor antigens; however, the tumor microenvironment prevents TCM from differentiating into effector cells [186,187]. To check the inhibitory state of lymph nodes, in another study, we assessed the expression of PD-1 and its ligands, as immune checkpoint inhibitors on different memory subsets and found that while PD-1 expression was very low on TSCM, it was highly expressed on TEM and TCM (unpublished data). Although PD-1 was generally considered a hallmark of T-cells' exhaustion, it was shown that the expression levels of this inhibitory molecule change depending on the stage of the T-cell differentiation [188], suggesting that PD-1 expression on memory T-cells is related to the activation and differentiation status of the subsets rather than their exhaustion. However, more functional studies are needed to elucidate the regulatory role of the PD-1 molecule in memory cell development and function in BC.

11. Immune Suppression in Breast Carcinoma: A Role for Regulatory T-Cells

Substantial evidence from functional studies demonstrated that the microenvironment of breast tumors contains a large number of infiltrating leukocytes that do not expand or function normally [17,189]. It is now well-documented that the growth of most invasive carcinomas including BC not only depends on regulatory factors derived from malignant cells but is also affected by those from nearby stromal cells [9,190,191]. In this regard, a large number of studies have indicated that the frequency of regulatory cells increases in the peripheral blood, TDLNs, and more greatly in different tumor tissues including breast adenocarcinomas [22,70,192]. However, Foxp3$^+$ regulatory T-cells have low frequency in BC tissues, and their densities and localizations in a large series of patients with primary BC showed a significant association with higher tumor grade, HER2 positivity, ER negativity, and poor prognosis [11,193–195]. Infiltration of Foxp3$^+$ cells also predicted lymph node metastasis and was associated with a shorter RFS independent of other prognostic factors [193,194,196,197]. The higher frequency of Tregs was also observed in TDLNs of the patients with at least one involved node compared with those with the tumor-free nodes [22]. Interestingly, the proliferation potential of Tregs was shown to be enhanced during disease [159], while we observed that regulatory populations contained a lower percentage of IL-2 and IFNγ-producing cells compared with effector cells [198]. Many studies extend these findings in the animal models and show that Tregs increased in BC and their specific depletion markedly suppresses tumor growth and provokes strong and persistent anti-tumor immune responses [199,200]. The suppressive role of Tregs was further confirmed by their adoptive transfer into the depleted mice that abrogated the immune response [199,200].

Some other studies, however, did not find any significant predictive role for Tregs in BC [11,148] or intriguingly reported that Foxp3 expression in tumor tissue is an independent predictor of better outcomes in HER2$^+$ patients [201] or those who received chemotherapy [202]. Besides intrinsic differences among different patients, this discrepancy might be also attributed to the heterogeneity of Foxp3$^+$ cells, their functional status, and the expression of different immunosuppressive molecules used to phenotype Tregs. Consistently, phenotypic and transcriptional profiling of the tumor and normal-tissue Tregs exhibited that these cells were Foxp3$^+$Helios$^+$ cells with high levels of inhibitory molecules (PD-1 and CTLA-4) and were associated with poor prognosis [203]. It is also reported that CCR8 was upregulated in tumor-tissue Tregs and these CCR8$^+$ Tregs have high activity and proliferation potential [204]. Controversial reports could also be found regarding CCL22. While there are several studies indicating that CCL22 contributes to the recruitment of Tregs to BC tissues and worse clinical outcomes [205,206], Freier et al.

reported that this phenotype correlated with good prognosis. They intriguingly proposed that Tregs confine inflammation and tumor spreading, thereby attenuating nodal involvement [207]. In another study, immunochemistry analysis showed a positive relationship between CCL1 expression, but not CCL22, with Tregs frequency and improved survival in breast tumors [208]. In addition, finding a new subset of Foxp3-expressing CD4$^+$ cells with no expression of CD25 increases the Tregs' complexity. CD25 (IL-2 receptor alpha chain) is commonly considered an indicative marker for both murine and human CD4$^+$ Tregs. The presence of this subset has been introduced in different settings including cancer [209]. To the best of our knowledge, it was for the first time that we reported the presence of this subset in draining lymph nodes of patients with BC. This subset showed elevated frequency in node-positive patients with the invasive ductal carcinoma subtype of BC implying an inhibitory role for these cells in breast tumor immunity [22]. However, further evaluations of these cells revealed lower Foxp3 expression compared with the CD25$^+$ Tregs [192,198] and lower inflammatory and inhibitory cytokines, i.e., IFNγ, IL-2, IL-17, IL-22, and IL-10, compared with the effectors. We concluded that due to lacking the CD25, these cells probably do not respond to IL-2, and accordingly do not expand and produce effector cytokines. Our results introduced these cells as a heterogeneous exhausted population containing both effector and regulatory cells and/or a subset in the intermediate state [198]. Overall, to clarify the precise role of Tregs in BC, several factors including the heterogeneity of Foxp3$^+$ cells and the expression of various immunosuppressive molecules should be considered during studying these cells.

12. T-Cell and B-Cell Crosstalk

While the most of focuses are on T-cells in cancer, immunosurveillance is not just limited to T lymphocytes [210]. In fact, the crosstalk of T-cells with other innate and acquired immune cells finally determines the outcome. The interaction of B-cells, the humoral arm of the adaptive immune response, and T-cells becomes more pronounced in GCs where T-cells provide help for B-cells to produce more efficient antibodies [211]. In this regard, Gu-Trantien et al. showed that in BC tumors, inflammatory milieu including Tfh cells directed B-cell migration and subsequently induced lymphoid structure and GCs formation, and finally led to the generation of long-lived B-cells and plasma cells and the elimination of tumor residues [140]. Besides this traditional belief, the crosstalk of B-cells and T-cells is not limited to antibody production but it is a dual and more complex interaction [212].

B-cells constitute up to 40% of TILs in BC tissues and draining nodes [210,213], with dominant memory phenotype (CD24^{hi}CD27$^+$) in TDLNs [213]. Recent studies showed a strong association between B-cells and different helper and cytotoxic subsets of T-cells. Infiltrated B-cells or a higher ratio of CD20/Foxp3$^+$ were also associated with improved clinical prognosis and longer OS [214,215]. B-cells could facilitate tumor regression by presenting antigens to T-cells via MHC I/II, and the recruitment of other immune cells including T-cells to effector sites, secondary and tertiary lymphoid organs [210]. In addition, antibodies bind to tumor antigens and help macrophages and DCs to uptake, process, and present or cross-present them to CD4$^+$ and CD8$^+$ T-cells, respectively [210]. However, Deola et al. introduced a novel interaction between CTLs and B-cells independent of common antigen presentation, recommending a helper role for B-cells. They showed that engagement of CTLs with bystander B-lymphocytes through CD27/CD70 contact led to the increased proliferation and improved survival of CD8$^+$ T-cells [216]. Additionally, B-cells express TNF-α which can provide the co-stimulatory signal for either T effector or Tregs [217]. However, we observed that the frequency of B-cells expressing high levels of TNF-α negatively correlated with the frequency of Foxp3$^+$ Tregs in BC draining LNs [217].

On the dark side, it has been shown that B-cells could suppress the anti-tumor function of T-cells through different mechanisms. A subset of B-cells with CD27^{hi}CD25$^+$ phenotype has been shown to help Treg cell expansion [218]. Consistently, we showed that this subset had a higher frequency in non-metastatic nodes of LN$^+$ patients with BC [213]. Another subset of B-cells, IgA$^+$PD-L1$^+$IL10$^+$, impairs the maturation of DCs and the activation of T-cells. In addition, antibodies could bind to FC receptors on myeloid cells and induce their differentiation to MDSCs, and indirectly suppress CD4$^+$ and CD8$^+$ T-cell responses. B-cells also secret TGF-β and IL-10 that deviate T-cells toward Tregs and suppress T-cell function [210,219]. Based on above, while B-cells are generally known for their anti-tumor effects, primarily through producing antibodies and their interactions with T-cells, specific B-cell subtypes are also able to promote tumor progression.

13. Conclusions

Clinical and experimental studies have demonstrated that T-cells play a crucial, yet dual, role in breast tumor development and progression. On the one hand, they suppress breast tumor by destroying tumor cells or inhibiting their growth (Figure 1), and on the other hand, they can facilitate the progression of those tumors that evolved to evade immune surveillance, by inducing the expression of growth factors and regulatory molecules (Figure 2). Despite the presence of many investigations, it is not possible to reach a clear and uniform conclusion about the role of each T-cell subset in the breast tumor microenvironment or its association with BC outcome. In this connection, however, the data were not confirmed by all the investigations; the majority of reports have shown a link between CTLs and Th1 cells and good prognosis, whereas Th2 cells, Tregs, and their related cytokines or molecules correlate with tumor growth and poor prognosis. Information about the exact role of new T-cell subsets such as Th22, Th9, Th25, and ThGM in tumor immunity is still limited and controversial. One of the main reasons for the observed controversies is the use of different markers for the study of one immune cell type. In addition, the classification of T-cells into definite and committed subsets is not simple since it is recently well-appreciated that committed T-cell subsets can be differentiated from other subsets in the inflammatory condition of the tumor microenvironment. Therefore, it is too difficult to divide T-cell subsets into black and white categories in terms of their role in anti-tumor immunity or their association with cancer outcomes. Besides this, the role of T-cells in BC immunity depends on a variety of extrinsic factors, including tumor type or subtype, the stage of the disease, tumor immunogenicity, localization of cells in the tumor tissue and crosstalk with other cells or cytokines.

Author Contributions: All authors have made substantial contributions to the medical literature review. Conceptualization and Validation: Z.F. and A.G. Writing—Original Draft Preparation: M.Z., M.R.-Z. and Z.F. Writing—Review and Editing: Z.F., A.G. and F.M. Supervision: Z.F. and A.G. All authors have read and agreed to the published version of the manuscript.

Funding: This work was financially supported by the Shiraz Institute for Cancer Research (grant numbers: ICR-100-504 and ICR-100-508).

Institutional Review Board Statement: Not applicable.

Informed Consent Statement: Not applicable.

Data Availability Statement: Not applicable.

Conflicts of Interest: The authors declare no conflict of interest.

References

1. Sørlie, T.; Perou, C.M.; Tibshirani, R.; Aas, T.; Geisler, S.; Johnsen, H.; Hastie, T.; Eisen, M.B.; van de Rijn, M.; Jeffrey, S.S.; et al. Gene expression patterns of breast carcinomas distinguish tumor subclasses with clinical implications. *Proc. Natl. Acad. Sci. USA* **2001**, *98*, 10869–10874. [CrossRef] [PubMed]
2. Momenimovahed, Z.; Salehiniya, H. Epidemiological characteristics of and risk factors for breast cancer in the world. *Breast Cancer (Dove Med. Press)* **2019**, *11*, 151–164. [CrossRef]
3. Zardavas, D.; Irrthum, A.; Swanton, C.; Piccart, M. Clinical management of breast cancer heterogeneity. *Nat. Rev. Clin. Oncol.* **2015**, *12*, 381–394. [CrossRef] [PubMed]

4. Rody, A.; Holtrich, U.; Pusztai, L.; Liedtke, C.; Gaetje, R.; Ruckhaeberle, E.; Solbach, C.; Hanker, L.; Ahr, A.; Metzler, D.; et al. T-cell metagene predicts a favorable prognosis in estrogen receptor-negative and HER2-positive breast cancers. *Breast Cancer Res. BCR* **2009**, *11*, R15. [CrossRef]
5. Hanker, L.C.; Rody, A.; Holtrich, U.; Pusztai, L.; Ruckhaeberle, E.; Liedtke, C.; Ahr, A.; Heinrich, T.M.; Sanger, N.; Becker, S.; et al. Prognostic evaluation of the B cell/IL-8 metagene in different intrinsic breast cancer subtypes. *Breast Cancer Res. Treat.* **2013**, *137*, 407–416. [CrossRef] [PubMed]
6. Schmidt, M.; Bohm, D.; von Torne, C.; Steiner, E.; Puhl, A.; Pilch, H.; Lehr, H.A.; Hengstler, J.G.; Kolbl, H.; Gehrmann, M. The humoral immune system has a key prognostic impact in node-negative breast cancer. *Cancer Res.* **2008**, *68*, 5405–5413. [CrossRef]
7. Huang, E.; Cheng, S.H.; Dressman, H.; Pittman, J.; Tsou, M.H.; Horng, C.F.; Bild, A.; Iversen, E.S.; Liao, M.; Chen, C.M.; et al. Gene expression predictors of breast cancer outcomes. *Lancet* **2003**, *361*, 1590–1596. [CrossRef]
8. Pagès, F.; Galon, J.; Dieu-Nosjean, M.C.; Tartour, E.; Sautès-Fridman, C.; Fridman, W.H. Immune infiltration in human tumors: A prognostic factor that should not be ignored. *Oncogene* **2010**, *29*, 1093–1102. [CrossRef]
9. DeNardo, D.G.; Coussens, L.M. Inflammation and breast cancer. Balancing immune response: Crosstalk between adaptive and innate immune cells during breast cancer progression. *Breast Cancer Res. BCR* **2007**, *9*, 212. [CrossRef]
10. Rahir, G.; Moser, M. Tumor microenvironment and lymphocyte infiltration. *Cancer Immunol. Immunother. CII* **2012**, *61*, 751–759. [CrossRef]
11. Ahmadvand, S.; Faghih, Z.; Montazer, M.; Safaei, A.; Mokhtari, M.; Jafari, P.; Talei, A.R.; Tahmasebi, S.; Ghaderi, A. Importance of CD45RO+ tumor-infiltrating lymphocytes in post-operative survival of breast cancer patients. *Cell. Oncol.* **2019**, *42*, 343–356. [CrossRef] [PubMed]
12. Solinas, C.; Carbognin, L.; De Silva, P.; Criscitiello, C.; Lambertini, M. Tumor-infiltrating lymphocytes in breast cancer according to tumor subtype: Current state of the art. *Breast* **2017**, *35*, 142–150. [CrossRef] [PubMed]
13. Staaf, J.; Ringnér, M.; Vallon-Christersson, J.; Jönsson, G.; Bendahl, P.O.; Holm, K.; Arason, A.; Gunnarsson, H.; Hegardt, C.; Agnarsson, B.A.; et al. Identification of subtypes in human epidermal growth factor receptor 2—Positive breast cancer reveals a gene signature prognostic of outcome. *J. Clin. Oncol. Off. J. Am. Soc. Clin. Oncol.* **2010**, *28*, 1813–1820. [CrossRef] [PubMed]
14. Georgiannos, S.N.; Renaut, A.; Goode, A.W.; Sheaff, M. The immunophenotype and activation status of the lymphocytic infiltrate in human breast cancers, the role of the major histocompatibility complex in cell-mediated immune mechanisms, and their association with prognostic indicators. *Surgery* **2003**, *134*, 827–834. [CrossRef]
15. Wong, P.Y.; Staren, E.D.; Tereshkova, N.; Braun, D.P. Functional analysis of tumor-infiltrating leukocytes in breast cancer patients. *J. Surg. Res.* **1998**, *76*, 95–103. [CrossRef]
16. Helal, T.E.; Ibrahim, E.A.; Alloub, A.I. Immunohistochemical analysis of tumor-infiltrating lymphocytes in breast carcinoma: Relation to prognostic variables. *Indian J. Pathol. Microbiol.* **2013**, *56*, 89–93. [CrossRef]
17. Ruffell, B.; Au, A.; Rugo, H.S.; Esserman, L.J.; Hwang, E.S.; Coussens, L.M. Leukocyte composition of human breast cancer. *Proc. Natl. Acad. Sci. USA* **2012**, *109*, 2796–2801. [CrossRef]
18. Reome, J.B.; Hylind, J.C.; Dutton, R.W.; Dobrzanski, M.J. Type 1 and type 2 tumor infiltrating effector cell subpopulations in progressive breast cancer. *Clin. Immunol.* **2004**, *111*, 69–81. [CrossRef]
19. Gu-Trantien, C.; Loi, S.; Garaud, S.; Equeter, C.; Libin, M.; de Wind, A.; Ravoet, M.; Le Buanec, H.; Sibille, C.; Manfouo-Foutsop, G.; et al. $CD4^+$ follicular helper T cell infiltration predicts breast cancer survival. *J. Clin. Investig.* **2013**, *123*, 2873–2892. [CrossRef]
20. Faghih, Z.; Deihimi, S.; Talei, A.; Ghaderi, A.; Erfani, N. Analysis of T cell receptor repertoire based on Vβ chain in patients with breast cancer. *Cancer Biomark.* **2018**, *22*, 733–745. [CrossRef]
21. DeNardo, D.G.; Brennan, D.J.; Rexhepaj, E.; Ruffell, B.; Shiao, S.L.; Madden, S.F.; Gallagher, W.M.; Wadhwani, N.; Keil, S.D.; Junaid, S.A.; et al. Leukocyte complexity predicts breast cancer survival and functionally regulates response to chemotherapy. *Cancer Discov.* **2011**, *1*, 54–67. [CrossRef] [PubMed]
22. Faghih, Z.; Erfani, N.; Haghshenas, M.R.; Safaei, A.; Talei, A.R.; Ghaderi, A. Immune profiles of CD4+ lymphocyte subsets in breast cancer tumor draining lymph nodes. *Immunol. Lett.* **2014**, *158*, 57–65. [CrossRef] [PubMed]
23. Faghih, Z.; Rezaeifard, S.; Safaei, A.; Ghaderi, A.; Erfani, N. IL-17 and IL-4 producing CD8+ T cells in tumor draining lymph nodes of breast cancer patients: Positive association with tumor progression. *Iran. J. Immunol. IJI* **2013**, *10*, 193–204. [PubMed]
24. Knutson, K.L.; Disis, M.L. Augmenting T helper cell immunity in cancer. *Curr. Drug Targets. Immune Endocr. Metab. Disord.* **2005**, *5*, 365–371. [CrossRef] [PubMed]
25. Chen, Y.; Sun, J.; Luo, Y.; Liu, J.; Wang, X.; Feng, R.; Huang, J.; Du, H.; Li, Q.; Tan, J.; et al. Pharmaceutical targeting Th2-mediated immunity enhances immunotherapy response in breast cancer. *J. Transl. Med.* **2022**, *20*, 615. [CrossRef] [PubMed]
26. Matkowski, R.; Gisterek, I.; Halon, A.; Lacko, A.; Szewczyk, K.; Staszek, U.; Pudelko, M.; Szynglarewicz, B.; Szelachowska, J.; Zolnierek, A.; et al. The prognostic role of tumor-infiltrating CD4 and CD8 T lymphocytes in breast cancer. *Anticancer Res.* **2009**, *29*, 2445–2451.
27. Ehi, K.; Ishigami, S.; Masamoto, I.; Uenosono, Y.; Natsugoe, S.; Arigami, T.; Arima, H.; Kijima, Y.; Yoshinaka, H.; Yanagita, S.; et al. Analysis of T-helper type 1 and 2 cells and T-cytotoxic type 1 and 2 cells of sentinel lymph nodes in breast cancer. *Oncol. Rep.* **2008**, *19*, 601–607. [CrossRef]
28. Matsuura, K.; Yamaguchi, Y.; Ueno, H.; Osaki, A.; Arihiro, K.; Toge, T. Maturation of dendritic cells and T-cell responses in sentinel lymph nodes from patients with breast carcinoma. *Cancer* **2006**, *106*, 1227–1236. [CrossRef]

29. Caras, I.; Grigorescu, A.; Stavaru, C.; Radu, D.L.; Mogos, I.; Szegli, G.; Salageanu, A. Evidence for immune defects in breast and lung cancer patients. *Cancer Immunol. Immunother. CII* **2004**, *53*, 1146–1152. [CrossRef]
30. Fracol, M.; Datta, J.; Lowenfeld, L.; Xu, S.; Zhang, P.J.; Fisher, C.S.; Czerniecki, B.J. Loss of Anti-HER-3 CD4+ T-Helper Type 1 Immunity Occurs in Breast Tumorigenesis and is Negatively Associated with Outcomes. *Ann. Surg. Oncol.* **2017**, *24*, 407–417. [CrossRef]
31. Eftekhari, R.; Esmaeili, R.; Mirzaei, R.; Bidad, K.; de Lima, S.; Ajami, M.; Shirzad, H.; Hadjati, J.; Majidzadeh-A, K. Study of the tumor microenvironment during breast cancer progression. *Cancer Cell Int.* **2017**, *17*, 123. [CrossRef] [PubMed]
32. Finak, G.; Bertos, N.; Pepin, F.; Sadekova, S.; Souleimanova, M.; Zhao, H.; Chen, H.; Omeroglu, G.; Meterissian, S.; Omeroglu, A.; et al. Stromal gene expression predicts clinical outcome in breast cancer. *Nat. Med.* **2008**, *14*, 518–527. [CrossRef] [PubMed]
33. Manjili, M.H.; Najarian, K.; Wang, X.Y. Signatures of tumor-immune interactions as biomarkers for breast cancer prognosis. *Future Oncol.* **2012**, *8*, 703–711. [CrossRef]
34. Einav, U.; Tabach, Y.; Getz, G.; Yitzhaky, A.; Ozbek, U.; Amariglio, N.; Izraeli, S.; Rechavi, G.; Domany, E. Gene expression analysis reveals a strong signature of an interferon-induced pathway in childhood lymphoblastic leukemia as well as in breast and ovarian cancer. *Oncogene* **2005**, *24*, 6367–6375. [CrossRef] [PubMed]
35. Bos, R.; Sherman, L.A. CD4+ T-cell help in the tumor milieu is required for recruitment and cytolytic function of CD8+ T lymphocytes. *Cancer Res.* **2010**, *70*, 8368–8377. [CrossRef] [PubMed]
36. Nocera, N.F.; Lee, M.C.; De La Cruz, L.M.; Rosemblit, C.; Czerniecki, B.J. Restoring Lost Anti-HER-2 Th1 Immunity in Breast Cancer: A Crucial Role for Th1 Cytokines in Therapy and Prevention. *Front. Pharmacol.* **2016**, *7*, 356. [CrossRef]
37. Namjoshi, P.; Showalter, L.; Czerniecki, B.J.; Koski, G.K. T-helper 1-type cytokines induce apoptosis and loss of HER-family oncodriver expression in murine and human breast cancer cells. *Oncotarget* **2016**, *5*, 6006. [CrossRef]
38. Braumüller, H.; Wieder, T.; Brenner, E.; Aßmann, S.; Hahn, M.; Alkhaled, M.; Schilbach, K.; Essmann, F.; Kneilling, M.; Griessinger, C.; et al. T-helper-1-cell cytokines drive cancer into senescence. *Nature* **2013**, *494*, 361–365. [CrossRef]
39. Datta, J.; Xu, S.; Rosemblit, C.; Smith, J.B.; Cintolo, J.A.; Powell, D.J., Jr.; Czerniecki, B.J. CD4(+) T-Helper Type 1 Cytokines and Trastuzumab Facilitate CD8(+) T-cell Targeting of HER2/neu-Expressing Cancers. *Cancer Immunol. Res.* **2015**, *3*, 455–463. [CrossRef]
40. Zhao, X.; Liu, J.; Ge, S.; Chen, C.; Li, S.; Wu, X.; Feng, X.; Wang, Y.; Cai, D. Saikosaponin A Inhibits Breast Cancer by Regulating Th1/Th2 Balance. *Front. Pharmacol.* **2019**, *10*, 624. [CrossRef] [PubMed]
41. DeNardo, D.G.; Barreto, J.B.; Andreu, P.; Vasquez, L.; Tawfik, D.; Kolhatkar, N.; Coussens, L.M. CD4(+) T cells regulate pulmonary metastasis of mammary carcinomas by enhancing protumor properties of macrophages. *Cancer Cell* **2009**, *16*, 91–102. [CrossRef]
42. Park, J.M.; Terabe, M.; Donaldson, D.D.; Forni, G.; Berzofsky, J.A. Natural immunosurveillance against spontaneous, autochthonous breast cancers revealed and enhanced by blockade of IL-13-mediated negative regulation. *Cancer Immunol. Immunother. CII* **2008**, *57*, 907–912. [CrossRef] [PubMed]
43. Wu, W.J.; Wang, S.H.; Wu, C.C.; Su, Y.A.; Chiang, C.Y.; Lai, C.H.; Wang, T.H.; Cheng, T.L.; Kuo, J.Y.; Hsu, T.C.; et al. IL-4 and IL-13 Promote Proliferation of Mammary Epithelial Cells through STAT6 and IRS-1. *Int. J. Mol. Sci.* **2021**, *22*, 12008. [CrossRef] [PubMed]
44. Aspord, C.; Pedroza-Gonzalez, A.; Gallegos, M.; Tindle, S.; Burton, E.C.; Su, D.; Marches, F.; Banchereau, J.; Palucka, A.K. Breast cancer instructs dendritic cells to prime interleukin 13-secreting CD4+ T cells that facilitate tumor development. *J. Exp. Med.* **2007**, *204*, 1037–1047. [CrossRef]
45. Todaro, M.; Lombardo, Y.; Francipane, M.G.; Alea, M.P.; Cammareri, P.; Iovino, F.; Di Stefano, A.B.; Di Bernardo, C.; Agrusa, A.; Condorelli, G.; et al. Apoptosis resistance in epithelial tumors is mediated by tumor-cell-derived interleukin-4. *Cell Death Differ.* **2008**, *15*, 762–772. [CrossRef]
46. Pedroza-Gonzalez, A.; Xu, K.; Wu, T.C.; Aspord, C.; Tindle, S.; Marches, F.; Gallegos, M.; Burton, E.C.; Savino, D.; Hori, T.; et al. Thymic stromal lymphopoietin fosters human breast tumor growth by promoting type 2 inflammation. *J. Exp. Med.* **2011**, *208*, 479–490. [CrossRef] [PubMed]
47. Boieri, M.; Malishkevich, A.; Guennoun, R.; Marchese, E.; Kroon, S.; Trerice, K.E.; Awad, M.; Park, J.H.; Iyer, S.; Kreuzer, J.; et al. CD4+ T helper 2 cells suppress breast cancer by inducing terminal differentiation. *J. Exp. Med.* **2022**, *219*, e20201963. [CrossRef]
48. Gooch, J.L.; Christy, B.; Yee, D. STAT6 mediates interleukin-4 growth inhibition in human breast cancer cells. *Neoplasia* **2002**, *4*, 324–331. [CrossRef] [PubMed]
49. Nagai, S.; Toi, M. Interleukin-4 and breast cancer. *Breast Cancer* **2000**, *7*, 181–186. [CrossRef]
50. Haricharan, S.; Li, Y. STAT signaling in mammary gland differentiation, cell survival and tumorigenesis. *Mol. Cell. Endocrinol.* **2014**, *382*, 560–569. [CrossRef]
51. Bożek, A.; Jarzab, J.; Mielnik, M.; Bogacz, A.; Kozlowska, R.; Mangold, D. Can atopy have a protective effect against cancer? *PLoS ONE* **2020**, *15*, e0226950. [CrossRef] [PubMed]
52. Wang, H.; Diepgen, T.L. Is atopy a protective or a risk factor for cancer? A review of epidemiological studies. *Allergy* **2005**, *60*, 1098–1111. [CrossRef]
53. Dong, C. Defining the T(H)17 cell lineage. *Nat. Rev. Immunol.* **2021**, *21*, 618. [CrossRef] [PubMed]
54. Benevides, L.; da Fonseca, D.M.; Donate, P.B.; Tiezzi, D.G.; De Carvalho, D.D.; de Andrade, J.M.; Martins, G.A.; Silva, J.S. IL17 Promotes Mammary Tumor Progression by Changing the Behavior of Tumor Cells and Eliciting Tumorigenic Neutrophils Recruitment. *Cancer Res.* **2015**, *75*, 3788–3799. [CrossRef]

55. Chen, W.C.; Lai, Y.H.; Chen, H.Y.; Guo, H.R.; Su, I.J.; Chen, H.H. Interleukin-17-producing cell infiltration in the breast cancer tumour microenvironment is a poor prognostic factor. *Histopathology* **2013**, *63*, 225–233. [CrossRef] [PubMed]
56. Benevides, L.; Cardoso, C.R.; Tiezzi, D.G.; Marana, H.R.; Andrade, J.M.; Silva, J.S. Enrichment of regulatory T cells in invasive breast tumor correlates with the upregulation of IL-17A expression and invasiveness of the tumor. *Eur. J. Immunol.* **2013**, *43*, 1518–1528. [CrossRef]
57. Ji, Y.; Zhang, W. Th17 cells: Positive or negative role in tumor? *Cancer Immunol. Immunother. CII* **2010**, *59*, 979–987. [CrossRef]
58. Wilke, C.M.; Kryczek, I.; Wei, S.; Zhao, E.; Wu, K.; Wang, G.; Zou, W. Th17 cells in cancer: Help or hindrance? *Carcinogenesis* **2011**, *32*, 643–649. [CrossRef]
59. Wang, D.; Yu, W.; Lian, J.; Wu, Q.; Liu, S.; Yang, L.; Li, F.; Huang, L.; Chen, X.; Zhang, Z.; et al. Th17 cells inhibit CD8(+) T cell migration by systematically downregulating CXCR3 expression via IL-17A/STAT3 in advanced-stage colorectal cancer patients. *J. Hematol. Oncol.* **2020**, *13*, 68. [CrossRef]
60. Shibabaw, T.; Teferi, B.; Ayelign, B. The role of Th-17 cells and IL-17 in the metastatic spread of breast cancer: As a means of prognosis and therapeutic target. *Front. Immunol.* **2023**, *14*, 1094823. [CrossRef]
61. Welte, T.; Zhang, X.H. Interleukin-17 Could Promote Breast Cancer Progression at Several Stages of the Disease. *Mediat. Inflamm.* **2015**, *2015*, 804347. [CrossRef] [PubMed]
62. Kim, G.; Khanal, P.; Lim, S.C.; Yun, H.J.; Ahn, S.G.; Ki, S.H.; Choi, H.S. Interleukin-17 induces AP-1 activity and cellular transformation via upregulation of tumor progression locus 2 activity. *Carcinogenesis* **2013**, *34*, 341–350. [CrossRef] [PubMed]
63. Ma, K.; Yang, L.; Shen, R.; Kong, B.; Chen, W.; Liang, J.; Tang, G.; Zhang, B. Th17 cells regulate the production of CXCL1 in breast cancer. *Int. Immunopharmacol.* **2018**, *56*, 320–329. [CrossRef]
64. Fabre, J.A.S.; Giustinniani, J.; Garbar, C.; Merrouche, Y.; Antonicelli, F.; Bensussan, A. The Interleukin-17 Family of Cytokines in Breast Cancer. *Int. J. Mol. Sci.* **2018**, *19*, 3880. [CrossRef] [PubMed]
65. Su, X.; Ye, J.; Hsueh, E.C.; Zhang, Y.; Hoft, D.F.; Peng, G. Tumor microenvironments direct the recruitment and expansion of human Th17 cells. *J. Immunol.* **2010**, *184*, 1630–1641. [CrossRef]
66. Kim, H.; Kim, Y.; Bae, S.; Kong, J.M.; Choi, J.; Jang, M.; Choi, J.; Hong, J.M.; Hwang, Y.I.; Kang, J.S.; et al. Direct Interaction of CD40 on Tumor Cells with CD40L on T Cells Increases the Proliferation of Tumor Cells by Enhancing TGF-beta Production and Th17 Differentiation. *PLoS ONE* **2015**, *10*, e0125742. [CrossRef]
67. Novitskiy, S.V.; Pickup, M.W.; Gorska, A.E.; Owens, P.; Chytil, A.; Aakre, M.; Wu, H.; Shyr, Y.; Moses, H.L. TGF-beta receptor II loss promotes mammary carcinoma progression by Th17 dependent mechanisms. *Cancer Discov.* **2011**, *1*, 430–441. [CrossRef]
68. Yang, L.; Qi, Y.; Hu, J.; Tang, L.; Zhao, S.; Shan, B. Expression of Th17 cells in breast cancer tissue and its association with clinical parameters. *Cell Biochem. Biophys.* **2012**, *62*, 153–159. [CrossRef]
69. Faucheux, L.; Grandclaudon, M.; Perrot-Dockès, M.; Sirven, P.; Berger, F.; Hamy, A.S.; Fourchotte, V.; Vincent-Salomon, A.; Mechta-Grigoriou, F.; Reyal, F.; et al. A multivariate Th17 metagene for prognostic stratification in T cell non-inflamed triple negative breast cancer. *Oncoimmunology* **2019**, *8*, e1624130. [CrossRef]
70. Wang, J.; Cai, D.; Ma, B.; Wu, G.; Wu, J. Skewing the balance of regulatory T-cells and T-helper 17 cells in breast cancer patients. *J. Int. Med. Res.* **2011**, *39*, 691–701. [CrossRef]
71. Horlock, C.; Stott, B.; Dyson, P.J.; Morishita, M.; Coombes, R.C.; Savage, P.; Stebbing, J. The effects of trastuzumab on the CD4+CD25+FoxP3+ and CD4+IL17A+ T-cell axis in patients with breast cancer. *Br. J. Cancer* **2009**, *100*, 1061–1067. [CrossRef] [PubMed]
72. Hemdan, N.Y. Anti-cancer versus cancer-promoting effects of the interleukin-17-producing T helper cells. *Immunol. Lett.* **2013**, *149*, 123–133. [CrossRef] [PubMed]
73. Vitiello, G.A.; Miller, G. Targeting the interleukin-17 immune axis for cancer immunotherapy. *J. Exp. Med.* **2020**, *217*, e20190456. [CrossRef]
74. Ma, M.; Huang, W.; Kong, D. IL-17 inhibits the accumulation of myeloid-derived suppressor cells in breast cancer via activating STAT3. *Int. Immunopharmacol.* **2018**, *59*, 148–156. [CrossRef]
75. Tato, C.M.; Laurence, A.; O'Shea, J.J. Helper T cell differentiation enters a new era: Le Roi est mort; vive le Roi! *J. Exp. Med.* **2006**, *203*, 809–812. [CrossRef] [PubMed]
76. Vock, C.; Hauber, H.P.; Wegmann, M. The other T helper cells in asthma pathogenesis. *J. Allergy* **2010**, *2010*, 519298. [CrossRef]
77. Yuan, Q.; Peng, N.; Xiao, F.; Shi, X.; Zhu, B.; Rui, K.; Tian, J.; Lu, L. New insights into the function of Interleukin-25 in disease pathogenesis. *Biomark. Res.* **2023**, *11*, 36. [CrossRef]
78. de Sousa, J.R.; Quaresma, J.A.S. The role of T helper 25 cells in the immune response to Mycobacterium leprae. *J. Am. Acad. Dermatol.* **2018**, *78*, 1009–1011. [CrossRef]
79. Gowhari Shabgah, A.; Amir, A.; Gardanova, Z.R.; Olegovna Zekiy, A.; Thangavelu, L.; Ebrahimi Nik, M.; Ahmadi, M.; Gholizadeh Navashenaq, J. Interleukin-25: New perspective and state-of-the-art in cancer prognosis and treatment approaches. *Cancer Med.* **2021**, *10*, 5191–5202. [CrossRef]
80. Mombelli, S.; Cochaud, S.; Merrouche, Y.; Garbar, C.; Antonicelli, F.; Laprevotte, E.; Alberici, G.; Bonnefoy, N.; Eliaou, J.F.; Bastid, J.; et al. IL-17A and its homologs IL-25/IL-17E recruit the c-RAF/S6 kinase pathway and the generation of pro-oncogenic LMW-E in breast cancer cells. *Sci. Rep.* **2015**, *5*, 11874. [CrossRef]

81. Merrouche, Y.; Fabre, J.; Cure, H.; Garbar, C.; Fuselier, C.; Bastid, J.; Antonicelli, F.; Al-Daccak, R.; Bensussan, A.; Giustiniani, J. IL-17E synergizes with EGF and confers in vitro resistance to EGFR-targeted therapies in TNBC cells. *Oncotarget* **2016**, *7*, 53350–53361. [CrossRef]
82. Shuai, C.; Yang, X.; Pan, H.; Han, W. Estrogen Receptor Downregulates Expression of PD-1/PD-L1 and Infiltration of CD8(+) T Cells by Inhibiting IL-17 Signaling Transduction in Breast Cancer. *Front. Oncol.* **2020**, *10*, 582863. [CrossRef] [PubMed]
83. Jiang, Z.; Chen, J.; Du, X.; Cheng, H.; Wang, X.; Dong, C. IL-25 blockade inhibits metastasis in breast cancer. *Protein Cell* **2017**, *8*, 191–201. [CrossRef] [PubMed]
84. Huang, C.K.; Yang, C.Y.; Jeng, Y.M.; Chen, C.L.; Wu, H.H.; Chang, Y.C.; Ma, C.; Kuo, W.H.; Chang, K.J.; Shew, J.Y.; et al. Autocrine/paracrine mechanism of interleukin-17B receptor promotes breast tumorigenesis through NF-κB-mediated antiapoptotic pathway. *Oncogene* **2014**, *33*, 2968–2977. [CrossRef] [PubMed]
85. Barati, M.; Sinaeian, M.; Shokrollahi Barough, M.; Pak, F.; Semnani, V.; Kokhaei, P.; Momtazi-Borojeni, A.A. Evaluation of Interleukin 25 and Interleukin 25 Receptor Expression in Peripheral Blood Mononuclear Cells of Breast Cancer Patients and Normal Subjects. *J. Interferon Cytokine Res. Off. J. Int. Soc. Interferon Cytokine Res.* **2020**, *40*, 139–144. [CrossRef] [PubMed]
86. Yin, S.Y.; Jian, F.Y.; Chen, Y.H.; Chien, S.C.; Hsieh, M.C.; Hsiao, P.W.; Lee, W.H.; Kuo, Y.H.; Yang, N.S. Induction of IL-25 secretion from tumour-associated fibroblasts suppresses mammary tumour metastasis. *Nat. Commun.* **2016**, *7*, 11311. [CrossRef]
87. Furuta, S.; Jeng, Y.M.; Zhou, L.; Huang, L.; Kuhn, I.; Bissell, M.J.; Lee, W.H. IL-25 causes apoptosis of IL-25R-expressing breast cancer cells without toxicity to nonmalignant cells. *Sci. Transl. Med.* **2011**, *3*, 78ra31. [CrossRef]
88. Gelaleti, G.B.; Borin, T.F.; Maschio-Signorini, L.B.; Moschetta, M.G.; Jardim-Perassi, B.V.; Calvinho, G.B.; Facchini, M.C.; Viloria-Petit, A.M.; de Campos Zuccari, D.A.P. Efficacy of melatonin, IL-25 and siIL-17B in tumorigenesis-associated properties of breast cancer cell lines. *Life Sci.* **2017**, *183*, 98–109. [CrossRef]
89. Younesi, V.; Nejatollahi, F. Induction of anti-proliferative and apoptotic effects by anti-IL-25 receptor single chain antibodies in breast cancer cells. *Int. Immunopharmacol.* **2014**, *23*, 624–632. [CrossRef]
90. Tian, T.; Yu, S.; Ma, D. Th22 and related cytokines in inflammatory and autoimmune diseases. *Expert. Opin. Ther. Targets* **2013**, *17*, 113–125. [CrossRef]
91. Eyerich, K.; Dimartino, V.; Cavani, A. IL-17 and IL-22 in immunity: Driving protection and pathology. *Eur. J. Immunol.* **2017**, *47*, 607–614. [CrossRef]
92. Doulabi, H.; Masoumi, E.; Rastin, M.; Foolady Azarnaminy, A.; Esmaeili, S.A.; Mahmoudi, M. The role of Th22 cells, from tissue repair to cancer progression. *Cytokine* **2022**, *149*, 155749. [CrossRef]
93. Cui, G. T(H)9, T(H)17, and T(H)22 Cell Subsets and Their Main Cytokine Products in the Pathogenesis of Colorectal Cancer. *Front. Oncol.* **2019**, *9*, 1002. [CrossRef]
94. Qin, J.J.; Yan, L.; Zhang, J.; Zhang, W.D. STAT3 as a potential therapeutic target in triple negative breast cancer: A systematic review. *J. Exp. Clin. Cancer Res. CR* **2019**, *38*, 195. [CrossRef] [PubMed]
95. Salmanpour, A.; Rezaeifard, S.; Kiani, R.; Tahmasebi, S.; Faghih, Z.; Erfani, N. IFNγ-IL-17-IL-22+CD4+ subset and IL-22-producing cells in tumor draining lymph nodes of patients with breast cancer. *Breast Dis.* **2022**, *41*, 383–390. [CrossRef] [PubMed]
96. Abdulwahid, A.G.; Abdullah, H.N. Expression of Serum IL-22, IL-23, and TLR9 as Tumor Markers in Untreated Breast Cancer Patients. 2020. Available online: http://impactfactor.org/PDF/IJDDT/10/IJDDT,Vol10,Issue3,Article30.pdf (accessed on 18 August 2023).
97. Zhao, J.; Liu, H.; Zhang, X.; Zhang, W.; Liu, L.; Yu, Y.; Ren, S.; Yang, Q.; Liu, B.; Li, J.; et al. Tumor Cells Interleukin-22 Expression Associates with Elevated Tumor-Associated Macrophages Infiltrating and Poor Prognosis in Patients with Breast Cancer. *Cancer Biother. Radiopharm.* **2021**, *36*, 160–166. [CrossRef] [PubMed]
98. Zhang, Y.; Liu, C.; Gao, J.; Shao, S.; Cui, Y.; Yin, S.; Pan, B. IL-22 promotes tumor growth of breast cancer cells in mice. *Aging* **2020**, *12*, 13354–13364. [CrossRef]
99. Rasé, V.J.; Hayward, R.; Haughian, J.M.; Pullen, N.A. T(h)17, T(h)22, and Myeloid-Derived Suppressor Cell Population Dynamics and Response to IL-6 in 4T1 Mammary Carcinoma. *Int. J. Mol. Sci.* **2022**, *23*, 10299. [CrossRef]
100. Katara, G.K.; Kulshrestha, A.; Schneiderman, S.; Riehl, V.; Ibrahim, S.; Beaman, K.D. Interleukin-22 promotes development of malignant lesions in a mouse model of spontaneous breast cancer. *Mol. Oncol.* **2020**, *14*, 211–224. [CrossRef]
101. Kim, E.Y.; Choi, B.; Kim, J.E.; Park, S.O.; Kim, S.M.; Chang, E.J. Interleukin-22 Mediates the Chemotactic Migration of Breast Cancer Cells and Macrophage Infiltration of the Bone Microenvironment by Potentiating S1P/SIPR Signaling. *Cells* **2020**, *9*, 131. [CrossRef]
102. Rui, J.; Chunming, Z.; Binbin, G.; Na, S.; Shengxi, W.; Wei, S. IL-22 promotes the progression of breast cancer through regulating HOXB-AS5. *Oncotarget* **2017**, *8*, 103601–103612. [CrossRef] [PubMed]
103. Voigt, C.; May, P.; Gottschlich, A.; Markota, A.; Wenk, D.; Gerlach, I.; Voigt, S.; Stathopoulos, G.T.; Arendt, K.A.M.; Heise, C.; et al. Cancer cells induce interleukin-22 production from memory CD4(+) T cells via interleukin-1 to promote tumor growth. *Proc. Natl. Acad. Sci. USA* **2017**, *114*, 12994–12999. [CrossRef] [PubMed]
104. Weber, G.F.; Gaertner, F.C.; Erl, W.; Janssen, K.P.; Blechert, B.; Holzmann, B.; Weighardt, H.; Essler, M. IL-22-mediated tumor growth reduction correlates with inhibition of ERK1/2 and AKT phosphorylation and induction of cell cycle arrest in the G2-M phase. *J. Immunol.* **2006**, *177*, 8266–8272. [CrossRef] [PubMed]
105. Chen, T.; Guo, J.; Cai, Z.; Li, B.; Sun, L.; Shen, Y.; Wang, S.; Wang, Z.; Wang, Z.; Wang, Y.; et al. Th9 Cell Differentiation and Its Dual Effects in Tumor Development. *Front. Immunol.* **2020**, *11*, 1026. [CrossRef]

106. Fischer, M.; Bijman, M.; Molin, D.; Cormont, F.; Uyttenhove, C.; van Snick, J.; Sundstrom, C.; Enblad, G.; Nilsson, G. Increased serum levels of interleukin-9 correlate to negative prognostic factors in Hodgkin's lymphoma. *Leukemia* **2003**, *17*, 2513–2516. [CrossRef] [PubMed]
107. Elyaman, W.; Bradshaw, E.M.; Uyttenhove, C.; Dardalhon, V.; Awasthi, A.; Imitola, J.; Bettelli, E.; Oukka, M.; van Snick, J.; Renauld, J.C.; et al. IL-9 induces differentiation of TH17 cells and enhances function of FoxP3+ natural regulatory T cells. *Proc. Natl. Acad. Sci. USA* **2009**, *106*, 12885–12890. [CrossRef]
108. Fang, Y.; Chen, X.; Bai, Q.; Qin, C.; Mohamud, A.O.; Zhu, Z.; Ball, T.W.; Ruth, C.M.; Newcomer, D.R.; Herrick, E.J.; et al. IL-9 inhibits HTB-72 melanoma cell growth through upregulation of p21 and TRAIL. *J. Surg. Oncol.* **2015**, *111*, 969–974. [CrossRef]
109. Schlapbach, C.; Gehad, A.; Yang, C.; Watanabe, R.; Guenova, E.; Teague, J.E.; Campbell, L.; Yawalkar, N.; Kupper, T.S.; Clark, R.A. Human TH9 cells are skin-tropic and have autocrine and paracrine proinflammatory capacity. *Sci. Transl. Med.* **2014**, *6*, 219ra218. [CrossRef]
110. Lu, Y.; Hong, B.; Li, H.; Zheng, Y.; Zhang, M.; Wang, S.; Qian, J.; Yi, Q. Tumor-specific IL-9-producing CD8+ Tc9 cells are superior effector than type-I cytotoxic Tc1 cells for adoptive immunotherapy of cancers. *Proc. Natl. Acad. Sci. USA* **2014**, *111*, 2265–2270. [CrossRef]
111. Carlsson, A.; Wingren, C.; Kristensson, M.; Rose, C.; Fernö, M.; Olsson, H.; Jernström, H.; Ek, S.; Gustavsson, E.; Ingvar, C.; et al. Molecular serum portraits in patients with primary breast cancer predict the development of distant metastases. *Proc. Natl. Acad. Sci. USA* **2011**, *108*, 14252–14257. [CrossRef]
112. You, F.P.; Zhang, J.; Cui, T.; Zhu, R.; Lv, C.Q.; Tang, H.T.; Sun, D.W. Th9 cells promote antitumor immunity via IL-9 and IL-21 and demonstrate atypical cytokine expression in breast cancer. *Int. Immunopharmacol.* **2017**, *52*, 163–167. [CrossRef]
113. Ding, P.; Zhu, R.; Cai, B.; Zhang, J.; Bu, Q.; Sun, D.W. IL-9-producing CD8(+) T cells represent a distinctive subset with different transcriptional characteristics from conventional CD8(+) T cells, and partially infiltrate breast tumors. *Int. J. Biochem. Cell Biol.* **2019**, *115*, 105576. [CrossRef] [PubMed]
114. Hoelzinger, D.B.; Dominguez, A.L.; Cohen, P.A.; Gendler, S.J. Inhibition of adaptive immunity by IL-9 can be disrupted to achieve rapid T cell sensitization and rejection of progressive tumor challenges. *Cancer Res.* **2014**, *74*, 6845–6855. [CrossRef] [PubMed]
115. Zhang, J.; Roberts, A.I.; Liu, C.; Ren, G.; Xu, G.; Zhang, L.; Devadas, S.; Shi, Y. A novel subset of helper T cells promotes immune responses by secreting GM-CSF. *Cell Death Differ.* **2013**, *20*, 1731–1741. [CrossRef] [PubMed]
116. Shiomi, A.; Usui, T. Pivotal roles of GM-CSF in autoimmunity and inflammation. *Mediat. Inflamm.* **2015**, *2015*, 568543. [CrossRef]
117. Robinson, R.T. T Cell Production of GM-CSF Protects the Host during Experimental Tuberculosis. *mBio* **2017**, *8*, e02087-17. [CrossRef]
118. Ariafar, A.; Zareinejad, M.; Soltani, M.; Vahidi, Y.; Faghih, Z. GM-CSF-producing lymphocytes in tumor-draining lymph nodes of patients with bladder cancer. *Eur. Cytokine Netw.* **2021**, *32*, 1–7. [CrossRef]
119. Kumar, A.; Taghi Khani, A.; Sanchez Ortiz, A.; Swaminathan, S. GM-CSF: A Double-Edged Sword in Cancer Immunotherapy. *Front. Immunol.* **2022**, *13*, 901277. [CrossRef]
120. Chaubey, N.; Ghosh, S.S. Overexpression of granulocyte macrophage colony stimulating factor in breast cancer cells leads towards drug sensitization. *Appl. Biochem. Biotechnol.* **2015**, *175*, 1948–1959. [CrossRef]
121. Eubank, T.D.; Roberts, R.; Galloway, M.; Wang, Y.; Cohn, D.E.; Marsh, C.B. GM-CSF induces expression of soluble VEGF receptor-1 from human monocytes and inhibits angiogenesis in mice. *Immunity* **2004**, *21*, 831–842. [CrossRef]
122. Razmkhah, M.; Karimzadeh, P.; Eghbali, F.; Rezaeifard, S.; Faghih, Z. Serum level of colony stimulating factors, Granulocyte, Monocyte and Granulocyte-Monocyte, in peripheral blood of patients with breast cancer. *J. Sabzevar Univ. Med. Sci.* **2019**, *25*, 809–817.
123. Al-Rashed, F.; Thomas, R.; Al-Roub, A.; Al-Mulla, F.; Ahmad, R. LPS Induces GM-CSF Production by Breast Cancer MDA-MB-231 Cells via Long-Chain Acyl-CoA Synthetase 1. *Molecules* **2020**, *25*, 4709. [CrossRef]
124. Thomas, R.; Al-Rashed, F.; Akhter, N.; Al-Mulla, F.; Ahmad, R. ACSL1 Regulates TNFα-Induced GM-CSF Production by Breast Cancer MDA-MB-231 Cells. *Biomolecules* **2019**, *9*, 555. [CrossRef]
125. Yonemitsu, K.; Pan, C.; Fujiwara, Y.; Miyasato, Y.; Shiota, T.; Yano, H.; Hosaka, S.; Tamada, K.; Yamamoto, Y.; Komohara, Y. GM-CSF derived from the inflammatory microenvironment potentially enhanced PD-L1 expression on tumor-associated macrophages in human breast cancer. *Sci. Rep.* **2022**, *12*, 12007. [CrossRef] [PubMed]
126. Ghirelli, C.; Reyal, F.; Jeanmougin, M.; Zollinger, R.; Sirven, P.; Michea, P.; Caux, C.; Bendriss-Vermare, N.; Donnadieu, M.H.; Caly, M.; et al. Breast Cancer Cell-Derived GM-CSF Licenses Regulatory Th2 Induction by Plasmacytoid Predendritic Cells in Aggressive Disease Subtypes. *Cancer Res.* **2015**, *75*, 2775–2787. [CrossRef] [PubMed]
127. Su, X.; Xu, Y.; Fox, G.C.; Xiang, J.; Kwakwa, K.A.; Davis, J.L.; Belle, J.I.; Lee, W.C.; Wong, W.H.; Fontana, F.; et al. Breast cancer-derived GM-CSF regulates arginase 1 in myeloid cells to promote an immunosuppressive microenvironment. *J. Clin. Investig.* **2021**, *131*, e145296. [CrossRef]
128. Cho, H.; Seo, Y.; Loke, K.M.; Kim, S.W.; Oh, S.M.; Kim, J.H.; Soh, J.; Kim, H.S.; Lee, H.; Kim, J.; et al. Cancer-Stimulated CAFs Enhance Monocyte Differentiation and Protumoral TAM Activation via IL6 and GM-CSF Secretion. *Clin. Cancer Res. Off. J. Am. Assoc. Cancer Res.* **2018**, *24*, 5407–5421. [CrossRef]
129. Espinoza-Sánchez, N.A.; Vadillo, E.; Balandrán, J.C.; Monroy-García, A.; Pelayo, R.; Fuentes-Pananá, E.M. Evidence of lateral transmission of aggressive features between different types of breast cancer cells. *Int. J. Oncol.* **2017**, *51*, 1482–1496. [CrossRef]

130. Su, S.; Liu, Q.; Chen, J.; Chen, J.; Chen, F.; He, C.; Huang, D.; Wu, W.; Lin, L.; Huang, W.; et al. A positive feedback loop between mesenchymal-like cancer cells and macrophages is essential to breast cancer metastasis. *Cancer Cell* **2014**, *25*, 605–620. [CrossRef]
131. Chen, G.; Gupta, R.; Petrik, S.; Laiko, M.; Leatherman, J.M.; Asquith, J.M.; Daphtary, M.M.; Garrett-Mayer, E.; Davidson, N.E.; Hirt, K.; et al. A feasibility study of cyclophosphamide, trastuzumab, and an allogeneic GM-CSF-secreting breast tumor vaccine for HER2+ metastatic breast cancer. *Cancer Immunol. Res.* **2014**, *2*, 949–961. [CrossRef]
132. Mittendorf, E.A.; Lu, B.; Melisko, M.; Price Hiller, J.; Bondarenko, I.; Brunt, A.M.; Sergii, G.; Petrakova, K.; Peoples, G.E. Efficacy and Safety Analysis of Nelipepimut-S Vaccine to Prevent Breast Cancer Recurrence: A Randomized, Multicenter, Phase III Clinical Trial. *Clin. Cancer Res. Off. J. Am. Assoc. Cancer Res.* **2019**, *25*, 4248–4254. [CrossRef] [PubMed]
133. King, C.; Tangye, S.G.; Mackay, C.R. T follicular helper (TFH) cells in normal and dysregulated immune responses. *Annu. Rev. Immunol.* **2008**, *26*, 741–766. [CrossRef]
134. Liu, X.; Nurieva, R.I.; Dong, C. Transcriptional regulation of follicular T-helper (Tfh) cells. *Immunol. Rev.* **2013**, *252*, 139–145. [CrossRef] [PubMed]
135. Gutiérrez-Melo, N.; Baumjohann, D. T follicular helper cells in cancer. *Trends Cancer* **2023**, *9*, 309–325. [CrossRef] [PubMed]
136. Noël, G.; Fontsa, M.L.; Garaud, S.; De Silva, P.; de Wind, A.; Van den Eynden, G.G.; Salgado, R.; Boisson, A.; Locy, H.; Thomas, N.; et al. Functional Th1-oriented T follicular helper cells that infiltrate human breast cancer promote effective adaptive immunity. *J. Clin. Investig.* **2021**, *131*, e139905. [CrossRef]
137. Razis, E.; Kalogeras, K.T.; Kotoula, V.; Eleftheraki, A.G.; Nikitas, N.; Kronenwett, R.; Timotheadou, E.; Christodoulou, C.; Pectasides, D.; Gogas, H.; et al. Improved outcome of high-risk early HER2 positive breast cancer with high CXCL13-CXCR5 messenger RNA expression. *Clin. Breast Cancer* **2012**, *12*, 183–193. [CrossRef] [PubMed]
138. Panse, J.; Friedrichs, K.; Marx, A.; Hildebrandt, Y.; Luetkens, T.; Barrels, K.; Horn, C.; Stahl, T.; Cao, Y.; Milde-Langosch, K.; et al. Chemokine CXCL13 is overexpressed in the tumour tissue and in the peripheral blood of breast cancer patients. *Br. J. Cancer* **2008**, *99*, 930–938. [CrossRef]
139. Biswas, S.; Sengupta, S.; Roy Chowdhury, S.; Jana, S.; Mandal, G.; Mandal, P.K.; Saha, N.; Malhotra, V.; Gupta, A.; Kuprash, D.V.; et al. CXCL13-CXCR5 co-expression regulates epithelial to mesenchymal transition of breast cancer cells during lymph node metastasis. *Breast Cancer Res. Treat.* **2014**, *143*, 265–276. [CrossRef]
140. Gu-Trantien, C.; Migliori, E.; Buisseret, L.; de Wind, A.; Brohée, S.; Garaud, S.; Noël, G.; Dang Chi, V.L.; Lodewyckx, J.N.; Naveaux, C.; et al. CXCL13-producing TFH cells link immune suppression and adaptive memory in human breast cancer. *JCI Insight* **2017**, *2*, e91487. [CrossRef]
141. Zhu, S.; Lin, J.; Qiao, G.; Wang, X.; Xu, Y. Tim-3 identifies exhausted follicular helper T cells in breast cancer patients. *Immunobiology* **2016**, *221*, 986–993. [CrossRef]
142. Linterman, M.A.; Pierson, W.; Lee, S.K.; Kallies, A.; Kawamoto, S.; Rayner, T.F.; Srivastava, M.; Divekar, D.P.; Beaton, L.; Hogan, J.J.; et al. Foxp3+ follicular regulatory T cells control the germinal center response. *Nat. Med.* **2011**, *17*, 975–982. [CrossRef] [PubMed]
143. Li, L.; Ma, Y.; Xu, Y. Follicular regulatory T cells infiltrated the ovarian carcinoma and resulted in CD8 T cell dysfunction dependent on IL-10 pathway. *Int. Immunopharmacol.* **2019**, *68*, 81–87. [CrossRef] [PubMed]
144. Cha, Z.; Gu, H.; Zang, Y.; Wang, Z.; Li, J.; Huang, W.; Qin, A.; Zhu, L.; Tu, X.; Cheng, N.; et al. The prevalence and function of CD4(+)CXCR5(+)Foxp3(+) follicular regulatory T cells in diffuse large B cell lymphoma. *Int. Immunopharmacol.* **2018**, *61*, 132–139. [CrossRef]
145. Brady, M.T.; Hilchey, S.P.; Hyrien, O.; Spence, S.A.; Bernstein, S.H. Mesenchymal stromal cells support the viability and differentiation of follicular lymphoma-infiltrating follicular helper T-cells. *PLoS ONE* **2014**, *9*, e97597. [CrossRef] [PubMed]
146. Song, H.; Liu, A.; Liu, G.; Wu, F.; Li, Z. T follicular regulatory cells suppress Tfh-mediated B cell help and synergistically increase IL-10-producing B cells in breast carcinoma. *Immunol. Res.* **2019**, *67*, 416–423. [CrossRef]
147. Xie, M.M.; Dent, A.L. Unexpected Help: Follicular Regulatory T Cells in the Germinal Center. *Front. Immunol.* **2018**, *9*, 1536. [CrossRef]
148. Leong, P.P.; Mohammad, R.; Ibrahim, N.; Ithnin, H.; Abdullah, M.; Davis, W.C.; Seow, H.F. Phenotyping of lymphocytes expressing regulatory and effector markers in infiltrating ductal carcinoma of the breast. *Immunol. Lett.* **2006**, *102*, 229–236. [CrossRef]
149. Sun, Y.P.; Ke, Y.L.; Li, X. Prognostic value of CD8(+) tumor-infiltrating T cells in patients with breast cancer: A systematic review and meta-analysis. *Oncol. Lett.* **2023**, *25*, 39. [CrossRef]
150. Li, K.; Li, T.; Feng, Z.; Huang, M.; Wei, L.; Yan, Z.; Long, M.; Hu, Q.; Wang, J.; Liu, S.; et al. CD8(+) T cell immunity blocks the metastasis of carcinogen-exposed breast cancer. *Sci. Adv.* **2021**, *7*, eabd8936. [CrossRef]
151. Egelston, C.A.; Avalos, C.; Tu, T.Y.; Simons, D.L.; Jimenez, G.; Jung, J.Y.; Melstrom, L.; Margolin, K.; Yim, J.H.; Kruper, L.; et al. Human breast tumor-infiltrating CD8(+) T cells retain polyfunctionality despite PD-1 expression. *Nat. Commun.* **2018**, *9*, 4297. [CrossRef]
152. Savas, P.; Virassamy, B.; Ye, C.; Salim, A.; Mintoff, C.P.; Caramia, F.; Salgado, R.; Byrne, D.J.; Teo, Z.L.; Dushyanthen, S.; et al. Single-cell profiling of breast cancer T cells reveals a tissue-resident memory subset associated with improved prognosis. *Nat. Med.* **2018**, *24*, 986–993. [CrossRef] [PubMed]
153. Meng, S.; Li, L.; Zhou, M.; Jiang, W.; Niu, H.; Yang, K. Distribution and prognostic value of tumorinfiltrating T cells in breast cancer. *Mol. Med. Rep.* **2018**, *18*, 4247–4258. [CrossRef]

154. Ali, H.R.; Provenzano, E.; Dawson, S.J.; Blows, F.M.; Liu, B.; Shah, M.; Earl, H.M.; Poole, C.J.; Hiller, L.; Dunn, J.A.; et al. Association between CD8+ T-cell infiltration and breast cancer survival in 12,439 patients. *Ann. Oncol. Off. J. Eur. Soc. Med. Oncol.* **2014**, *25*, 1536–1543. [CrossRef]
155. Baker, K.; Lachapelle, J.; Zlobec, I.; Bismar, T.A.; Terracciano, L.; Foulkes, W.D. Prognostic significance of CD8+ T lymphocytes in breast cancer depends upon both oestrogen receptor status and histological grade. *Histopathology* **2011**, *58*, 1107–1116. [CrossRef] [PubMed]
156. Wang, K.; Shen, T.; Siegal, G.P.; Wei, S. The CD4/CD8 ratio of tumor-infiltrating lymphocytes at the tumor-host interface has prognostic value in triple-negative breast cancer. *Hum. Pathol.* **2017**, *69*, 110–117. [CrossRef] [PubMed]
157. St Paul, M.; Ohashi, P.S. The Roles of CD8(+) T Cell Subsets in Antitumor Immunity. *Trends Cell Biol.* **2020**, *30*, 695–704. [CrossRef]
158. Nam, J.S.; Terabe, M.; Kang, M.J.; Chae, H.; Voong, N.; Yang, Y.; Laurence, A.; Michalowska, A.; Mamura, M.; Lonning, S.; et al. TGF-β subverts the immune system into directly promoting tumor growth through IL-17. *Cancer Res.* **2008**, *68*, 3915–3923. [CrossRef] [PubMed]
159. Zhu, S.; Lin, J.; Qiao, G.; Xu, Y.; Zou, H. Differential regulation and function of tumor-infiltrating T cells in different stages of breast cancer patients. *Tumour Biol. J. Int. Soc. Oncodev. Biol. Med.* **2015**, *36*, 7907–7913. [CrossRef]
160. Schule, J.M.; Bergkvist, L.; Hakansson, L.; Gustafsson, B.; Hakansson, A. CD28 expression in sentinel node biopsies from breast cancer patients in comparison with CD3-zeta chain expression. *J. Transl. Med.* **2004**, *2*, 45. [CrossRef] [PubMed]
161. Dolina, J.S.; Van Braeckel-Budimir, N.; Thomas, G.D.; Salek-Ardakani, S. CD8(+) T Cell Exhaustion in Cancer. *Front. Immunol.* **2021**, *12*, 715234. [CrossRef]
162. Shariati, S.; Ghods, A.; Zohouri, M.; Rasolmali, R.; Talei, A.R.; Mehdipour, F.; Ghaderi, A. Significance of TIM-3 expression by CD4(+) and CD8(+) T lymphocytes in tumor-draining lymph nodes from patients with breast cancer. *Mol. Immunol.* **2020**, *128*, 47–54. [CrossRef] [PubMed]
163. Du, H.; Yi, Z.; Wang, L.; Li, Z.; Niu, B.; Ren, G. The co-expression characteristics of LAG3 and PD-1 on the T cells of patients with breast cancer reveal a new therapeutic strategy. *Int. Immunopharmacol.* **2020**, *78*, 106113. [CrossRef] [PubMed]
164. Chauvin, J.M.; Pagliano, O.; Fourcade, J.; Sun, Z.; Wang, H.; Sander, C.; Kirkwood, J.M.; Chen, T.H.; Maurer, M.; Korman, A.J.; et al. TIGIT and PD-1 impair tumor antigen-specific $CD8^+$ T cells in melanoma patients. *J. Clin. Investig.* **2015**, *125*, 2046–2058. [CrossRef]
165. Canale, F.P.; Ramello, M.C.; Nunez, N.; Araujo Furlan, C.L.; Bossio, S.N.; Gorosito Serran, M.; Tosello Boari, J.; Del Castillo, A.; Ledesma, M.; Sedlik, C.; et al. CD39 Expression Defines Cell Exhaustion in Tumor-Infiltrating CD8(+) T Cells. *Cancer Res.* **2018**, *78*, 115–128. [CrossRef] [PubMed]
166. Tallón de Lara, P.; Castañón, H.; Vermeer, M.; Núñez, N.; Silina, K.; Sobottka, B.; Urdinez, J.; Cecconi, V.; Yagita, H.; Movahedian Attar, F.; et al. CD39(+)PD-1(+)CD8(+) T cells mediate metastatic dormancy in breast cancer. *Nat. Commun.* **2021**, *12*, 769. [CrossRef]
167. Cortesini, R.; LeMaoult, J.; Ciubotariu, R.; Cortesini, N.S.F. CD8+ CD28− T suppressor cells and the induction of antigen-specific, antigen-presenting cell-mediated suppression of Th reactivity. *Immunol. Rev.* **2001**, *182*, 201–206. [CrossRef]
168. Liu, Q.; Sun, Z.; Chen, L. Memory T cells: Strategies for optimizing tumor immunotherapy. *Protein Cell* **2020**, *11*, 549–564. [CrossRef]
169. Coventry, B.J.; Weightman, M.J.; Bradley, J.; Skinner, J.M. Immune profiling in human breast cancer using high-sensitivity detection and analysis techniques. *JRSM Open* **2015**, *6*, 2054270415603909. [CrossRef]
170. Buisseret, L.; Garaud, S.; de Wind, A.; Van den Eynden, G.; Boisson, A.; Solinas, C.; Gu-Trantien, C.; Naveaux, C.; Lodewyckx, J.-N.; Duvillier, H. Tumor-infiltrating lymphocyte composition, organization and PD-1/PD-L1 expression are linked in breast cancer. *Oncoimmunology* **2017**, *6*, e1257452. [CrossRef]
171. Savas, P.; Salgado, R.; Denkert, C.; Sotiriou, C.; Darcy, P.K.; Smyth, M.J.; Loi, S. Clinical relevance of host immunity in breast cancer: From TILs to the clinic. *Nat. Rev. Clin. Oncol.* **2016**, *13*, 228. [CrossRef]
172. Schnellhardt, S.; Erber, R.; Büttner-Herold, M.; Rosahl, M.C.; Ott, O.J.; Strnad, V.; Beckmann, M.W.; King, L.; Hartmann, A.; Fietkau, R.; et al. Tumour-Infiltrating Inflammatory Cells in Early Breast Cancer: An Underrated Prognostic and Predictive Factor? *Int. J. Mol. Sci.* **2020**, *21*, 8238. [CrossRef] [PubMed]
173. Yajima, R.; Yajima, T.; Fujii, T.; Yanagita, Y.; Fujisawa, T.; Miyamoto, T.; Tsutsumi, S.; Iijima, M.; Kuwano, H. Tumor-infiltrating CD45RO(+) memory cells are associated with a favorable prognosis breast cancer. *Breast. Cancer* **2016**, *23*, 668–674. [CrossRef] [PubMed]
174. Fu, P.; Li, X.; Shi, P. CD45RO positive expression correlates with lower histological grade, less lymph node metastasis and prolonged overall survival in surgical patients with breast cancer. *Int. J. Clin. Exp. Med.* **2017**, *10*, 16336–16343.
175. Naylor, K.; Li, G.; Vallejo, A.N.; Lee, W.-W.; Koetz, K.; Bryl, E.; Witkowski, J.; Fulbright, J.; Weyand, C.M.; Goronzy, J.J. The influence of age on T cell generation and TCR diversity. *J. Immunol.* **2005**, *174*, 7446–7452. [CrossRef] [PubMed]
176. Dieci, M.; Mathieu, M.; Guarneri, V.; Conte, P.; Delaloge, S.; Andre, F.; Goubar, A. Prognostic and predictive value of tumor-infiltrating lymphocytes in two phase III randomized adjuvant breast cancer trials. *Ann. Oncol.* **2015**, *26*, 1698–1704. [CrossRef]
177. Loi, S.; Michiels, S.; Salgado, R.; Sirtaine, N.; Jose, V.; Fumagalli, D.; Kellokumpu-Lehtinen, P.-L.; Bono, P.; Kataja, V.; Desmedt, C. Tumor infiltrating lymphocytes are prognostic in triple negative breast cancer and predictive for trastuzumab benefit in early breast cancer: Results from the FinHER trial. *Ann. Oncol.* **2014**, *25*, 1544–1550. [CrossRef] [PubMed]

178. Adams, S.; Goldstein, L.J.; Sparano, J.A.; Demaria, S.; Badve, S.S. Tumor infiltrating lymphocytes (TILs) improve prognosis in patients with triple negative breast cancer (TNBC). *Oncoimmunology* **2015**, *4*, e985930. [CrossRef]
179. Cejalvo, J.M.; Pascual, T.; Fernández-Martínez, A.; Brasó-Maristany, F.; Gomis, R.R.; Perou, C.M.; Muñoz, M.; Prat, A. Clinical implications of the non-luminal intrinsic subtypes in hormone receptor-positive breast cancer. *Cancer Treat. Rev.* **2018**, *67*, 63–70. [CrossRef]
180. Bardou, V.-J.; Arpino, G.; Elledge, R.M.; Osborne, C.K.; Clark, G.M. Progesterone receptor status significantly improves outcome prediction over estrogen receptor status alone for adjuvant endocrine therapy in two large breast cancer databases. *J. Clin. Oncol.* **2003**, *21*, 1973–1979. [CrossRef]
181. Järvinen, T.A.; Pelto-Huikko, M.; Holli, K.; Isola, J. Estrogen receptor β is coexpressed with ERα and PR and associated with nodal status, grade, and proliferation rate in breast cancer. *Am. J. Pathol.* **2000**, *156*, 29–35. [CrossRef]
182. Triulzi, T.; Forte, L.; Regondi, V.; Di Modica, M.; Ghirelli, C.; Carcangiu, M.L.; Sfondrini, L.; Balsari, A.; Tagliabue, E. HER2 signaling regulates the tumor immune microenvironment and trastuzumab efficacy. *Oncoimmunology* **2019**, *8*, e1512942. [CrossRef] [PubMed]
183. Jameson, S.C.; Masopust, D. Understanding Subset Diversity in T Cell Memory. *Immunity* **2018**, *48*, 214–226. [CrossRef] [PubMed]
184. Wang, Z.Q.; Milne, K.; Derocher, H.; Webb, J.R.; Nelson, B.H.; Watson, P.H. CD103 and Intratumoral Immune Response in Breast Cancer. *Clin. Cancer Res. Off. J. Am. Assoc. Cancer Res.* **2016**, *22*, 6290–6297. [CrossRef] [PubMed]
185. Lee, Y.J.; Kim, J.Y.; Jeon, S.H.; Nam, H.; Jung, J.H.; Jeon, M.; Kim, E.S.; Bae, S.J.; Ahn, J.; Yoo, T.K.; et al. CD39(+) tissue-resident memory CD8(+) T cells with a clonal overlap across compartments mediate antitumor immunity in breast cancer. *Sci. Immunol.* **2022**, *7*, eabn8390. [CrossRef]
186. Vahidi, Y.; Faghih, Z.; Talei, A.R.; Doroudchi, M.; Ghaderi, A. Memory CD4(+) T cell subsets in tumor draining lymph nodes of breast cancer patients: A focus on T stem cell memory cells. *Cell. Oncol.* **2018**, *41*, 1–11. [CrossRef]
187. Vahidi, Y.; Bagheri, M.; Ghaderi, A.; Faghih, Z. CD8-positive memory T cells in tumor-draining lymph nodes of patients with breast cancer. *BMC Cancer* **2020**, *20*, 257. [CrossRef]
188. Legat, A.; Speiser, D.E.; Pircher, H.; Zehn, D.; Fuertes Marraco, S.A. Inhibitory Receptor Expression Depends More Dominantly on Differentiation and Activation than "Exhaustion" of Human CD8 T Cells. *Front. Immunol.* **2013**, *4*, 455. [CrossRef]
189. Kohrt, H.E.; Nouri, N.; Nowels, K.; Johnson, D.; Holmes, S.; Lee, P.P. Profile of immune cells in axillary lymph nodes predicts disease-free survival in breast cancer. *PLoS Med.* **2005**, *2*, e284. [CrossRef]
190. Razmkhah, M.; Abedi, N.; Hosseini, A.; Imani, M.T.; Talei, A.R.; Ghaderi, A. Induction of T regulatory subsets from naive CD4+ T cells after exposure to breast cancer adipose derived stem cells. *Iran. J. Immunol. IJI* **2015**, *12*, 1–15.
191. Razmkhah, M.; Jaberipour, M.; Erfani, N.; Habibagahi, M.; Talei, A.R.; Ghaderi, A. Adipose derived stem cells (ASCs) isolated from breast cancer tissue express IL-4, IL-10 and TGF-beta1 and upregulate expression of regulatory molecules on T cells: Do they protect breast cancer cells from the immune response? *Cell. Immunol.* **2011**, *266*, 116–122. [CrossRef]
192. Jafarinia, M.; Mehdipour, F.; Hosseini, S.V.; Ghahramani, L.; Hosseinzadeh, M.; Ghaderi, A. Determination of a CD4(+)CD25(-)FoxP3(+) T cells subset in tumor-draining lymph nodes of colorectal cancer secreting IL-2 and IFN-γ. *Tumour Biol. J. Int. Soc. Oncodev. Biol. Med.* **2016**, *37*, 14659–14666. [CrossRef] [PubMed]
193. Zhou, Y.; Shao, N.; Aierken, N.; Xie, C.; Ye, R.; Qian, X.; Hu, Z.; Zhang, J.; Lin, Y. Prognostic value of tumor-infiltrating Foxp3+ regulatory T cells in patients with breast cancer: A meta-analysis. *J. Cancer* **2017**, *8*, 4098–4105. [CrossRef] [PubMed]
194. Shou, J.; Zhang, Z.; Lai, Y.; Chen, Z.; Huang, J. Worse outcome in breast cancer with higher tumor-infiltrating FOXP3+ Tregs: A systematic review and meta-analysis. *BMC Cancer* **2016**, *16*, 687. [CrossRef] [PubMed]
195. Mahmoud, S.M.; Paish, E.C.; Powe, D.G.; Macmillan, R.D.; Lee, A.H.; Ellis, I.O.; Green, A.R. An evaluation of the clinical significance of FOXP3+ infiltrating cells in human breast cancer. *Breast Cancer Res. Treat.* **2011**, *127*, 99–108. [CrossRef]
196. Jiang, D.; Gao, Z.; Cai, Z.; Wang, M.; He, J. Clinicopathological and prognostic significance of FOXP3+ tumor infiltrating lymphocytes in patients with breast cancer: A meta-analysis. *BMC Cancer* **2015**, *15*, 727. [CrossRef]
197. Bates, G.J.; Fox, S.B.; Han, C.; Leek, R.D.; Garcia, J.F.; Harris, A.L.; Banham, A.H. Quantification of regulatory T cells enables the identification of high-risk breast cancer patients and those at risk of late relapse. *J. Clin. Oncol. Off. J. Am. Soc. Clin. Oncol.* **2006**, *24*, 5373–5380. [CrossRef]
198. Niakan, A.; Faghih, Z.; Talei, A.R.; Ghaderi, A. Cytokine profile of CD4(+)CD25(-)FoxP3(+) T cells in tumor-draining lymph nodes from patients with breast cancer. *Mol. Immunol.* **2019**, *116*, 90–97. [CrossRef]
199. Bos, P.D.; Plitas, G.; Rudra, D.; Lee, S.Y.; Rudensky, A.Y. Transient regulatory T cell ablation deters oncogene-driven breast cancer and enhances radiotherapy. *J. Exp. Med.* **2013**, *210*, 2435–2466. [CrossRef]
200. Chen, L.; Huang, T.G.; Meseck, M.; Mandeli, J.; Fallon, J.; Woo, S.L. Rejection of metastatic 4T1 breast cancer by attenuation of Treg cells in combination with immune stimulation. *Mol. Ther. J. Am. Soc. Gene Ther.* **2007**, *15*, 2194–2202. [CrossRef]
201. Ladoire, S.; Arnould, L.; Mignot, G.; Coudert, B.; Rebe, C.; Chalmin, F.; Vincent, J.; Bruchard, M.; Chauffert, B.; Martin, F.; et al. Presence of Foxp3 expression in tumor cells predicts better survival in HER2-overexpressing breast cancer patients treated with neoadjuvant chemotherapy. *Breast Cancer Res. Treat.* **2011**, *125*, 65–72. [CrossRef]
202. Ladoire, S.; Mignot, G.; Dalban, C.; Chevriaux, A.; Arnould, L.; Rebe, C.; Apetoh, L.; Boidot, R.; Penault-Llorca, F.; Fumoleau, P.; et al. FOXP3 expression in cancer cells and anthracyclines efficacy in patients with primary breast cancer treated with adjuvant chemotherapy in the phase III UNICANCER-PACS 01 trial. *Ann. Oncol. Off. J. Eur. Soc. Med. Oncol.* **2012**, *23*, 2552–2561. [CrossRef]

203. Syed Khaja, A.S.; Toor, S.M.; El Salhat, H.; Faour, I.; Ul Haq, N.; Ali, B.R.; Elkord, E. Preferential accumulation of regulatory T cells with highly immunosuppressive characteristics in breast tumor microenvironment. *Oncotarget* **2017**, *8*, 33159–33171. [CrossRef]
204. Plitas, G.; Konopacki, C.; Wu, K.; Bos, P.; Morrow, M.; Putintseva, E.V.; Chudakov, D.M.; Rudensky, A.Y. Regulatory T cells exhibit distinct features in human breast cancer. *Immunity* **2016**, *45*, 1122–1134. [CrossRef]
205. Li, Y.Q.; Liu, F.F.; Zhang, X.M.; Guo, X.J.; Ren, M.J.; Fu, L. Tumor secretion of CCL22 activates intratumoral Treg infiltration and is independent prognostic predictor of breast cancer. *PLoS ONE* **2013**, *8*, e76379. [CrossRef]
206. Gobert, M.; Treilleux, I.; Bendriss-Vermare, N.; Bachelot, T.; Goddard-Leon, S.; Arfi, V.; Biota, C.; Doffin, A.C.; Durand, I.; Olive, D.; et al. Regulatory T cells recruited through CCL22/CCR4 are selectively activated in lymphoid infiltrates surrounding primary breast tumors and lead to an adverse clinical outcome. *Cancer Res.* **2009**, *69*, 2000–2009. [CrossRef] [PubMed]
207. Freier, C.P.; Kuhn, C.; Endres, S.; Mayr, D.; Friese, K.; Jeschke, U.; Anz, D. FOXP3+ Cells Recruited by CCL22 into Breast Cancer Correlates with Less Tumor Nodal Infiltration. *Anticancer Res.* **2016**, *36*, 3139–3145. [PubMed]
208. Kuehnemuth, B.; Piseddu, I.; Wiedemann, G.M.; Lauseker, M.; Kuhn, C.; Hofmann, S.; Schmoeckel, E.; Endres, S.; Mayr, D.; Jeschke, U.; et al. CCL1 is a major regulatory T cell attracting factor in human breast cancer. *BMC Cancer* **2018**, *18*, 1278. [CrossRef] [PubMed]
209. Zohouri, M.; Mehdipour, F.; Razmkhah, M.; Faghih, Z.; Ghaderi, A. CD4(+)CD25(-)FoxP3(+) T cells: A distinct subset or a heterogeneous population? *Int. Rev. Immunol.* **2021**, *40*, 307–316. [CrossRef] [PubMed]
210. Yuen, G.J.; Demissie, E.; Pillai, S. B lymphocytes and cancer: A love-hate relationship. *Trends Cancer* **2016**, *2*, 747–757. [CrossRef]
211. Li, M.; Quintana, A.; Alberts, E.; Hung, M.S.; Boulat, V.; Ripoll, M.M.; Grigoriadis, A. B Cells in Breast Cancer Pathology. *Cancers* **2023**, *15*, 1517. [CrossRef]
212. Linnebacher, M.; Maletzki, C. Tumor-infiltrating B cells: The ignored players in tumor immunology. *Oncoimmunology* **2012**, *1*, 1186–1188. [CrossRef]
213. Mehdipour, F.; Razmkhah, M.; Hosseini, A.; Bagheri, M.; Safaei, A.; Talei, A.R.; Ghaderi, A. Increased B Regulatory Phenotype in Non-Metastatic Lymph Nodes of Node-Positive Breast Cancer Patients. *Scand. J. Immunol.* **2016**, *83*, 195–202. [CrossRef] [PubMed]
214. Kuroda, H.; Jamiyan, T.; Yamaguchi, R.; Kakumoto, A.; Abe, A.; Harada, O.; Masunaga, A. Tumor-infiltrating B cells and T cells correlate with postoperative prognosis in triple-negative carcinoma of the breast. *BMC Cancer* **2021**, *21*, 286. [CrossRef]
215. Garaud, S.; Buisseret, L.; Solinas, C.; Gu-Trantien, C.; de Wind, A.; Van den Eynden, G.; Naveaux, C.; Lodewyckx, J.N.; Boisson, A.; Duvillier, H.; et al. Tumor infiltrating B-cells signal functional humoral immune responses in breast cancer. *JCI Insight* **2019**, *5*, e129641. [CrossRef]
216. Deola, S.; Panelli, M.C.; Maric, D.; Selleri, S.; Dmitrieva, N.I.; Voss, C.Y.; Klein, H.; Stroncek, D.; Wang, E.; Marincola, F.M. Helper B cells promote cytotoxic T cell survival and proliferation independently of antigen presentation through CD27/CD70 interactions. *J. Immunol.* **2008**, *180*, 1362–1372. [CrossRef]
217. Mehdipour, F.; Razmkhah, M.; Faghih, Z.; Bagheri, M.; Talei, A.R.; Ghaderi, A. The significance of cytokine-producing B cells in breast tumor-draining lymph nodes. *Cell. Oncol.* **2019**, *42*, 381–395. [CrossRef] [PubMed]
218. Kessel, A.; Haj, T.; Peri, R.; Snir, A.; Melamed, D.; Sabo, E.; Toubi, E. Human CD19(+)CD25(high) B regulatory cells suppress proliferation of CD4(+) T cells and enhance Foxp3 and CTLA-4 expression in T-regulatory cells. *Autoimmun. Rev.* **2012**, *11*, 670–677. [CrossRef] [PubMed]
219. Wang, J.Z.; Zhang, Y.H.; Guo, X.H.; Zhang, H.Y.; Zhang, Y. The double-edge role of B cells in mediating antitumor T-cell immunity: Pharmacological strategies for cancer immunotherapy. *Int. Immunopharmacol.* **2016**, *36*, 73–85. [CrossRef]

cancers

Article

AKT2 Loss Impairs BRAF-Mutant Melanoma Metastasis

Siobhan K. McRee [1,2], Abraham L. Bayer [3,4], Jodie Pietruska [2], Philip N. Tsichlis [5] and Philip W. Hinds [1,2,*]

1 Program in Genetics, Graduate School of Biomedical Sciences, Tufts University, Boston, MA 02111, USA; skmcree@gmail.com
2 Department of Developmental, Molecular and Chemical Biology, Tufts University School of Medicine, Boston, MA 02111, USA; jodie.pietruska@gmail.com
3 Program in Immunology, Graduate School of Biomedical Sciences, Tufts University, Boston, MA 02111, USA; abraham.bayer@tufts.edu
4 Department of Immunology, Tufts University School of Medicine, Boston, MA 02111, USA
5 Department of Cancer Biology and Genetics, The Ohio State University, Columbus, OH 43210, USA; tsichlis.1@osu.edu
* Correspondence: phil.hinds@tufts.edu

Simple Summary: Skin cancer, such as melanoma, is often treatable but becomes deadly when it expands to other places in the body, a process called metastasis. While many current drugs target melanoma tumor growth, no drugs yet exist to specifically prevent metastasis. Improving treatment outcomes, therefore, will require an understanding of the molecular mechanisms leading to metastasis. The AKT family of proteins are important regulators of cellular growth and signaling, which play differing roles in cancer. This work sought to study each of the AKT isoforms and their contributions to melanoma cell migration and metastasis. We found that AKT2 specifically regulates melanoma cell metastasis through effects on metabolism and melanoma cell properties, while AKT1 is involved in cellular proliferation and growth. This study suggests that specifically targeting AKT2 and AKT1 represents novel therapeutic strategies for different-stage melanoma patients.

Abstract: Despite recent advances in treatment, melanoma remains the deadliest form of skin cancer due to its highly metastatic nature. Melanomas harboring oncogenic BRAF^{V600E} mutations combined with PTEN loss exhibit unrestrained PI3K/AKT signaling and increased invasiveness. However, the contribution of different AKT isoforms to melanoma initiation, progression, and metastasis has not been comprehensively explored, and questions remain about whether individual isoforms play distinct or redundant roles in each step. We investigate the contribution of individual AKT isoforms to melanoma initiation using a novel mouse model of AKT isoform-specific loss in a murine melanoma model, and we investigate tumor progression, maintenance, and metastasis among a panel of human metastatic melanoma cell lines using AKT isoform-specific knockdown studies. We elucidate that AKT2 is dispensable for primary tumor formation but promotes migration and invasion in vitro and metastatic seeding in vivo, whereas AKT1 is uniquely important for melanoma initiation and cell proliferation. We propose a mechanism whereby the inhibition of AKT2 impairs glycolysis and reduces an EMT-related gene expression signature in PTEN-null BRAF-mutant human melanoma cells to limit metastatic spread. Our data suggest that the elucidation of AKT2-specific functions in metastasis might inform therapeutic strategies to improve treatment options for melanoma patients.

Keywords: melanoma; metastasis; cancer; AKT; signaling

1. Introduction

Metastasis is responsible for the vast majority of all cancer deaths, yet no therapies preferentially target the metastatic process. A key hallmark of cutaneous melanoma is its rapid and efficient metastasis, facilitated, in part, by the PI3K/AKT pathway [1]. PI3K/AKT signaling is hyperactivated in late-stage melanomas [2], often through the loss of its negative regulator, PTEN. PTEN inactivation leads to unrestrained activity regarding the

PI3K effector kinase, AKT, which co-operates with pre-existing oncogenic BRAF mutations to promote metastasis [3]. Indeed, while the BRAF mutation V600E hyperactivates the MAPK signaling pathway and is a very common early mutation, it alone is insufficient for tumorigenesis, and subsequent genetic and/or epigenetic alterations are required [4,5]. Alterations in PTEN, including inactivating mutations, epigenetic silencing, protein destabilization, or genetic loss, occur in as many as 50% of all melanomas and correlate with advanced disease, including brain metastasis [6,7]. Accordingly, AKT is hyperactivated in melanoma brain metastases [8], and the inhibition of the PI3K pathway effectively reduces metastasis in murine melanoma models [9]. Despite this, pan-PI3K inhibitors have not yet demonstrated clinical efficacy [10,11]. While six PI3K inhibitors have been approved for clinical use against breast cancer, hematologic cancers, and other diseases, only one exhibits pan-PI3K inhibition, whereas others are isoform-specific. [12] A similar spectrum of clinical efficacy is likely from the AKT inhibitors currently being used in clinical trials, with isoform-specific inhibition better tolerated compared to pan-inhibition. This suggests that the additional isoform-specific PI3K/AKT inhibitors under development might be more therapeutically effective than pan-inhibition strategies. [13,14].

The AKT kinase family comprises three highly homologous yet functionally distinct isoforms (AKT1, AKT2, and AKT3) [15–18]. They differentially contribute to tumorigenesis in the breast, colon, and prostate [19,20], among other tissues, but isoform-specific inhibition has yet to be exploited as a therapeutic strategy. AKT isoforms achieve their specificity through differential activation, preferential substrate phosphorylation, and tissue distribution [19], but how these differential properties contribute to distinct cell transformation events is still being elucidated. While the isoform-specific effects of AKTs have been demonstrated in numerous other cancers [21–23] and in melanoma [24–27], many questions remain regarding the contribution of individual AKT isoforms to melanoma progression or metastasis, and few studies have interrogated the contribution of individual AKT isoforms to melanoma initiation. Large genomic datasets have provided some insights as to the relative frequency of AKT-isoform-specific alterations; the activation of mutations or gene amplification of AKT1 and AKT2 are more common in BRAF^{V600E} mutant and PTEN-null melanomas [28,29], whereas AKT3 amplification is more common in BRAF^{WT}PTEN^{WT} tumors [26], although AKT3 has also demonstrated co-operativity with BRAF^{V600E} to induce murine melanoma development [30]. With regard to melanoma metastasis, multiple and differing roles for AKT isoforms have been described in both mouse and human models. For example, upon the loss of PTEN, the invasiveness of human melanoma cell lines was enhanced through the preferential phosphorylation of AKT2, whereas AKT3 phosphorylation reduced the invasive potential of PTEN-null melanoma cells [31]. Similarly, it has been shown that the PHLPP1 phosphatase suppressed melanoma metastasis through the dephosphorylation of AKT2 or AKT3 but not AKT1 in human melanoma cells [25]. In Braf^{V600E}-driven murine melanomas lacking Cdkn2a, AKT1 activation was shown to enhance brain metastasis, whereas AKT2 and AKT3 had a less pronounced effect [27,32]. While the totality of data is suggestive of AKT isoforms playing differing roles in melanoma initiation, progression, and metastasis, further clarification through additional studies is needed, especially given the importance of metastasis to melanoma outcomes.

Here, we employ inducible AKT-isoform-specific shRNAs and CRISPR/Cas9-mediated gene editing in human melanoma cell lines to outline the role of AKT2 in the metastatic cell seeding of oncogenic BRAF-driven PTEN-null melanoma through the enhancement of cell migration and invasiveness. As such, AKT2 depletion impairs and prevents metastasis, improving overall survival in mice. We further investigate AKT isoform phosphorylation in primary spontaneous murine melanomas and human metastatic melanoma cell lines, finding that AKT2 phosphorylation specifically increases in metastatic lesions. We further show that while AKT2 is dispensable for melanoma initiation, AKT1 contributes to BRAF-driven murine melanoma initiation and cell proliferation, which is in line with our previous results [33]. Mechanistically, we find AKT2 promotes metastasis and regulates glycolysis through a PDHK1-PDHE1α axis. This work supports the continued investigation of targeted AKT inhibition as an anti-melanoma therapy and presents a novel potential target for preventing metastatic spread.

2. Materials and Methods

2.1. Mouse Strains

All mice were maintained in a heat- and humidity-controlled AAALAC-accredited vivarium operating under a standard light-dark cycle. All protocols have been approved by the Institutional Animal Care and Use Committee (IACUC) at Tufts University School of Medicine, where the mice were housed and the experiments were conducted. $BRAF^{V600E}$; $Arf^{-/-}$ mice were bred in-house, whereas the AKT isoform knockout mice have been described previously [23]. NOD/SCID and C57Bl6/J mice were purchased from the Jackson Laboratory (Bar Harbor, ME, USA).

2.2. Xenografts

Male NOD/SCID mice (Jackson Laboratory, 6–10 weeks old) were injected subcutaneously with 2×10^6 human melanoma cells, according to approved protocols. Once palpable (after roughly 3–4 weeks, and ~200 mm^3), by using light touch compared to non-inoculated flank, the tumors were measured 3× weekly using calipers, and tumor volume was calculated using the formula $[(\pi/6) \times L \times W^2]$. Tumors were allowed to grow until a limit of 1500 mm^3 or 2 cm in any single direction or until mice became moribund. Doxycycline chow (200 mg/kg, Teklad, Chicago, IL, USA) was introduced when tumors were palpable when indicated.

2.3. Luciferase Imaging

Mice were anesthetized with isofluorane and injected intraperitoneally with 10 µL/g of body weight of Luciferin (Thermofisher, Waltham, MA, USA) 5 min prior to imaging. Imaging was performed using an IVIS SpectrumCT in vivo imaging system and analyzed using Living Image® Software, version 4.7.

2.4. Metastasis Assays

Human (1×10^6) or mouse melanoma cells (0.5×10^6) were injected into a lateral tail vein of NOD/SCID or C57Bl6/J mice, respectively. Mice were maintained on regular or doxycycline chow (200 mg/kg, Teklad), imaged at 2 weeks post-injection and weekly thereafter. Mice were euthanized when moribund.

2.5. Tumor Cell Isolation and Tissue Preparation

Tumors were minced and digested with 3 mg/mL collagenase and 250 U/mL hyaluronidase for 2–4 h at 37 °C. The contaminating red blood cells were lysed with RBC Lysis Buffer (Sigma), and the organoids were triturated using an 18 G syringe needle, incubated in 0.05% Trypsin/0.53 mM EDTA, passed through a 40 µM cell strainer, and cultured in RPMI1640/10% FBS/1% penicillin/streptomycin/fungizone.

2.6. Cell Lines/Tissue Culture

SM1 cells were a generous gift from Antoni Ribas (UCLA, Los Angeles, CA, USA), and the TUMM cell lines were generated as described above. Human melanoma cell lines were generous gifts from Frank Haluska (Tufts Medical Center, Boston, MA, USA) and were routinely validated for melanocytic identity by the RNA or protein expression of MITF-M and pigment enzymes TYR and DCT and tested for mycoplasma contamination. Human cells were maintained in DMEM (Invitrogen, Cambridge, MA, USA) with 10% FBS (Atlanta Biologicals, Flowery Branch, GA, USA) and 1% penicillin/streptomycin (Invitrogen). The SM1 and TUMM murine cell lines were cultured in RPMI1640/10% FBS/1%penicillin/streptomycin/fungizone. Knockdown experiments utilized doxycycline (Sigma, St. Oouis, MO, USA) at concentrations of 0.5–1 µg/mL. Doxycycline-inducible shAKT plasmids were generous gifts from Drs. Alex Toker and Rebecca Chin (Beth Israel Deaconess Medical Center, Boston, MA, USA) [34]. A nontargeting hairpin scramble sequence (Sigma) was cloned into the Tet-pLKO-puro backbone, a gift from Dmitri Wiederschain (Addgene plasmid #21915). The 293T cells were transfected with shRNA constructs and packaging plasmids (psPAX2 and VSV-G) using PEI (polyethylenimine MW25,000, Polysciences, Warrington, PA, USA). Viral supernatants were collected at 48 h and 72 h post-transfection and mixed. Stably transduced cell lines were generated by infection overnight in the presence of 8 μg/mL polybrene, and 48 h after infection, they were selected with 1 μg/mL Puromycin (Gibco, Cambridge, MA, USA) for three days. Cells infected with pLENTI-Luciferase-expressing virus (generous gift of Charlotte Kuperwasser, Tufts University) were selected with neomycin (G418, 500 μg/mL, Gibco) for 2–3 weeks. CRISPR knockout WM1799 cells were previously generated and cultured, as described [33].

2.7. Immunoblot and Immunoprecipitation Analysis

Cells were lysed in RIPA buffer containing protease and phosphatase inhibitors (Roche, Basel Switzerland) and cleared by centrifugation. Protein concentration was determined by DC Protein Assay (BioRad, Hercules, CA, USA), and equivalent masses of protein were resolved using SDS-PAGE and transferred to 0.2 μm PVDF membranes (BioRad) for immunoblotting with indicated antibodies (see Supplementary Table S1). Membranes were incubated with horseradish peroxidase-conjugated secondary antibodies and visualized using enhanced chemiluminescence (Pierce, Waltham, MA, USA). For immunoprecipitation, samples were lysed in RIPA or CST lysis buffer (Cell Signaling Technologies, Danvers, MA, USA), and protein concentrations were normalized to 1 mg/mL and precleared with Protein A magnetic beads (Pierce). The antibody or equivalent control IgG was incubated overnight at 4 °C, then with Protein A beads for 60–90 min. Antibody/bead complexes were washed extensively in lysis buffer, eluted by boiling in Laemmli buffer, and resolved by immunoblotting, as described.

2.8. Wound Healing Assay

Cells were pre-incubated with doxycycline (DOX, 1 μg/mL) or DMSO for 24 h; then, a wound was made vertically across the growth area using a P1000 pipette tip. The cells were then washed with PBS, and the media was replaced with DOX or DMSO, as indicated. The cells were immediately imaged using an inverted scope at 4× magnification, aligned to a horizontally drawn guide perpendicular to the wound to ensure consistent imaging. After 16 h, the plates were imaged again using the same guide. Scratch distances were quantified using ImageJ and expressed as a percentage of wound closure relative to the DMSO-treated cells.

2.9. Migration/Invasion Assays

Cells were pre-incubated with doxycycline (DOX, 1 μg/mL) or DMSO for 24 h, and then seeded (25,000 cells) in the upper chamber of trans-wells (8 μm pores, Corning, Corning, NY, USA) in serum-free DMEM with DOX or DMSO, using 10% FBS as a chemoattractant and incubated overnight. Inserts were washed with PBS, scrubbed, fixed in ice-cold methanol, and stained with DAPI. Invasion assays utilized growth factor-reduced (GFR) Matrigel-coated trans-wells (Corning) and a 36 h incubation time.

2.10. Anchorage-Independent Growth (Soft Agar) Assay

Sterile low melting agarose (SeaPlaque, Lonza, Basel Switzerland) was prepared at a 5% stock concentration, with a 1% final concentration (bottom layer) prepared by dilution in an appropriate cell culture medium; this was allowed to solidify at room temperature for 30 min and was overlaid with 10,000 cells per 6 wells in 0.5% agarose. After solidifying for 30 min, the wells were overlaid with 0.5 mL of media containing 0.5 μg/mL of DMSO or doxycycline and allowed to incubate at 37 °C for 2–3 weeks or until macroscopic colonies were visible. The media was refreshed every 2–3 days. The plates were then fixed with 10% neutral buffered formalin, washed 1× with PBS, stained with 0.05% crystal violet overnight, and washed with PBS until clear. Images were taken at 10× magnification and quantified using ImageJ.

2.11. Seahorse Glycolytic Rate Assay

The WM1799 cells were pre-incubated with DMSO/DOX for 48 h, and 21,000 cells were seeded into microtiter plates (Agilent, Santa Clara, CA, USA) 1 day prior to the assay. One hour prior to the assay, the cells were washed and incubated with RPMI at pH 7.4 (Agilent) and placed in a non-CO_2 incubator. The assays were performed using a Seahorse XFe96 Analyzer and Wave Software 2.4.0.

2.12. Quantitative RT-PCR

The cells were lysed in TRIzol (Invitrogen), and the RNA was isolated by phenol-chloroform extraction according to the manufacturer's instructions, followed by lithium chloride/isopropanol precipitation. cDNA synthesis was performed using SMARTScribe™ Reverse Transcriptase (Takara, Inc., Kusatsu, Japan), and qPCR was performed on a CFX96 real-time thermal cycler (Bio-Rad). See Supplementary Table S1 for the primer sequences.

2.13. Cell Cycle Analysis

The cells were cultured in DMSO or DOX, as indicated, collected and washed with cold PBS, and then fixed in ice-cold 70% EtOH for 30 min at 4 °C. The cell pellets were washed in staining buffer (PBS no Ca/Mg, 3% FBS, +1 mM EDTA), then incubated in staining buffer with 80 μg/mL Propidium Iodide and 0.125 mg/mL RNAse A (Thermofisher) for 40 min at 30 °C. The samples were run on a BD LSRII (BD Biosciences) and analyzed using FlowJo software (version 10.8.2).

2.14. Statistical Analysis

Statistics were performed using GraphPad Prism 5.02, utilizing the Student's unpaired t-test for two means, one-way ANOVA with Tukey post-test for multiple means, or Kaplan–Meyer survival analysis with Mantel–Cox tests, as indicated. The significant p-values are listed or noted as * $p < 0.05$, ** $p < 0.01$, or *** $p < 0.001$. Error bars represent standard error means.

3. Results

3.1. AKT2 Depletion Impairs Cell Migration and Invasion in Human Melanoma Cells

It is well-known that AKT phosphorylation, a surrogate marker for active AKT, increases with disease stage in clinical samples [35], but there is a paucity of information regarding the relative contributions of individual isoforms to disease progression. In order to investigate this, we first characterized total AKT phosphorylation (Ser473 and Thr308) across a panel of human melanoma cell lines. Immunoblotting revealed a wide range of AKT phosphorylation levels and several cell lines that showed significant AKT2 phosphorylation (Supplementary Figure S1A). Furthermore, as expected, there was an inverse correlation between PTEN expression and total AKT phosphorylation (Supplementary Figure S1B).

In order to investigate the possible contribution of AKT isoforms to metastatic potential in human cell lines, we generated luciferized doxycycline-inducible shRNA hairpins to AKT1, AKT2, and AKT3 [34], as well as a nontargeting hairpin (shNT) that efficiently reduced protein expression in the majority of melanoma cell lines from our panel, (Figures 1A,B and S1C–E) without affecting viability (Figure 1C). We focused on cell lines in which there was significant AKT2 phosphorylation and robust knockdown efficiency (WM1799, UACC903, and WM455) and then sought to test the individual functions contributing to metastasis in these cells. By using a conventional in vitro wound healing assay, it was found that only AKT2 depletion and not AKT1 or AKT3 depletion inhibited cell migration in three different human melanoma cell lines (Figures 1D,E and S2A–D). Next, we utilized a complementary trans-well assay and observed that AKT2 depletion reduced cell migration in response to a serum gradient in multiple cell lines (Supplementary Figure S2E,G). Further, invasion through Matrigel was impaired by AKT2 depletion in the same human melanoma cell lines (Figures 1F,G and S2F,H). This reduction was not due to defects in cellular proliferation, as AKT2 did not impair cell proliferation in any cell line, as assessed by cell counting with trypan blue exclusion (Supplementary Figure S5). When taken together, these data suggest that AKT2 depletion in vitro impairs the functions required for melanoma metastasis, such as migration and invasion.

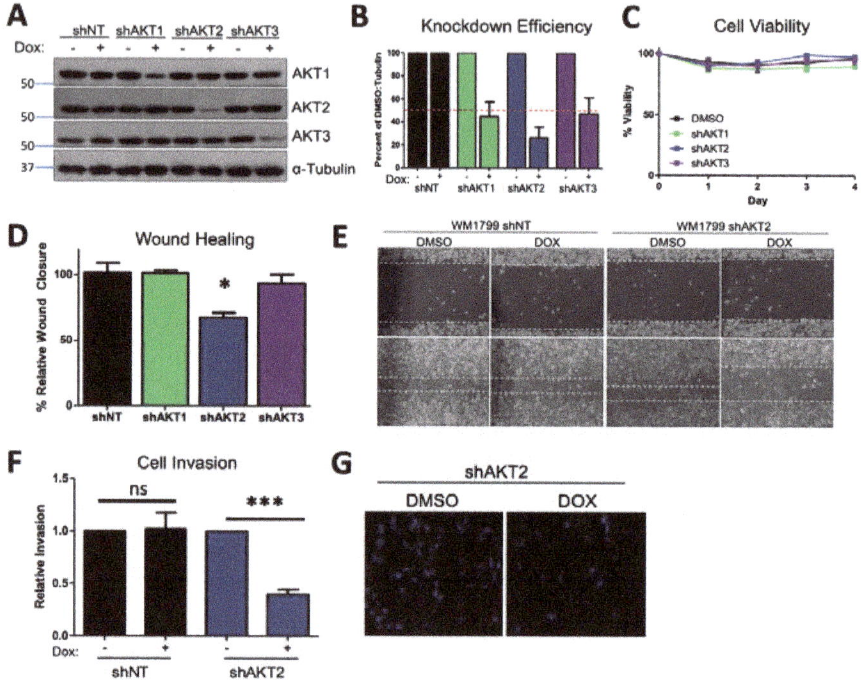

Figure 1. AKT2 depletion impairs migration and invasion in WM1799 human melanoma cell lines. (**A**) A representative immunoblot of WM1799 human melanoma cells after stable cell line generation. Dox-inducible AKT-isoform knockdown (shAKT1/2/3) or nontargeting hairpin (shNT) expressing cell lines are shown after 72 h of treatment with 0.5 ug/mL doxycycline. (**B**) Quantitation of KD efficiency from three independent experiments using ImageJ to quantify the intensity of the bands normalized to percent of DMSO-treated total protein and loading control. (**C**) The cell lines from (**A**) were grown in the presence of DOX or DMSO and collected at indicated times for cell counting with trypan blue exclusion to assess cell viability from two independent experiments. (**D**) Quantitation of wound closure in shWM1799 cells at 0 h and 16 h post-wounding, comparing DMSO- and DOX-treated cells from three independent experiments, with representative images at 4× magnification (**E**) of shNT or shAKT2 WM1799 cells treated with DOX relative to DMSO vehicle-treated cells. (**F**) Quantitation of invasion ability of control (NT) or AKT2 KD (shAKT2) cells with or without DOX treatment, as indicated from three independent experiments. (**G**) Representative images of DAPI-stained AKT2 KD cells on the underside of a Matrigel-coated membrane (20× magnification) following treatment with DMSO or DOX (left) from n = 3 total experiments. ns: no significant difference, *: $p < 0.05$, ***: $p < 0.001$.

3.2. AKT2 Depletion Restricts Anchorage-Independent Growth In Vitro and In Vivo

Anchorage-independent growth is required for the growth of metastatic cells; therefore, we interrogated the role of AKT2 in this process. By focusing our efforts on WM1799 cells, we seeded shAKT2 transduced WM1799 cells in soft agar, overlaid with either DMSO- or doxycycline-containing media to knock down AKT2. We observed that AKT2 depletion reduced the colony number (Figure 2A,B), with only a minor trend in decreased colony size (Figure 2C), indicating that AKT2 depletion limits the ability of cells to grow in a 3D culture. Next, we assessed the effect of AKT2 depletion on the growth of WM1799 cells as subcutaneous tumors in vivo, which similarly requires anchorage-independent growth in early tumorigenesis. We injected WM1799 shAKT2 cells subcutaneously into immunodeficient NOD/SCID mice (to exclude changes in melanoma cell growth due to an adaptive immune response) and allowed palpable tumors to form before transitioning a subset of these mice to doxycycline chow (Figure 2D). AKT2 knockdown significantly slowed tumor growth relative to the control mice fed with regular chow (Figure 2E), but tumors eventually grew in both groups despite persistent AKT2 knockdown (Figure 2F) in the majority of tumors. Further, we observed no change in total AKT phosphorylation across those tumors in which AKT2 was depleted, suggesting AKT1 or AKT3 activity may compensate over time to promote tumor growth.

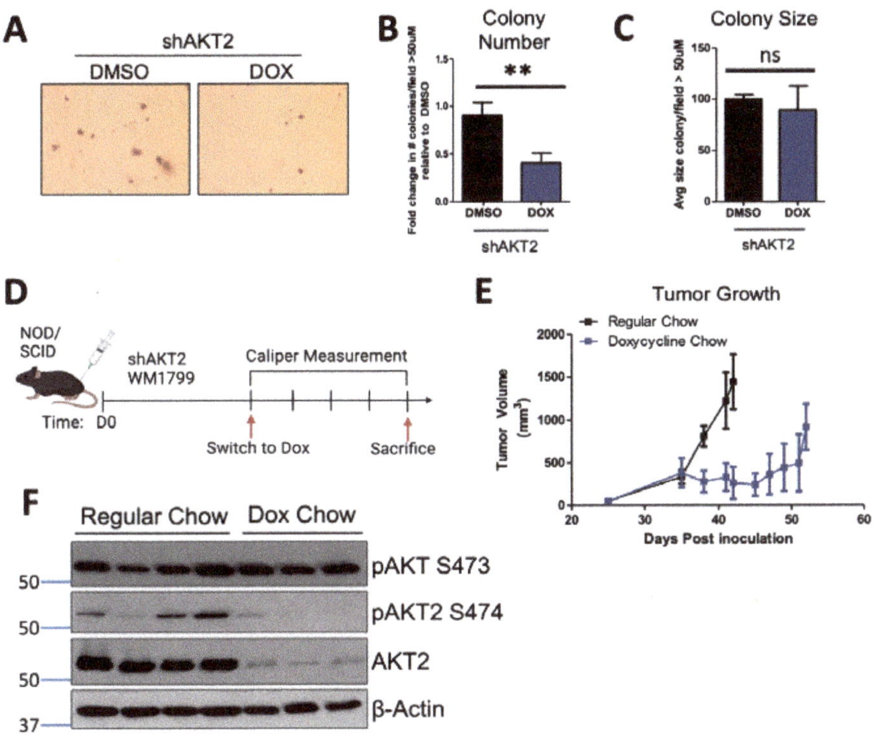

Figure 2. AKT2 depletion restricts anchorage-independent growth. (**A**) Anchorage-independent growth was assessed by seeding WM1799 AKT2 KD cells in soft agar and culturing in DMSO- or DOX-containing media for three weeks, after which the cells were fixed and stained with crystal violet (left; representative image at 4× magnification). (**B**,**C**) Colonies greater than 50 μM were counted and quantified using ImageJ from three independent experiments. (**D**) Schematic of experimental design to assess tumor growth potential. WM1799 shAKT2 cells were injected subcutaneously into NOD/SCID mice and allowed to form palpable tumors; then, the mice were randomized into groups receiving either regular or DOX-containing chow to induce AKT2 KD (n = 3–4 mice per group). (**E**) Tumor volumes were determined using calipers at the indicated times, and the mice were sacrificed when tumors reached 1500 m^3 in size. (**F**) The tumors isolated from individual mice receiving

either DMSO or DOX chow were subjected to immunoblotting for AKT2 protein and phosphorylation using beta-actin as a loading control. ns: no significant difference, **: $p < 0.01$.

3.3. AKT2 Depletion Delays Metastatic Onset and Extends the Survival of Melanoma-Bearing Mice

In order to determine whether AKT2 is important for the process of metastatic seeding and metastatic nodule growth, we performed a tail vein metastasis assay in which WM1799 shAKT2-Luc melanoma cells were allowed to seed the lungs. This was followed by initiating AKT2 depletion 24 h later by switching some mice to doxycycline-containing chow compared to mice fed with regular chow (Figure 3A). By performing immunoblot analysis using lysate derived from isolated pulmonary metastases, we confirmed that stable and robust AKT2 depletion occurred only in the doxycycline-fed mice (Figure 3B). By using weekly imaging, we found that the control mice fed regular chow displayed advanced metastatic disease at 6 weeks, with tumors observed at multiple sites; however, the mice fed doxycycline chow displayed only occasional tumor nodules at 6 weeks, albeit with significantly reduced size and frequency compared to the control mice (Figure 3C,D). While AKT2 depletion after metastatic seeding significantly improved overall survival in the doxycycline chow-fed mice versus regular chow mice, the doxycycline-fed mice did succumb to eventual metastatic disease, suggesting that AKT2 reduction only delayed the growth of the metastatic lesions, and other AKT activity might compensate over the long term (Figure 3E).

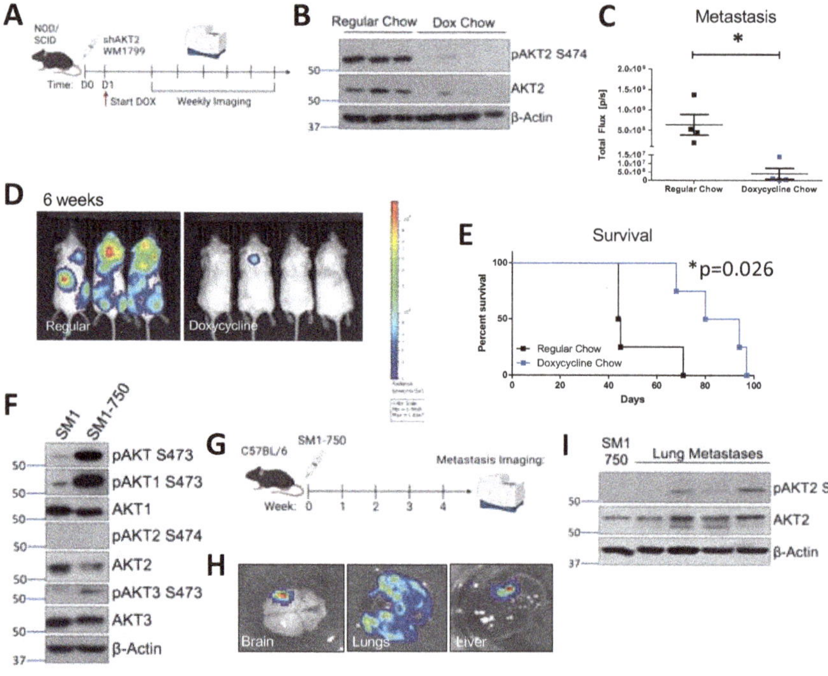

Figure 3. AKT2 depletion delays metastatic onset and extends the overall survival of melanoma-bearing mice. (**A**) Schematic representation of experimental design. Luciferized WM1799 cells expressing DOX-inducible shAKT2 hairpins were injected into the tail vein of NOD/SCID mice. The mice were started on DOX chow 1 d post-injection, and were maintained on regular chow or DOX chow and monitored weekly using IVIS SpectrumCT imaging. (**B**) Immunoblot analysis of total and phospho-S474 AKT2 (P-AKT2) levels in the metastatic tumors isolated from the mice fed regular or doxycycline chow at the time of death. (**C**) Quantification of mice bearing metastases at 6 weeks

post-injection, and the representative images (**D**) of luminescence 6 weeks after cell injection in the mice fed doxycycline chow compared to regular chow (N = 3–4 mice per group). (**E**) Percent survival of mice bearing metastases. (**F**) SM1 and SM1-750 cell lines were established from spontaneous primary murine melanomas expressing BRAF^{V600E} and passaged in the C57Bl6/J mice; this was then analyzed by immunoblotting. (**G**) Schematic of experimental design of in vivo metastasis assays. SM1-750 cells expressing luciferase were injected into the tail vein of C57Bl6/J mice, and the tissues were analyzed using IVIS SpectrumCT imaging at 4 weeks post-injection. (**H**) The lungs, liver, and brains collected from the mice were imaged ex vivo, and the representative images from n = 4 mice are shown. (**I**) The lung metastases were excised from four individual mice (at least 3–4 nodules per mouse were combined) and homogenized to generate protein lysates; this was analyzed by immunoblotting. *: $p < 0.05$.

3.4. AKT2 Phosphorylation Occurs in Metastatic Mouse Melanoma Lesions

In order to study AKT2 in the context of metastatic murine melanoma, we generated an aggressive murine melanoma cell line, SM1-750, which exhibits high metastatic potential. The SM1 cell line was previously derived from melanomas arising in a BRAFV600E mouse and is tumorigenic in syngeneic mice [36]. In our hands, the subcutaneous tumor formation by SM1 cells in the C57BL6/J strain occurred in only a small subset of injected mice. In order to increase the tumor "take rate", we passaged the SM1 cell line in tumor-forming C57BL6/J mice through several rounds of injection and tumor formation, resulting in the isolation of the SM1-750 line, which displays a nearly 100% take rate. Subsequently, the SM1-750 line was engineered to express luciferase, and the maintenance of the ability of these cells to form tumors was confirmed (Figures 3F and S3).

The SM1 cells exhibited significant AKT1 and AKT3 phosphorylation but nearly undetectable AKT2 phosphorylation and the newly derived SM1-750 cells showed further elevated phosphorylation in terms of only AKT1 and AKT3. The SM1-750 cells were injected into the tail vein of syngeneic mice, and the injected mice were imaged after luciferin injection to visualize metastatic progression (Figure 3G). The mice readily developed metastatic lesions, which could be visualized in the lungs, brain, and liver (Figure 3H). Spontaneous brain metastases are relatively rare in murine melanomas and have previously been linked to AKT1 activation [27]. Interestingly, the immunoblotting of AKT isoform phosphorylation in discrete lung tumor metastases from individual mice revealed that AKT2 activation (indicated by S474 phosphorylation) dramatically increased when compared to that of primary tumors (Figure 3I). We further found that AKT2 upregulation in metastatic lesions was not due to adaptive immunity-dependent selection, as we also observed this in the SM1-750 metastases isolated from immune-deficient Rag2$^{-/-}$ mice (Supplementary Figure S3A).

We then sought to characterize AKT isoform phosphorylation in spontaneously arising primary melanomas relative to murine melanomas with metastatic potential. We previously developed a mouse model of BRAF^{V600E}-driven spontaneous melanoma, in which both melanocyte-targeted human BRAF^{V600E} co-operates with tumor suppressor p19^{ARF} loss (hereafter referred to as Arf$^{-/-}$) to facilitate melanoma formation and AKT phosphorylation is observed in tumors but not normal skin [37]. The tumor suppressor Arf is commonly lost in human melanomas, and melanoma penetrance in our model increased on an Arf$^{-/-}$ background [38]. We isolated the cell lines from the spontaneously arising primary tumors in these mice and investigated their isoform-specific phosphorylation patterns (Tufts University Mouse Melanoma, TUMM, see Supplementary Figure S3B,C). This analysis revealed ubiquitous total AKT phosphorylation on the activating residue serine 473 in the TUMM cell lines, as expected, as well as readily detectable AKT1 phosphorylation, with rare AKT3 phosphorylation and almost no AKT2 phosphorylation, which is in line with our observations of SM1-750 cells (Supplementary Figure S3C). Further, neither the TUMM nor SM1-750 cells showed AKT2 phosphorylation in low-attachment cell culture plates or in the primary tumors produced following subcutaneous injection into NOD/SCID mice (Supplementary Figure S3D,E), indicating that while AKT1 or AKT3 phosphorylation can drive melanoma cell growth and primary tumor formation, the activation of AKT2 correlates strongly with an ability to grow in the metastatic niche.

3.5. Prophylactic AKT2 Depletion Prevents Metastatic Cell Seeding

In order to determine if AKT2 plays a role in cellular migration and invasion, as well as in metastatic progression that is further extended to initial metastatic seeding, we pre-treated the WM1799 shAKT2 cells with doxycycline-containing media (compared to DMSO-containing media alone) for 72 h prior to tail vein injection into the NOD/SCID mice; we then maintained the mice on doxycycline-containing food or regular chow for 6 weeks. Metastatic progression was then monitored by using luminescence (Figure 4A). The mice maintained on regular chow displayed luminescent tumor nodules within 3 weeks, which progressed to advanced metastatic disease by 6 weeks, including multiple distant metastases under the forelimbs, the head, neck, and mesenteric lining (determined at autopsy, not shown), which is consistent with lymphatic dissemination (Figure 4B,C). Contrary to the control mice, those mice that received the shAKT2 cells and doxycycline chow remained healthy, with no detectable tumors after 6 weeks (Figure 4B,C). Additionally, when maintained on doxycycline chow, these mice were protected from detectable metastasis, even for up to 12 weeks (Figure 4D,E). These findings suggest that AKT2 is either required for metastatic seeding or the growth of seeded cells but do not distinguish whether AKT2 KD cells were eliminated from circulation or were simply latent in the mice fed with the doxycycline chow. In order to address this question, we asked if tumors would emerge after the removal of doxycycline chow. A subset of mice injected with WM1799 shAKT2 Luc cells and fed doxycycline chow were switched to regular chow after the first 6 weeks, at which time they still did not have detectable metastases. The mice were monitored weekly, and after an additional 6 weeks, none of the mice that were removed from doxycycline chow developed metastases (Figure 4F,G). These results suggest that the prophylactic targeting of AKT2 impairs the seeding of invasive cells in the metastatic niche, fully preventing metastatic formation rather than restraining the growth of dormant but intact metastatic tumor cells.

3.6. AKT2 Deletion Impairs Melanoma Migration, Invasion, and Metastasis

In order to confirm these findings with a more robust depletion of AKT2, we utilized isoform-specific CRISPR/Cas9 knockout in WM1799 cells for which we previously confirmed stable isoform-specific knockout (Figures 5A and S4A) [33]. We first characterized the properties of AKT2 knockout (KO) WM1799 cells compared to nontargeting (NT) cells in vitro, as was carried out for the inducible knockdown cell lines (Figures 1 and 2). As expected, and in line with our results using inducible knockdown, we found that the AKT2 KO cells had impaired wound healing (scratch assay), migration (trans-wells), and Matrigel invasion when compared to the NT cells (Figure 5B,D), which are all indicative of impaired metastatic properties. In order to test this, we further engineered NT and AKT2 cells to express luciferase and injected them into the tail veins of NOD SCID mice (Figure 5E). At 8 weeks post-inoculation, the NOD SCID mice that received AKT2 KO cells had significantly lower metastatic burden than those injected with NT cells (Figure 5F,G). Further, the mice injected with AKT2 KO cells showed a survival benefit relative to NT-injected mice (Figure 5H), which is consistent with the previous results, indicating that AKT2 depletion delays the onset of metastatic disease and overall survival.

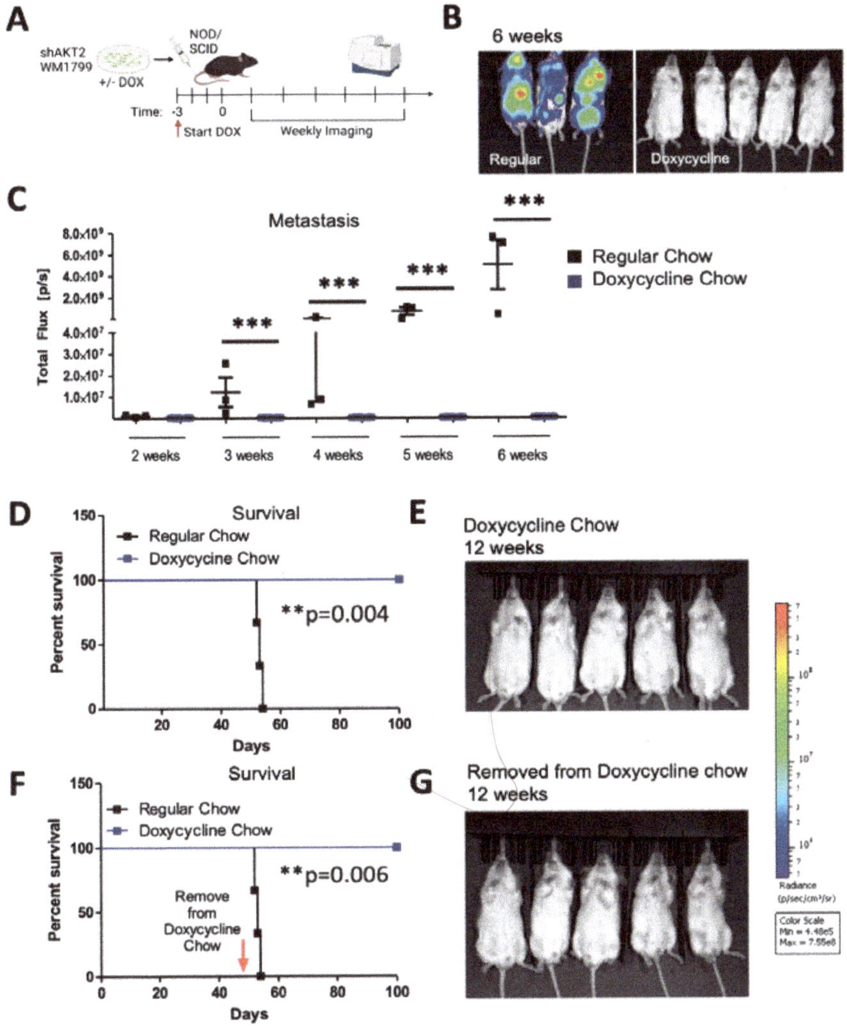

Figure 4. Prophylactic AKT2 depletion prevents metastatic cell seeding. (**A**) Schematic representation of experimental design. AKT2-depleted (3 days of DOX treatment) or control (DMSO treated) WM1799 shAKT2 Luc cells were injected into the tail vein of the NOD/SCID mice fed DOX chow or regular chow, respectively, for 3 days prior to injection (N = 3–5 mice per group). Mice were provided with regular or DOX chow and monitored weekly by IVIS SpectrumCT imaging. (**B**) Image of mice 6 weeks post-injection for DOX-treated versus regular chow mice. (**C**) Quantification of luminescence from mice in (**B**) using LivingImage software at indicated times. (**D**) Survival of regular-chow-fed (control) and DOX-fed mice injected with DMSO- or DOX-treated shAKT2 Luc cells, respectively (n = 3–5 mice per group). Mice were sacrificed when moribund. (**E**) Luminescence assayed at 12 weeks for mice injected with AKT2 KD cells and fed Dox chow. (**F**,**G**) A subset of mice treated as per D and E that were switched to regular chow after 6 weeks on DOX chow; these were assessed for the presence of metastases by IVIS SpectrumCT imaging after an additional 6 weeks and showed no evidence of progressing metastatic lesions (n = 3–5 mice per group). **: $p < 0.01$, ***: $p < 0.001$.

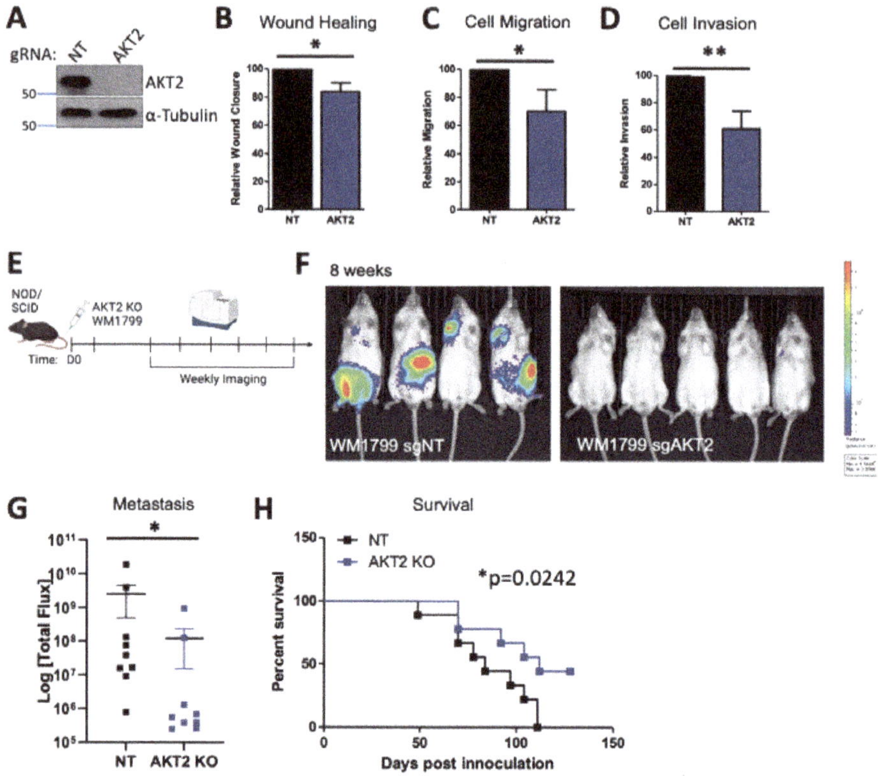

Figure 5. AKT2 Knockout impairs human melanoma cell migration, invasion, and metastasis. (**A**) Representative immunoblot of KO WM1799 cells for AKT2. (**B–D**) Ability of WT or AKT2 KO WM1799 cells in wound closure using scratch assay (**B**), cell migration using trans-well assay (**C**), or cell invasion through a Matrigel-coated membrane (**D**), all from n = 3 independent experiments. (**E**) Experimental schematic, in which AKT2 KO WM1799 cells were engineered to express luciferase and were injected into the tail veins of NOD/SCID mice; metastasis was monitored using IVIS SpectrumCT imaging; representative mice at 8 weeks are shown (**F**) with quantification (**G**). (**H**) Overall percent survival of mice injected with WM1799 AKT2 KO cells compared to NT cells. *: $p < 0.05$, **: $p < 0.01$.

3.7. AKT1 Deletion Impairs Primary Tumor Formation

We next sought to test if AKT2 knockout impaired tumor initiation when compared to AKT1 and AKT3 deletion, as our CRISPR-edited WM1799 cells exhibited more robust AKT2 deletion compared to the inducible knockdown lines, which did not impact tumor initiation. We injected each isoform-specific knockout into the flanks of the NOD/SCID mice and monitored tumor formation and growth using caliper measurements (Supplementary Figure S4A,B). Interestingly, we found that while the AKT2 and AKT3 KO WM1799 cells developed tumors similarly to NT cells, only the AKT1 KO cells had significantly delayed tumor growth (Supplementary Figure S4C).

In order to further investigate the impact of the genetic loss of each AKT isoform on melanoma progression, we crossed melanoma-prone BRAF^{V600E}; Arf$^{-/-}$ mice with AKT isoform knockout mice [23] to generate a BRAF^{V600E}; Arf$^{-/-}$; AKT$^{-/-}$ compound mutant mouse (Breeding scheme; Supplementary Figure S4D). As the BRAF^{V600E}; Arf$^{-/-}$ mice demonstrate significant tumor formation and overall reduced survival [38], we analyzed the contribution of each AKT to long-term survival. Again, AKT1 loss in the context of oncogenic BRAF and Arf loss provided a survival benefit for these melanoma-prone mice, which was not observed for AKT2 or AKT3 loss (Supplementary Figure S4E,F). This is in line with our data showing only strong AKT1 phosphorylation in primary mouse tumors, distinguishing the role of AKT2 in metastatic seeding from the role of AKT1 in primary murine melanoma formation.

3.8. AKT1 Knockdown Impairs Cellular Proliferation and Anchorage-Independent Growth

In order to investigate why AKT1 but not AKT2 impaired primary tumor growth, we analyzed cellular proliferation in our inducible AKT1 and AKT2 knockdown melanoma cell lines. We found that in the WM1799, UACC903, and WM1158 cells, only AKT1 but not AKT2 knockdown impaired cellular proliferation after 4 days in culture with doxycycline-containing media (Supplementary Figure S5A). We went on to perform cell cycle analysis on the shWM1799 cells and discovered that doxycycline-treated AKT1 knockdown cells exhibited increased G1 arrested cells and fewer G2/M phase cells, whereas the NT and AKT2 knockdown cells showed no cell cycle defects in the presence of doxycycline (Supplementary Figure S5B,C). The AKT1 knockdown cells further showed significantly reduced BrdU incorporation in the presence of doxycycline when compared to the NT and AKT2 cells (Supplementary Figure S5D), confirming a role for AKT1 in cellular proliferation that is in line with the existing literature [17] and one mechanism of delayed primary tumor formation.

We further tested whether AKT1 played a role in anchorage-independent growth, as the AKT1 KO WM1799 cells showed not only delayed tumor formation but also a reduction in tumor growth. Indeed, the knockdown of AKT1 reduced colony formation in soft agar with a trend in decreased colony size (Supplementary Figure S6A,C). We went on to test this observation in vivo by implanting shAKT1 WM1799 cells into NOD/SCID mice and switching the mice to doxycycline-containing chow compared to regular chow (Supplementary Figure S6D). The mice on the doxycycline chow had significantly reduced tumor growth compared to the regular chow-fed mice, together showing that AKT1 drives tumor cell proliferation, whereas AKT1 and AKT2 both contribute to anchorage-independent growth.

3.9. AKT2 Depletion Inhibits EMT and Impairs Glycolysis through PDHK1 Activity in Melanoma Cells

In order to investigate the possible mechanisms whereby AKT2 could support metastatic growth and survival, we first interrogated the impact of AKT2 depletion on the epithelial-mesenchymal transition (EMT), an early step in the acquisition of invasive and metastatic capability [39]. Only AKT2 knockdown but not AKT1 or AKT3 knockdown in WM1799 cells reduced the expression of pro-metastatic transcription factors ZEB1 and Snail and the matrix-remodeling enzyme MMP2, which is concomitant with a trend in increased E-cadherin expression (Figure 6A). Additionally, only AKT2 knockdown decreased the expression of TEA domain (TEAD) genes (Figure 6A), which are invasion-associated genes previously implicated in melanoma metastasis [40]. These data are consistent with an isoform-specific role for AKT2 in programming melanoma cells for EMT transition and subsequent metastasis.

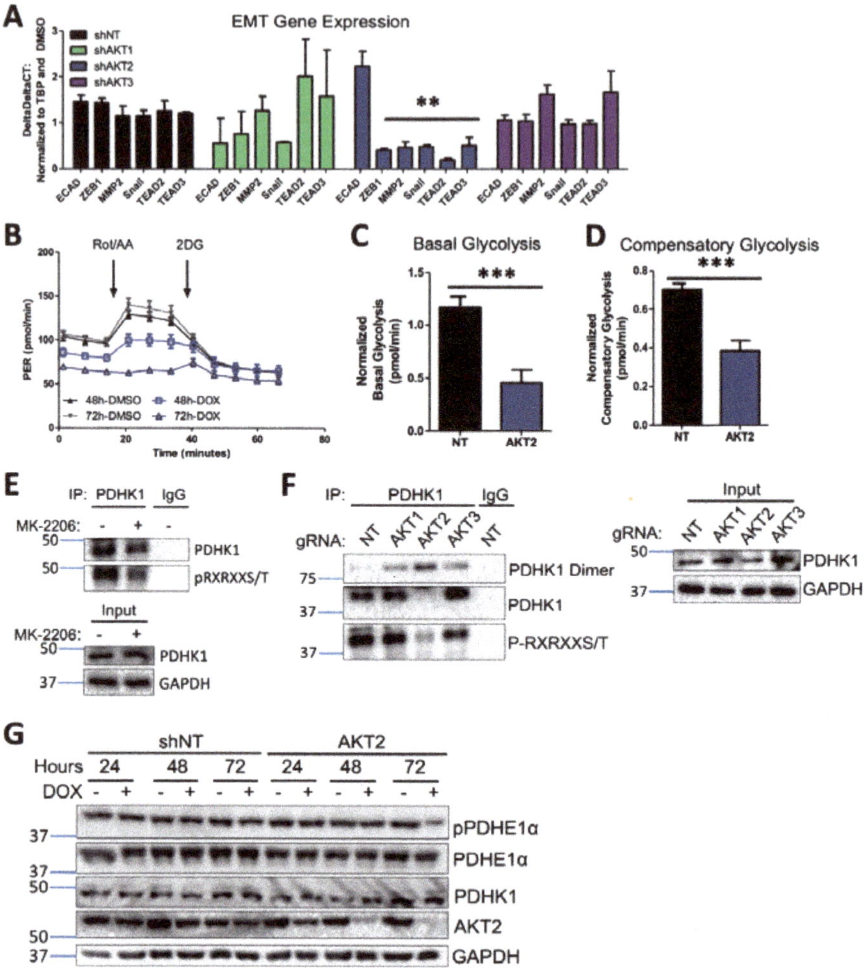

Figure 6. AKT2 depletion alters EMT and impairs glycolysis in melanoma cells. (**A**) RT-qPCR analysis of EMT and invasion-associated transcripts in WM1799 shNT, AKT1 KD, AKT2 KD, or AKT3 KD cells after 48 h of DOX treatment from three independent experiments. Expression levels were normalized to shNT cells. (**B**) Representative GRA plot shows the proton efflux rate (PER) of WM1799 shAKT2 cells cultured in DMSO- or DOX-containing media for 48 or 72 h, which was used to quantify basal glycolysis (**C**) and compensatory glycolysis (**D**) in DOX-treated WM1799 shNT cells from three independent experiments. (**E**) PDHK1 immunoprecipitation was performed after 24 h of MK-2206 treatment, and then immunoblotting was performed for AKT consensus site phosphorylation, with the input immunoblot for PDHK1 shown in the lower panel. (**F**) PDHK1 immunoprecipitation was performed on isoform-specific CRISPR KO WM1799 cells, then immunoblotting was performed for AKT consensus site phosphorylation. (**G**) Immunoblot analysis of PDHK1 expression or PDHE1a expression and phosphorylation in WM1799 shNT or AKT2 KD after treatment with DOX for indicated time points. ns: no significant difference. **: $p < 0.01$, ***: $p < 0.001$.

It has long been appreciated that tumor cells undergo metabolic reprogramming during tumor progression and especially during metastasis, with an increased reliance on glycolysis over oxidative phosphorylation, a phenomenon known as the Warburg effect [41]. In order to determine whether AKT2 knockdown directly disrupts glycolysis, we employed a Seahorse glycolytic rate assay (GRA), which quantitatively measures the proton efflux rate (PER) in real time under standard conditions disruptive to mitochondrial metabolism. Using shNT or shAKT2 WM1799 cells, we performed the glycolytic rate assay after 48 or 72 h of incubation with DMSO- or DOX-containing media (Figure 6B). A comparison of basal glycolytic metabolism in shNT and shAKT2 cells revealed that AKT2 knockdown suppressed basal glycolytic metabolism (Figure 6B,C). After 72 h of AKT2 knockdown, glycolysis that was independent of mitochondrial respiration was also suppressed since Rot/AA injection was unable to increase PER (Figure 6B). Compensatory glycolysis was also significantly reduced in the AKT2 knockdown cells compared to the DOX-treated WM1799 shNT cells (Figure 6C). Together, these results strongly suggest that selective AKT2 knockdown inhibits glycolysis in WM1799 human melanoma cells.

In order to explore the potential mechanisms underlying the observed AKT2-dependent glycolytic defect, we evaluated the impact of AKT2 suppression on key enzymatic regulators of glycolysis. Pyruvate dehydrogenase kinases (PDKs) are critical regulators of the pyruvate dehydrogenase complex (PDC) through inactivating phosphorylation events, with subsequent decreases in TCA cycle activity shifting towards anaerobic respiration [42]. Of the four PDK isoforms (PDHK1-4), PDHK1 is the most frequently studied in the context of tumorigenesis and has previously been shown to be a direct target of AKT2 in prostate adenocarcinoma cells [42]. We first confirmed that PDHK1 was an AKT target in WM1799 cells by performing the immunoprecipitation of PDHK1 in the presence or absence of the AKT inhibitor MK2206 and probing for the phosphorylation of the AKT consensus site, showing AKT-specific phosphorylation on isolated PDHK1 that was significantly decreased by MK2206 treatment (Figure 6E). In order to confirm the isoform specificity of PDHK1 phosphorylation, we performed PDHK1 immunoprecipitation in our CRISPR KO WM1799 cell lines and found that only AKT2 KO cells showed decreased PDHK1 phosphorylation when compared to NT, AKT1, or AKT3 KO cells (Figure 6F). Further, we found that the majority of the immunoprecipitated PDHK1 was in the dimerized state, which occurs when PDHK1 is dephosphorylated and, therefore, inactive [43], supporting that the activity of AKT2 is specifically required for PDHK1-induced glycolytic activity.

We further tested the effect of AKT2 knockdown on PDHK1 levels and found that while shAKT2 cells treated with doxycycline resulted in similar levels of PDHK1 (Figure 6G), a major target of PDHK1 kinase activity, pyruvate dehydrogenase E1 component subunit alpha (PDEH1α), had significantly lower PDEH1α phosphorylation in only those shAKT2 cells treated with doxycycline for 72 h. Decreased PDEH1α phosphorylation as a result of impaired PDK1 activity would be expected to shift metabolic flux away from glycolysis back to the TCA cycle, which is one mechanistic explanation for the loss of glycolytic activity in shAKT2 cells. Overall, our findings are consistent with a role for AKT2 in the regulation of not just EMT but also glycolytic versus aerobic respiration, promoting melanoma invasiveness and metastatic activity.

4. Discussion

Despite clear evidence that the PI3K/AKT pathway is important for tumor progression and metastasis in melanomas, the effective therapeutic inhibition of the AKT pathway has been challenging [10]. A potential pitfall in targeting the AKT pathway may be the strategy of pan-inhibition despite strong evidence of AKT isoform specificity in cancer. This study sought to investigate the contribution of AKT isoforms to melanoma initiation and metastasis, utilizing both murine and human melanoma models.

Although it is well-known that AKT activation drives tumor progression and metastasis in melanoma [44–47], the contributing AKT isoform(s) were not determined. Our data support an isoform-specific role for AKT2 in melanoma metastasis, with a shared role for AKT1 and AKT2 in tumorigenesis. Initially, we crossed mice with BRAF-driven murine melanoma previously developed in our lab [38] with mice lacking different AKT isoforms [23] and monitored the mice derived from these crosses for melanoma development. We observed that AKT1 loss extended overall survival in melanoma-prone mice, while

AKT2 and AKT3 loss did not (Figure 1C). This suggests that AKT1 promotes melanoma initiation and growth, which is consistent with a well-described role for AKT1 in tumor promotion [15,21] and cell proliferation and survival [17]. This is also consistent with our observation that tumor-derived cell lines consistently show robust phosphorylation of AKT1 when compared with other AKT isoforms (Figures 3F and S3), suggesting that AKT1 but not AKT2 or AKT3 plays a critical role in melanoma initiation. One caveat to these findings is that 20% of $BRAF^{V600E}/Arf^{-/-}$ mice die for reasons unrelated to melanoma [38], and determining the exact cause of death was not technically feasible in our initial studies. Additionally, while the effect of AKT1 on survival implies PI3K-AKT activation in this model, we did not explicitly test the extent of this within the resulting murine tumors. As such, melanoma-free survival or the exclusion of animals that both lacked melanoma and also died for reasons unrelated to melanoma would more accurately reflect the relative contributions of AKT isoforms to melanoma initiation and tumor promotion. However, because early primary melanomas pose a relatively low clinical risk, we focused our continuing studies on metastatic disease progression, which still presents the greatest treatment challenge for patients.

In order to address the potential involvement of AKT isoforms in metastasis, we chose to examine how the loss of AKT isoforms affects the metastatic potential of a melanoma cell line we established from a $BRAF^{V600E}/Arf^{-/-}$ mouse melanoma. The cell line was engineered to express luciferase, and following IV inoculation in syngeneic immunocompetent mice, it could be monitored by bioluminescence. By using this approach, we discovered that increased phosphorylation of AKT2 was present in metastatic lesions despite the lack of such phosphorylation in primary tumors and parental cell lines, suggesting that AKT2 activation may facilitate the metastatic seeding, survival, and/or growth of the cells seeded in metastatic sites. We have previously reported a low but significant rate of spontaneous lung metastasis in the $BRAF^{V600E}/Arf^{-/-}$ melanoma model [38]. Based on this observation, future studies could use this model to determine whether the AKT isoforms differentially control the rate of metastatic melanomas. This could be addressed by monitoring the metastatic burden of $BRAF^{V600E}/Arf^{-/-}/AKT^{-/-}$ mice lacking individual AKT isoforms.

The preceding data suggest that AKT2 differentially promotes the metastatic potential of melanomas, but to address the mechanism, we utilized a set of human melanoma cell lines driven by mutation in BRAF, as well as by PTEN loss. The functional loss of the tumor suppressor PTEN leading to AKT activation occurs in a large proportion of melanomas [7], and frequently co-occurs with oncogenic BRAF mutations [26]. However, the isoform specificity of AKT in the context of PTEN loss has remained largely unexplored. Therefore, using several $BRAF^{V600E}$ metastatic human melanoma cell lines with PTEN loss and robust AKT phosphorylation, we investigated the consequence of isoform-specific knockdown on cell migration and invasion. Our results show that AKT2 depletion consistently inhibits invasive and migratory behaviors, while AKT1 depletion reduces cellular proliferation and tumor growth. We postulated that AKT2-dependent invasion and migration were regulated through the AKT2-specific transcriptional changes in EMT-related genes E-cadherin, ZEB1, and Snail (Figure 6A). These data are consistent with findings in breast cancer, in which AKT2 but not AKT1 promotes cell migration and EMT [21,22,48,49] and the well-recognized role of EMT in metastatic promotion [50–52]. While our studies using shRNA-mediated AKT depletion used only one hairpin per isoform, the similar phenotypes between shRNA depletion and CRISPR knockout cells support the conclusion that impaired metastatic capability depends more specifically on AKT2.

While a consistent and explicit role for AKT3 was not identified in our study, this may be partially explained by AKT3-specific knockdown efficiency, which varied greatly among the cell lines (Supplementary Figure S1C). AKT3 amplification is known to occur in melanoma [26] but was not characterized across our panel and could attenuate knockdown efficiency if present, which is consistent with reports in the literature showing robust AKT3 phosphorylation in melanoma [24]. Nevertheless, our finding that AKT2 knockdown greatly attenuates the metastatic potential of human melanoma cell lines is consistent with previous data, suggesting that PHLPP1 inhibits melanoma metastasis by suppressing the phosphorylation of AKT2 and AKT3 but not AKT1 [25].

Once cells become migratory and invasive, extravasation and anchorage-independent growth are two necessary requirements for metastatic colonization [50]. Previous work in PTEN-deficient prostate tumors demonstrated their dependence on AKT2 for both maintenance and survival, but AKT1 in the same tumors was dispensable [53]. Our xenograft studies suggest that PTEN-deficient melanomas are initially sensitive to AKT2 inhibition but ultimately do not depend on AKT2 over the long term, given our observation of delayed tumor outgrowth in the AKT2 knockdown mice, as well as unmitigated tumor growth of AKT2 KO cells (Supplemental Figure S4C). In PTEN null melanomas and in the human cell lines in our study, a key difference is significant AKT1 phosphorylation, which could compensate to drive tumor cell growth and survival, given the cellular sensitivity to AKT1 depletion in cell proliferation and tumor growth that we observed. We hypothesize that potential compensation by AKT1 might explain why the CRIPSR KO AKT2 cells did not recapitulate the decreased subcutaneous tumor growth or the full prophylactic protection observed for the shRNA AKT2 KD cells, although we have not explicitly tested this hypothesis. Despite the apparent inability of AKT1 to fully support metastasis in AKT2-depleted human melanoma cells, hyperactive AKT1 contributes to the metastatic potential of the mouse SM1-750 model. By utilizing an autochthonous mouse model of melanoma, Kircher and colleagues showed that hyperactive, ectopic AKT1(E17K) resulted in an enhanced level of brain metastases and reduced overall survival compared to hyperactivating mutations in AKT2 or AKT3, mediated through a FAK an AKT1-specific substrate [32]. Indeed, we observed a significant increase in the phosphorylation of both AKT1 and AKT3 after passage in vivo to generate the brain metastatic-competent SM1-750 cell line (Figure 3F), in addition to increased AKT2 phosphorylation despite the presence of intact PTEN protein (Supplementary Figure S3A). Whether this increased phosphorylation is due to hyperactivating mutations arising in AKT1 remains to be determined. Interestingly, Kircher and colleagues also showed that mice with hyperactive AKT2 but not hyperactive AKT3 mutations still developed brain metastases, albeit to a lesser extent than AKT1. We are currently unable to perform AKT2 overexpression in our model systems due to the inability to interpret isoform-specific effects in the context of constitutively active AKT; however, future studies will aim to utilize this approach to test if AKT2 overactivity increases metastatic capability. These observations, when taken together with our data, suggest that AKT2 promotes melanoma metastasis in murine model systems, and further studies may elucidate the mechanistic differences in metastatic promotion between AKT isoforms in murine melanomas.

In order to understand the potential stages of metastatic dissemination for which AKT2 may be most critical, we explored extravasation from blood vessels and tumor cell proliferation at seeded distant sites as two candidate stages for the study of metastasis. The data presented in this report provide strong evidence that the key limiting event regulated by AKT2 was tumor cell extravasation. By pre-incubating cells with doxycycline to knock down AKT2 and then injecting them into the tail vein of mice, we observed the complete lack of metastatic dissemination despite the presence of phosphorylated AKT1 and AKT3. We ruled out the possibility that resident cells were dormant at distant metastatic sites by removing mice from doxycycline chow at 6 weeks. No tumors appeared after an additional 6-week observation period, suggesting that the disseminated cells may have been removed from circulation. It is also possible that 6 weeks was not sufficient time or that the dormant cells were too small to observe by our methods, as melanoma cells disseminating to the brain have been known to exist as single cells or small clusters barely visible by histology that can then regrow when the conditions are optimal [54]. Nevertheless, when cells expressing AKT2 were injected into the tail vein, and then mice were fed doxycycline chow to knock down AKT2 24 h later, there was a partial inhibition of metastasis. However, these mice were not fully protected from metastatic disease, unlike the mice inoculated with AKT2-depleted cells, suggesting AKT2 activity may be most impactful at the stage of extravasation, and its effect on tumor growth at the metastatic site was moderate. These results were phenocopied with AKT2 KO cells, in which AKT2 KO provided a survival benefit, perhaps by partially inhibiting the onset of metastatic disease.

Lastly, we investigated the mechanisms whereby AKT2 might mediate pro-metastatic behaviors, hypothesizing that AKT2 KD cells might be glycolytically impaired. Further, the relationship between impaired glycolysis and impaired EMT is well-established in the literature [55,56]. AKT2-specific roles in maintaining glucose homeostasis are well-documented [18], and metabolic rewiring in melanoma can facilitate metastatic dissemination [57]. We observed that AKT2 KD suppressed basal glycolytic metabolism and reduced compensatory glycolysis. In malignant glioma, the AKT2-specific phosphorylation of PDHK1 at Thr346 was shown to increase the phosphorylation of PDHE1α. This interferes with the entry of pyruvate into the TCA cycle, resulting in the stimulation of glycolysis, the maintenance of cell proliferation, and the inhibition of autophagy and apoptosis during severe hypoxia [42]. Earlier studies have shown that AKT2 selectively promotes the expression of miR-21 during hypoxia and renders cells resistant to hypoxia-induced cell death [58] Importantly, recent studies have shown that miR-21 promotes glycolysis by targeting pyruvate dehydrogenase A1 (PDHA1) [59] Therefore, AKT2 may inhibit the activity of pyruvate dehydrogenase by increasing the activity of PDHK1 and inhibiting the expression of PDHA1. In agreement with a significant role for AKT2-mediated regulation of PDHK1 and subsequent PDHE1α activity, we observed reductions in PDHK1 activity as well as phosphorylated PDHE1α with AKT2 KD. When taken together, the totality of our data suggests the AKT2-mediated regulation of PDHK1 and other components of pyruvate metabolism are important in both normal and hypoxic conditions, although the specific regulation may differ. Future studies should further define the role of PDHK1 in glycolytic maintenance that mediates melanoma metastasis.

5. Conclusions

In summary, by using genetically engineered human cell lines and novel syngeneic mouse models, we reveal multiple roles for AKT2 in melanoma metastasis. Indeed, as AKTs are central pleiotropic signaling hubs [60], it is not surprising that the AKT2-mediated cellular changes facilitating melanoma metastasis are manifold, including glycolytic changes and the enhancement of factors promoting epithelial-to-mesenchymal transition. This study reinforces the need for improved development in terms of clinically relevant small molecules for the selective targeting of AKT isoforms as a melanoma therapy.

Supplementary Materials: The following are available online at https://www.mdpi.com/article/10.3390/cancers15204958/s1. Supplementary Table S1: Primer Sequences and Hairpin Sequences, Supplementary Table S2: An-tibodies Used, Supplementary Figure S1: Characterization of AKT Phosphorylation and Knock-down Cell Lines, Supplementary Figure S2: AKT2 knockdown impairs wound healing, migration and invasion in WM455 and UACC903 cells, Supplementary Figure S3: Activation state of AKT isoforms in murine melanoma cells in vitro and in vivo, Supplementary Figure S4: AKT1 Deletion Delays Primary Melanoma Growth and Improves Survival, Supplementary Figure S5: AKT1 Knockdown impairs Melanoma Cell Proliferation, Supplementary Figure S6: AKT1 Knockdown Restricts An-chorage Independent Growth.

Author Contributions: S.K.M. designed and performed experiments, analyzed the data, and drafted the manuscript. J.P. performed experiments and edited the manuscript. A.L.B. performed the experiments and drafted the manuscript. P.N.T. assisted with the experimental design and interpretation of the results. P.W.H. initiated the project, designed the experiments, and helped write and edit the manuscript. All authors have read and agreed to the published version of the manuscript.

Funding: This work was supported by grants from the NSF (DGE-0806676, S. McRee), Tufts Collaborative Cancer Biology Award (S. McRee, A. Bayer), NIH (F31CA210312; S. McRee, F30HL162200), and the American Cancer Society (120825-PF-11-281-01-CCG, J. Pietruska).

Institutional Review Board Statement: These studies were conducted in accordance with the Tufts University Animal Care and Use Committee, as approved in protocol B2018-150 on 1/23/2019.

Informed Consent Statement: This study did not include humans.

Data Availability Statement: Data are available within the article or upon request from the authors.

Acknowledgments: We gratefully acknowledge all collaborators that provided the reagents and materials (as indicated in Materials and Methods section). We thank Gary Sahagian, Min Fang of the Small Animal Imaging Facility, and Alessandra Cecchini for their help with in vivo imaging. The model figures were made with Biorendr.

Conflicts of Interest: P.W.H. reports personal funds from Incyte and Leidos, Inc. that are outside the submitted work. No disclosures were reported by the other authors.

References

1. Fruman, D.A.; Chiu, H.; Hopkins, B.D.; Bagrodia, S.; Cantley, L.C.; Abraham, R.T. The PI3K Pathway in Human Disease. *Cell* **2017**, *170*, 605–635. [CrossRef]
2. Dai, D.L.; Martinka, M.; Li, G. Prognostic Significance of Activated Akt Expression in Melanoma: A Clinicopathologic Study of 292 Cases. *J. Clin. Oncol.* **2005**, *23*, 1473–1482. [CrossRef]
3. Dankort, D.; Curley, D.P.; Cartlidge, R.A.; Nelson, B.; Karnezis, A.N.; Damsky, W.E., Jr.; You, M.J.; DePinho, R.A.; McMahon, M.; Bosenberg, M. BrafV600E Cooperates with Pten Loss to Induce Metastatic Melanoma. *Nat. Genet.* **2009**, *41*, 544–552. [CrossRef]
4. Davies, H.; Bignell, G.R.; Cox, C.; Stephens, P.; Edkins, S.; Clegg, S.; Teague, J.; Woffendin, H.; Garnett, M.J.; Bottomley, W.; et al. Mutations of the BRAF Gene in Human Cancer. *Nature* **2002**, *417*, 949–954. [CrossRef]
5. Ibrahim, N.; Haluska, F.G. Molecular Pathogenesis of Cutaneous Melanocytic Neoplasms. *Annu. Rev. Pathol. Mech. Dis.* **2009**, *4*, 551–579. [CrossRef]
6. Bucheit, A.D.; Chen, G.; Siroy, A.; Tetzlaff, M.; Broaddus, R.; Milton, D.; Fox, P.; Bassett, R.; Hwu, P.; Gershenwald, J.E.; et al. Complete Loss of PTEN Protein Expression Correlates with Shorter Time to Brain Metastasis and Survival in Stage IIIB/C Melanoma Patients with BRAFV600 Mutations. *Clin. Cancer Res.* **2014**, *20*, 5527–5536. [CrossRef]
7. Giles, K.M.; Rosenbaum, B.E.; Berger, M.; Izsak, A.; Li, Y.; Bochaca, I.I.; Vega-Saenz de Miera, E.; Wang, J.; Darvishian, F.; Zhong, H.; et al. Revisiting the Clinical and Biologic Relevance of Partial PTEN Loss in Melanoma. *J. Investig. Dermatol.* **2019**, *139*, 430–438. [CrossRef]
8. Davies, M.A.; Katherine, S.-H.; Lin, E.; Tellez, C.; Deng, W.; Gopal, Y.N.; Woodman, S.E.; Calderone, T.C.; Ju, Z.; Lazar, A.J.; et al. Integrated Molecular and Clinical Analysis of AKT Activation in Metastatic Melanoma. *Clin. Cancer Res.* **2009**, *15*, 7538–7546. [CrossRef]
9. Niessner, H.; Schmitz, J.; Tabatabai, G.; Schmid, A.M.; Calaminus, C.; Sinnberg, T.; Weide, B.; Eigentler, T.K.; Garbe, C.; Schittek, B.; et al. PI3K Pathway Inhibition Achieves Potent Antitumor Activity in Melanoma Brain Metastases In Vitro and In Vivo. *Clin. Cancer Res.* **2016**, *22*, 5818–5828. [CrossRef]
10. Amaral, T.; Niessner, H.; Sinnberg, T.; Thomas, I.; Meiwes, A.; Garbe, C.; Garzarolli, M.; Rauschenberg, R.; Eigentler, T.; Meier, F. An Open-Label, Single-Arm, Phase II Trial of Buparlisib in Patients with Melanoma Brain Metastases Not Eligible for Surgery or Radiosurgery—The BUMPER Study. *Neurooncol Adv.* **2020**, *2*, vdaa140. [CrossRef]
11. Kuzu, O.F.; Gowda, R.; Sharma, A.; Noory, M.A.; Dinavahi, S.S.; Kardos, G.; Drabick, J.J.; Robertson, G.P. Improving Pharmacological Targeting of AKT in Melanoma. *Cancer Lett.* **2017**, *404*, 29–36. [CrossRef]
12. Vanhaesebroeck, B.; Perry, M.W.D.; Brown, J.R.; André, F.; Okkenhaug, K. PI3K Inhibitors Are Finally Coming of Age. *Nat. Rev. Drug Discov.* **2021**, *20*, 741–769. [CrossRef]
13. Halder, A.K.; Cordeiro, M.N.D.S. AKT Inhibitors: The Road Ahead to Computational Modeling-Guided Discovery. *Int. J. Mol. Sci.* **2021**, *22*, 3944. [CrossRef]
14. Song, M.; Bode, A.M.; Dong, Z.; Lee, M.-H. AKT as a Therapeutic Target for Cancer. *Cancer Res.* **2019**, *79*, 1019–1031. [CrossRef]
15. Dummler, B.; Hemmings, B. Physiological Roles of PKB/Akt Isoforms in Development and Disease. *Biochem. Soc. Trans.* **2007**, *35*, 231–235. [CrossRef]
16. Dummler, B.; Tschopp, O.; Hynx, D.; Yang, Z.-Z.; Dirnhofer, S. Hemmings Life with a Single Isoform of Akt: Mice Lacking Akt2 and Akt3 Are Viable but Display Impaired Glucose Homeostasis and Growth Deficiencies. *Mol. Cell. Biol.* **2006**, *26*, 8042–8051. [CrossRef]
17. Cho, H.; Thorvaldsen, J.L.; Chu, Q.; Feng, F.; Birnbaum, M.J. Akt1/PKBα Is Required for Normal Growth but Dispensable for Maintenance of Glucose Homeostasis in Mice. *J. Biol. Chem.* **2001**, *276*, 38349–38352. [CrossRef]
18. Cho, H.; Mu, J.; Kim, J.; Thorvaldsen, J.; Chu, Q.; Crenshaw, E.; Kaestner, K.; Bartolomei, M.; Shulman, G.; Birnbaum, M. Insulin Resistance and a Diabetes Mellitus-like Syndrome in Mice Lacking the Protein Kinase Akt2 (PKB Beta). *Science* **2001**, *292*, 1728–1731. [CrossRef]
19. Gonzalez, E.; McGraw, T.E. The Akt Kinases: Isoform Specificity in Metabolism and Cancer. *Cell Cycle* **2009**, *8*, 2502–2508. [CrossRef]
20. Testa, J.R.; Tsichlis, P.N. AKT Signaling in Normal and Malignant Cells. *Oncogene* **2005**, *24*, 7391–7393. [CrossRef]
21. Dillon, R.L.; Muller, W.J. Distinct Biological Roles for the Akt Family in Mammary Tumor Progression. *Cancer Res.* **2010**, *70*, 4260–4264. [CrossRef]

22. Irie, H.Y.; Pearline, R.V.; Grueneberg, D.; Hsia, M.; Ravichandran, P.; Kothari, N.; Natesan, S.; Brugge, J.S. Distinct Roles of Akt1 and Akt2 in Regulating Cell Migration and Epithelial-Mesenchymal Transition. *J. Cell Biol.* **2005**, *171*, 1023–1034. [CrossRef] [PubMed]
23. Maroulakou, I.G.; Oemler, W.; Naber, S.P.; Tsichlis, P.N. Akt1 Ablation Inhibits, Whereas Akt2 Ablation Accelerates, the Development of Mammary Adenocarcinomas in Mouse Mammary Tumor Virus (MMTV)-ErbB2/Neu and MMTV-Polyoma Middle T Transgenic Mice. *Cancer Res.* **2007**, *67*, 167–177. [CrossRef] [PubMed]
24. Stahl, J.M.; Sharma, A.; Cheung, M.; Zimmerman, M.; Cheng, J.Q.; Bosenberg, M.W.; Kester, M.; Sandirasegarane, L.; Robertson, G.P. Deregulated Akt3 Activity Promotes Development of Malignant Melanoma. *Cancer Res.* **2004**, *64*, 7002–7010. [CrossRef] [PubMed]
25. Yu, Y.; Dai, M.; Lu, A.; Yu, E.; Merlino, G. PHLPP1 Mediates Melanoma Metastasis Suppression through Repressing AKT2 Activation. *Oncogene* **2018**, *37*, 2225–2236. [CrossRef]
26. The Cancer Genome Atlas Network. Genomic Classification of Cutaneous Melanoma. *Cell* **2015**, *161*, 1681–1696. [CrossRef]
27. Cho, J.H.; Robinson, J.P.; Arave, R.A.; Burnett, W.J.; Kircher, D.A.; Chen, G.; Davies, M.A.; Grossmann, A.H.; VanBrocklin, M.W.; McMahon, M.; et al. AKT1 Activation Promotes Development of Melanoma Metastases. *Cell Rep.* **2015**, *13*, 898–905. [CrossRef]
28. Sanchez-Vega, F.; Mina, M.; Armenia, J.; Chatila, W.K.; Luna, A.; La, K.C.; Dimitriadoy, S.; Liu, D.L.; Kantheti, H.S.; Saghafinia, S.; et al. Oncogenic Signaling Pathways in The Cancer Genome Atlas. *Cell* **2018**, *173*, 321–337.e10. [CrossRef]
29. Zhang, Y.; Kwok-Shing Ng, P.; Kucherlapati, M.; Chen, F.; Liu, Y.; Tsang, Y.H.; de Velasco, G.; Jeong, K.J.; Akbani, R.; Hadjipanayis, A.; et al. A Pan-Cancer Proteogenomic Atlas of PI3K/AKT/mTOR Pathway Alterations. *Cancer Cell* **2017**, *31*, 820–832.e3. [CrossRef]
30. Cheung, M.; Sharma, A.; Madhunapantula, S.V.; Robertson, G.P. Akt3 and Mutant V600EB-Raf Cooperate to Promote Early Melanoma Development. *Cancer Res.* **2008**, *68*, 3429–3439. [CrossRef]
31. Nogueira, C.; Kim, K.-H.H.; Sung, H.; Paraiso, K.; Dannenberg, J.-H.H.; Bosenberg, M.; Chin, L.; Kim, M. Cooperative Interactions of PTEN Deficiency and RAS Activation in Melanoma Metastasis. *Oncogene* **2010**, *29*, 6222–6232. [CrossRef] [PubMed]
32. Kircher, D.A.; Trombetti, K.A.; Silvis, M.R.; Parkman, G.L.; Fischer, G.M.; Angel, S.N.; Stehn, C.M.; Strain, S.C.; Grossmann, A.H.; Duffy, K.L.; et al. AKT1E17K Activates Focal Adhesion Kinase and Promotes Melanoma Brain Metastasis. *Mol. Cancer Res.* **2019**, *17*, 1787–1800. [CrossRef]
33. Bayer, A.L.; Pietruska, J.; Farrell, J.; McRee, S.; Alcaide, P.; Hinds, P.W. AKT1 Is Required for a Complete Palbociclib-Induced Senescence Phenotype in BRAF-V600E-Driven Human Melanoma. *Cancers* **2022**, *14*, 572. [CrossRef]
34. Chin, Y.; Yoshida, T.; Marusyk, A.; Beck, A.H.; Polyak, K.; Toker, A. Targeting Akt3 Signaling in Triple-Negative Breast Cancer. *Cancer Res.* **2014**, *74*, 964–973. [CrossRef]
35. Davies, M.A. The Multi-Faceted Roles of the PI3K-AKT Pathway in Melanoma. *J. Transl. Med.* **2015**, *13*, 2039. [CrossRef] [PubMed]
36. Koya, R.C.; Mok, S.; Otte, N.; Blacketor, K.J.; Begonya, C.-A.; Tumeh, P.C.; Minasyan, A.; Graham, N.A.; Graeber, T.G.; Chodon, T.; et al. BRAF Inhibitor Vemurafenib Improves the Antitumor Activity of Adoptive Cell Immunotherapy. *Cancer Res.* **2012**, *72*, 3928–3937. [CrossRef]
37. Goel, V.; Ibrahim, N.; Jiang, G.; Singhal, M.; Fee, S.; Flotte, T.; Westmoreland, S.; Haluska, F.; Hinds, P.; Haluska, F. Melanocytic Nevus-like Hyperplasia and Melanoma in Transgenic BRAFV600E Mice. *Cancer Res.* **2009**, *28*, 2289–2298. [CrossRef] [PubMed]
38. Luo, C.; Sheng, J.; Hu, M.G.; Haluska, F.G.; Cui, R.; Xu, Z.; Tsichlis, P.N.; Hu, G.-F.F.; Hinds, P.W. Loss of ARF Sensitizes Transgenic BRAFV600E Mice to UV-Induced Melanoma via Suppression of XPC. *Cancer Res.* **2013**, *73*, 4337–4348. [CrossRef]
39. Pedri, D.; Karras, P.; Landeloos, E.; Marine, J.-C.; Rambow, F. Epithelial-to-Mesenchymal-like Transition Events in Melanoma. *FEBS J.* **2022**, *289*, 1352–1368. [CrossRef]
40. Verfaillie, A.; Imrichova, H.; Atak, Z.K.; Dewaele, M.; Rambow, F.; Hulselmans, G.; Christiaens, V.; Svetlichnyy, D.; Luciani, F.; Van den Mooter, L.; et al. Decoding the Regulatory Landscape of Melanoma Reveals TEADS as Regulators of the Invasive Cell State. *Nat. Commun.* **2015**, *6*, 6683. [CrossRef]
41. Lu, J. The Warburg Metabolism Fuels Tumor Metastasis. *Cancer Metastasis Rev.* **2019**, *38*, 157–164. [CrossRef] [PubMed]
42. Chae, Y.C.; Vaira, V.; Caino, M.C.; Tang, H.-Y.; Seo, J.H.; Kossenkov, A.V.; Ottobrini, L.; Martelli, C.; Lucignani, G.; Bertolini, I.; et al. Mitochondrial Akt Regulation of Hypoxic Tumor Reprogramming. *Cancer Cell* **2016**, *30*, 257–272. [CrossRef]
43. Masters, T.A.; Calleja, V.; Armoogum, D.A.; Marsh, R.J.; Applebee, C.J.; Laguerre, M.; Bain, A.J.; Larijani, B. Regulation of 3-Phosphoinositide–Dependent Protein Kinase 1 Activity by Homodimerization in Live Cells. *Sci. Signal.* **2010**, *3*, ra78. [CrossRef]
44. Dhawan, P.; Singh, A.B.; Ellis, D.L.; Richmond, A. Constitutive Activation of Akt/Protein Kinase B in Melanoma Leads to up-Regulation of Nuclear Factor-kappaB and Tumor Progression. *Cancer Res.* **2002**, *62*, 7335–7342. [PubMed]
45. Schadendorf, D.; Fisher, D.E.; Garbe, C.; Gershenwald, J.E.; Grob, J.-J.J.; Halpern, A.; Herlyn, M.; Marchetti, M.A.; Grant, M.; Ribas, A.; et al. Melanoma. *Nat. Rev. Dis. Primers* **2015**, *1*, 15003. [CrossRef]
46. Luo, C.; Pietruska, J.R.; Sheng, J.; Bronson, R.T.; Hu, M.G.; Cui, R.; Hinds, P.W. Expression of Oncogenic BRAFV600E in Melanocytes Induces Schwannian Differentiation in Vivo. *Pigment. Cell Melanoma Res.* **2015**, *28*, 603–606. [CrossRef]
47. Chen, G.; Chakravarti, N.; Aardalen, K.; Lazar, A.J.; Tetzlaff, M.; Wubberhorst, B.; Kim, S.-B.B.; Kopetz, S.; Ledoux, A.; Nanda, V.G.; et al. Molecular Profiling of Patient-Matched Brain and Extracranial Melanoma Metastases Implicates the PI3K Pathway as a Therapeutic Target. *Clin. Cancer Res.* **2014**, *20*, 5537–5546. [CrossRef] [PubMed]
48. Cheng, G.Z.; Chan, J.; Wang, Q.; Zhang, W.; Sun, C.D.; Wang, L.-H.H. Twist Transcriptionally Up-Regulates AKT2 in Breast Cancer Cells Leading to Increased Migration, Invasion, and Resistance to Paclitaxel. *Cancer Res.* **2007**, *67*, 1979–1987. [CrossRef]

49. Iliopoulos, D.; Polytarchou, C.; Hatziapostolou, M.; Kottakis, F.; Maroulakou, I.G.; Struhl, K.; Tsichlis, P.N. MicroRNAs Differentially Regulated by Akt Isoforms Control EMT and Stem Cell Renewal in Cancer Cells. *Sci. Signal.* **2009**, *2*, ra62. [CrossRef]
50. Ganesh, K.; Massagué, J. Targeting Metastatic Cancer. *Nat. Med.* **2021**, *27*, 34–44. [CrossRef]
51. Alonso, S.R.; Tracey, L.; Ortiz, P.; Pérez-Gómez, B.; Palacios, J.; Pollán, M.; Linares, J.; Serrano, S.; Sáez-Castillo, A.I.; Sánchez, L.; et al. A High-Throughput Study in Melanoma Identifies Epithelial-Mesenchymal Transition as a Major Determinant of Metastasis. *Cancer Res.* **2007**, *67*, 3450–3460. [CrossRef]
52. Caramel, J.; Papadogeorgakis, E.; Hill, L.; Browne, G.J.; Richard, G.; Wierinckx, A.; Saldanha, G.; Osborne, J.; Hutchinson, P.; Tse, G.; et al. A Switch in the Expression of Embryonic EMT-Inducers Drives the Development of Malignant Melanoma. *Cancer Cell* **2013**, *24*, 466–480. [CrossRef] [PubMed]
53. Chin, Y.; Yuan, X.; Balk, S.P.; Toker, A. PTEN-Deficient Tumors Depend on AKT2 for Maintenance and Survival. *Cancer Discov.* **2014**, *4*, 942–955. [CrossRef]
54. Izraely, S.; Sagi-Assif, O.; Klein, A.; Meshel, T.; Tsarfaty, G.; Pasmanik-Chor, M.; Nahmias, C.; Couraud, P.-O.; Ateh, E.; Bryant, J.L.; et al. The Metastatic Microenvironment: Brain-Residing Melanoma Metastasis and Dormant Micrometastasis. *Int. J. Cancer* **2012**, *131*, 1071–1082. [CrossRef] [PubMed]
55. Marcucci, F.; Rumio, C. Tumor Cell Glycolysis—At the Crossroad of Epithelial–Mesenchymal Transition and Autophagy. *Cells* **2022**, *11*, 1041. [CrossRef] [PubMed]
56. Bhattacharya, D.; Azambuja, A.P.; Simoes-Costa, M. Metabolic Reprogramming Promotes Neural Crest Migration via Yap/Tead Signaling. *Dev. Cell* **2020**, *53*, 199–211.e6. [CrossRef]
57. Fischer, G.M.; Vashisht Gopal, Y.; McQuade, J.L.; Peng, W.; DeBerardinis, R.J.; Davies, M.A. Metabolic Strategies of Melanoma Cells: Mechanisms, Interactions with the Tumor Microenvironment, and Therapeutic Implications. *Pigment. Cell Melanoma Res.* **2017**, *31*, 11–30. [CrossRef]
58. Polytarchou, C.; Iliopoulos, D.; Hatziapostolou, M.; Kottakis, F.; Maroulakou, I.; Struhl, K.; Tsichlis, P.N. Akt2 Regulates All Akt Isoforms and Promotes Resistance to Hypoxia through Induction of miR-21 upon Oxygen Deprivation. *Cancer Res.* **2011**, *71*, 4720–4731. [CrossRef]
59. Liu, Z.; Yu, M.; Fei, B.; Fang, X.; Ma, T.; Wang, D. miR-21-5p Targets PDHA1 to Regulate Glycolysis and Cancer Progression in Gastric Cancer. *Oncol. Rep.* **2018**, *40*, 2955–2963. [CrossRef]
60. Manning, B.D.; Cantley, L.C. AKT/PKB Signaling: Navigating Downstream. *Cell* **2007**, *129*, 1261–1274. [CrossRef]

cancers

MDPI

Article

Fasting-Mimicking Diet Inhibits Autophagy and Synergizes with Chemotherapy to Promote T-Cell-Dependent Leukemia-Free Survival

Roberta Buono [1,2], Jonathan Tucci [3], Raffaello Cutri [1], Novella Guidi [1], Serghei Mangul [4,5], Franca Raucci [6], Matteo Pellegrini [5,7], Steven D. Mittelman [3,8] and Valter D. Longo [1,6,9,*]

1 Department of Biological Sciences, Longevity Institute, School of Gerontology, University of Southern California, 3715 McClintock Avenue, Los Angeles, CA 90089, USA
2 Department of Molecular Biology and Biochemistry, University of California, Irvine, CA 92697, USA
3 Center for Endocrinology, Diabetes & Metabolism, Children's Hospital Los Angeles, 4650 Sunset Blvd, Los Angeles, CA 90027, USA
4 Department of Computer Science, University of California Los Angeles, 580 Portola Plaza, Los Angeles, CA 90095, USA
5 Institute for Quantitative and Computational Biosciences, Boyer Hall, 611 Charles Young Drive, University of California Los Angeles, Los Angeles, CA 90095, USA
6 IFOM AIRC Institute of Molecular Oncology, Via Adamello 16, 20139 Milan, Italy
7 Department of Molecular, Cell and Developmental Biology, University of California Los Angeles, 801 Hilgard Avenue, Los Angeles, CA 90095, USA
8 Division of Pediatric Endocrinology, UCLA Mattel Children's Hospital, 10833 Le Conte Avenue, MDCC 22-315, Los Angeles, CA 90095, USA
9 Eli and Edythe Broad Center for Regenerative Medicine and Stem Cell Research at USC, Keck School of Medicine, University of Southern California, Los Angeles, CA 90033, USA
* Correspondence: vlongo@usc.edu

Simple Summary: Despite the advances in the treatment of pre-B-ALL leukemia in children, adult pre-B-ALL continues to represent a major challenge. This work focuses on the use of differential responses to fasting conditions between normal and cancer cells to achieve cancer-free survival. We show that a fasting-mimicking diet in combination with vincristine causes a synergistic increase in the toxicity to pre-B-ALL cells resulting in high cancer cell death. While fasting is not sufficient to promote cancer-free survival, the combination of fasting/FMD and vincristine promotes autophagy inhibition, which is at the center of the high toxicity phenotype specific to leukemia cells, possibly through a mechanism involving immune cells.

Abstract: Fasting mimicking diets (FMDs) are effective in the treatment of many solid tumors in mouse models, but their effect on hematologic malignancies is poorly understood, particularly in combination with standard therapies. Here we show that cycles of a 3-day FMD given to high-fat-diet-fed mice once a week increased the efficacy of vincristine to improve survival from BCR-ABL B acute lymphoblastic leukemia (ALL). In mice fed a standard diet, FMD cycles in combination with vincristine promoted cancer-free survival. RNA seq and protein assays revealed a vincristine-dependent decrease in the expression of multiple autophagy markers, which was exacerbated by the fasting/FMD conditions. The autophagy inhibitor chloroquine could substitute for fasting/FMD to promote cancer-free survival in combination with vincristine. In vitro, targeted inhibition of autophagy genes *ULK1* and *ATG9a* strongly potentiated vincristine's toxicity. Moreover, anti-CD8 antibodies reversed the effects of vincristine plus fasting/FMD in promoting leukemia-free survival in mice, indicating a central role of the immune system in this response. Thus, the inhibition of autophagy and enhancement of immune responses appear to be mediators of the fasting/FMD-dependent cancer-free survival in ALL mice.

Keywords: Leukemia; pre-B-ALL; fasting-mimicking diet; autophagy; cancer treatment

1. Introduction

Pre-B cell acute lymphoblastic leukemia (ALL) is the most common childhood cancer but can also occur in older adults. It is a hematologic malignancy characterized by impaired differentiation and aggressive proliferation of clonal lymphoblasts in the bone marrow, spleen and blood [1,2]. The standard of care for B-ALL is chemotherapy, which includes vincristine, cyclophosphamide, anthracyclines, corticosteroids, L-asparaginase and other drugs [3]. The overall cure rate is about 90% in children but only 50% in adults, for whom the relapse occurrence is high due to drug resistance [2]. There is still no consensus regarding the best treatment approach for newly diagnosed adults and especially older patients.

It is becoming increasingly recognized that weight and nutritional status can impact cancer survival and that obesity is a risk factor for poor outcomes not only for B-ALL but also for other cancers [4–6]. Using a murine model of diet-induced obesity, obese mice with syngeneic BCR-ABL B-ALL were shown to have poorer response to chemotherapy than control mice [7–9]. Clinical studies have shown similar negative effects of obesity on cancer mortality [10].

We and other groups have previously demonstrated that periodic fasting or FMDs reduce chemotherapy side effects, improve cancer treatment efficacy and delay cancer progression in mice [11–13]. Here, we have investigated the effects of FMD cycles in combination with or without chemotherapy (vincristine) in both obese and normal weight syngeneic in vivo mouse models as well as in vitro models for mouse and human B-ALL.

2. Materials and Methods

2.1. Cell Culture and Treatment

BCR-ABL syngeneic leukemia cells ("8093 cells") were generated using BCR/ABL transgenic mouse [16] M-ALL were cultured at a density of 2.5×10^5 cells/mL in a standard condition McCoy's 5A media supplemented with 10% FBS, murine IL-3, beta-mercapto ethanol and gentamycin or in the STS condition (same as above but with 0.5 gr/liter of Glucose; 2% FBS).

Human leukemia cells H-ALL (BV173) were obtained from ATCC and were cultured in RPMI media supplemented with 10% FBS and gentamycin or in the STS condition.

M-ALL and H-ALL were treated for 24 h or 48 h with or without VC 5nM (Sigma Aldrich, Saint Louis, MO, USA Cat#V8388).

2.2. Cell Viability

Cells viability was measured by Mini Automated Cell Counter Moxi (Orflo, Ketchum, ID, USA) or by Trypan Blue exclusion dye (Corning, Glendale, AZ, USA Cat# MT25900CI).

2.3. ULK1 and ATG9a Silencing

Cells were seeded at 60–80% of confluence and then transfected for 48 h with 30 pM ULK1 and ATG9a siRNA (Life Technologies, Carlsbad, CA, USA ID: 65268, 125425) using Lipofectamine RNAiMAx Reagent (Thermo Fisher Scientific, Waltham, MA, USA Cat#13778100) according to the manufacturer's instructions.

2.4. LDH Assay

Cell cytotoxicity was measured using the colorimetric CytoTox 96® Non-Radioactive Cytotoxicity Assay (Promega, Madison, WI, USA, Cat# G1780) following the manufacturer's instructions.

2.5. CYTO-ID Staining

Autophagy was measured by the CYTO-ID® Autophagy detection kit (Enzo Life Sciences, Farmingdale, NY, USA Cat# ENZ-51031-0050) according to the manufacturer's protocols.

2.6. FACS Analyses

Flow cytometry analyses of mouse BM and SP GFP$^+$ leukemia cells were performed to assess the engraftment. Flow cytometry staining of BM, spleen and blood were performed using APC-CD3 (Thermo Fisher Scientific Cat# 17-0032-82), PerCP-eFluor710-CD8 (Thermo Fisher Scientific Cat# 46-0081-80), PE-CD4 (Thermo Fisher Scientific Cat# 12-0041-85), APC-eFluor780-CD25 (Thermo Fisher Scientific Cat# 47-0251-82) and PE-Cy7-PD1 antibodies (Thermo Fisher Scientific Cat# 25-9985-82). FlowJo 10 (Becton Dickinson, Franklin Lakes, NJ, USA) was used to analyze data and to prepare figures.

2.7. Annexin V Staining

BCR-ABL syngeneic leukemia cells and BV173 were stained with eFLuor780 Fixable Viability Dye (Thermo Fisher Scientific Cat# 65-0865) and PE-Cy7 Annexin V (Thermo Fisher Scientific Cat# 88-8103-72) according to the manufacturer's instructions. Analyses were performed with BD FACS diva on LSR II (Becton Dickinson, Franklin Lakes, NJ, USA).

2.8. Western Blotting

Total cell lysates were prepared using the RIPA buffer (Thermo Fisher Scientific Cat# 89900) according to the manufacturer's instructions. Protein concentration was measured with the BCA assay (Thermo Fisher Scientific Cat#23227). Equal amounts of protein (30 µg) were heat-denaturized in a lane marker-reducing sample buffer (Thermo Fisher Scientific Cat#39000), resolved by SDS-PAGE using Novex 4–20% Tris-Glycine MiniProtein Gels (Thermo Fisher Scientific) and transferred to PVDF membranes (Millipore, Darmstadt, Germany). The filters were blocked in 5% BSA for 1 h at room temperature and then incubated O.N at 4° with a primary antibody directed against cleaved caspase 3 (1:1000), phosphorylated p53 (1:1000), beclin1 (1:1000), p62 (1:1000), LC3B (1:1000), ULK1 (1:1000), vinculin (1:2000) (Cell Signaling, Danvers, MA, USA, rabbit mAb #9664, rabbit mAb #9284, rabbit mAb #3495, rabbit mAb #5114, rabbit mAb #2775, rabbit mAb #8054, rabbit mAb #18799), ATG9 (1:1000) (GeneTex, Irvine, CA, USA, rabbit mAb # GTX128427) and tubulin (1:2500) (Millipore, Burlington, MA, USA, #05-661).

Peroxidase-conjugated IgG (Santa Cruz, CA, USA) was used as a secondary antibody. ImageJ software version 1.52a was used to analyze western blot data.

2.9. RNA-seq Library Preparation and Data Analysis

(Supplementary methods): RNAseq data are deposited in the GEO database GSE212918.

2.10. In Vivo B-ALL Model

All animal protocols were approved by the Institutional Animal Care and Use Committee (IACUC) of the University of Southern California. All mice were maintained in a pathogen-free environment and housed in clear shoebox cages in groups of five animals per cage with a constant temperature and humidity, a 12 h/12 h light/dark cycle and unlimited access to water.

Obese model: Diet-induced obese C57BL/6J mice (Jackson Laboratories, Bar Harbor, ME, USA, Cat# 380050) were weaned onto 60% calories from fat diet (Research Diets, New Brunswick, NJ, USA, D12492) and maintained on the diet until dietary intervention. At 20 weeks of age, mice were injected retro-orbitally with 1×10^4 GFP-expressing BCR-ABL syngeneic leukemia cancer cells. One week later, mice were divided into 4 groups: HFD + vehicle (HFD $n = 5$), HFD + FMD ($n = 5$) 4 cycles + vehicle, HFD + chemo drugs (vincristine (HFD + VC $n = 5$)) I.P. 0.5 mg/kg once a week and HFD + FMD +VC ($n = 5$) (VC Sigma Aldrich, St. Louis, MO, USA, Cat#V8388).

Normal diet model: 50 C57BL/6J mice (Jackson Laboratories Cat# 000664) (20 weeks old) were injected as described above. One week later, the mice were divided into 4 groups: ad lib+ vehicle (AL $n = 10$), FMD ($n = 12$) 4 cycles + vehicle, ad lib+ vincristine (AL + VC $n = 14$) I.P. 0.5 mg/kg once a week and FMD +VC ($n = 14$).

2.11. Mouse FMD Diet

The mouse version of the FMD was a 3-day regimen [14,15] (Table S1 and supplementary methods).

2.12. Vincristine and Chloroquine in an In Vivo Model of BCR-ABL B-ALL

40 mice C57BL/6J (20 weeks old) were injected with GFP-expressing BCR-ABL syngeneic leukemia cancer cells. One week later after tumor injection, they were divided into 5 groups: ad lib+ vehicle (AL), ad lib+ chloroquine once a week I.P. 50 mg/kg/day (AL + CQ $n = 8$), fasting-mimicking diet (FMD + CQ $n = 8$) 4 cycles + CQ, ad lib+ VC + CQ once a week (AL + VC + CQ $n = 8$) and FMD +VC + CQ ($n = 8$) (CQ Sigma Aldrich Cat# PHR1258).

2.13. CD8+ Cells In Vivo Depletion

Complete depletion of CD8+ CTL was achieved by intraperitoneal administration of neutralizing monoclonal antibody (αCD8; clone YTS 169.4 BioXCell Lebanon, NH, USA, Cat# BP0117) or rat IgG (BioXCell, Cat# BP0090) every 4 days after 1 week of tumor implantation. The depletion of circulating CD8+ CTL over time was confirmed by FACS analysis.

2.14. Statistics and Experimental Design

For statistical significance of the differences between the means of the two groups, we used two-tailed Student's t-tests. The statistical significance of differences among multiple groups (≥3) was calculated by performing ANOVA multiple comparisons of the means for each group. The survival rates of the two groups were analyzed using a log–rank test and were considered to be statistically significant if $p < 0.05$.

No samples or animals were excluded from analysis, and sample size estimates were not used. Animals were randomly assigned to groups. Studies were not conducted blinded.

3. Results

3.1. FMD Cycles Promote Cancer-Free Survival in Obese Leukemic Mice in Combination with Vincristine

We investigated whether FMD cycles, in combination with vincristine therapy, could improve the outcome of BCR-ABL B-ALL in an obese murine model.

C57BL/6J (20 weeks old) obese mice raised on a high-fat (60%) diet were implanted with GFP-expressing BCR-ABL syngeneic leukemia cancer cells [16]. Seven days post-implantation, obese mice were randomized into four groups: high-fat diet (HFD), HFD + FMD (FMD 3–4 days a week for 4 weeks), HFD mice treated with vincristine and HFD mice treated with FMD plus VC (Figure 1a). Mice were routinely examined for tumor progression and body weight (Figure 1b). FMD alone did not improve ALL survival but did improve survival in combination with VC; 40% of HFD+FMD+VC mice showed long-term cancer-free survival versus 20% in the HFD+VC group and none in the HFD or HFD+FMD only groups (Figure 1c). In a separate experiment, flow cytometric analyses showed a major reduction of GFP^+ tumor cells in HFD+FMD+VC vs. HFD+VC spleens isolated from obese mice sacrificed after three weeks of treatment (Figure 1d,e).

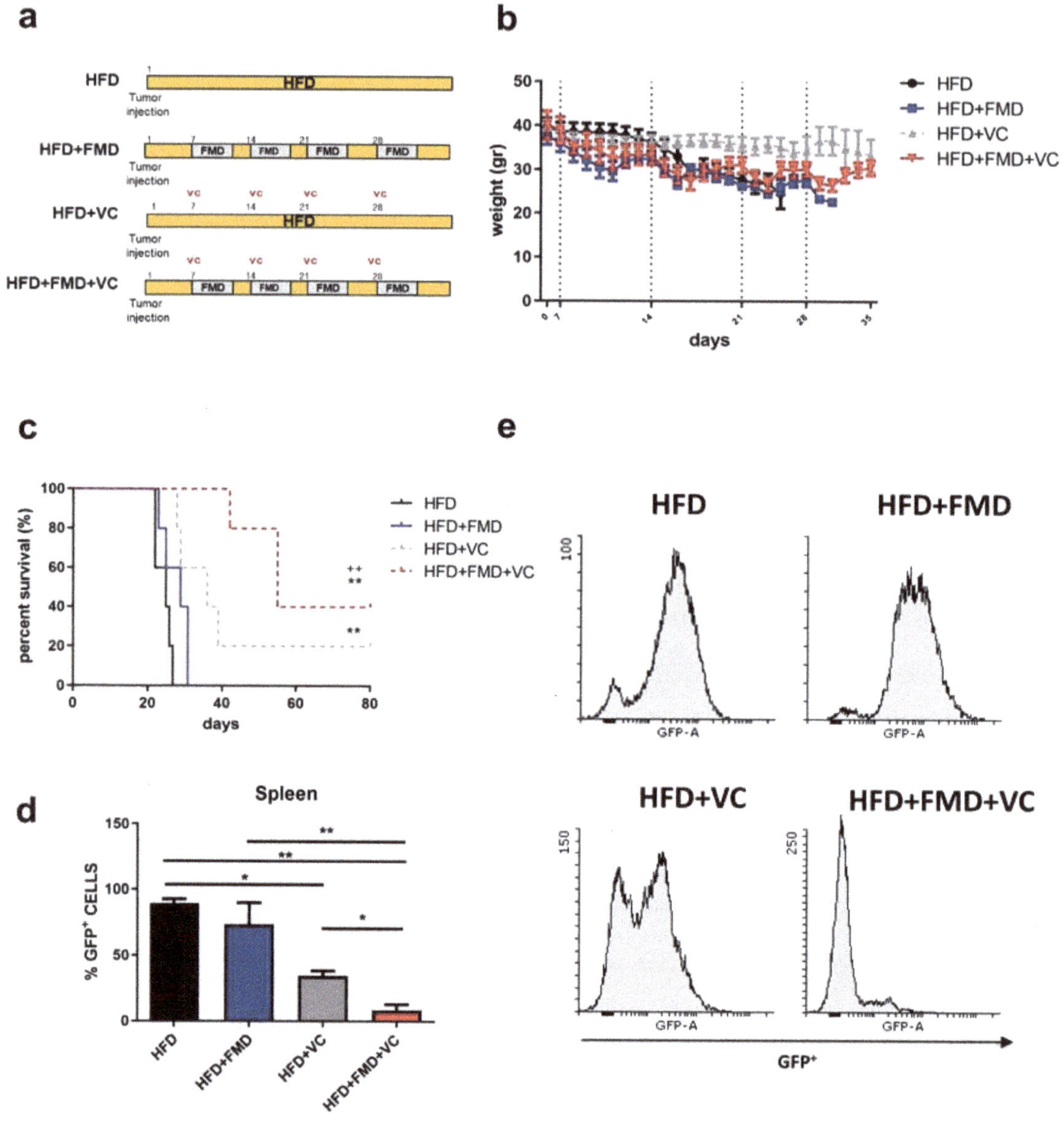

Figure 1. FMD cycles improve obese leukemic C57BL/6J mouse survival after vincristine treatment. (**a**) Experimental scheme, (**b**) body weight (gr) and (**c**) survival curve of periodical FMD in obese ALL murine model (n = 20) (p < 0.01 ** vs. HFD, ++ vs. FMD). A total of 20 diet-induced obese C57BL/6J mice were implanted with 1×10^4 GFP-expressing BCR-ABL syngeneic leukemia cancer cells on day 0. Mice were randomized into groups of 5 mice with the following conditions: obese mice remaining on high-fat diet without vincristine (HFD) and with vincristine (HFD + VC) and obese mice receiving intermittent FMD without vincristine (HFD + FMD) and with vincristine (HFD + FMD + VC). VC (0.5 mg/kg/wk × 4 wks) was delivered I.P. beginning on day 7 (n = 5/group). (**d**,**e**) FACS analyses and representative histograms of GFP+ ALL cells in spleens taken from obese mice in the third week of treatment with or without FMD and VC (n = 3/group). Data are expressed as mean ± s.e.m. * p < 0.05, ** p < 0.01, one-way ANOVA.

3.2. FMD Cycles in Combination with Vincristine Promote Cancer-Free Survival in a Normal-Weight Syngeneic Mouse Model of Pre-B-ALL

In a second in vivo model, we investigated the effect of periodic cycles of FMD + VC to treat ALL in mice maintained on a standard chow diet and of normal weight. C57BL/6J mice were injected with GFP-expressing BCR-ABL syngeneic leukemia cancer cells and one week later divided into four groups: regular chow ad libitum (ad lib+ vehicle (AL $n = 10$)), FMD ($n = 12$) + vehicle, ad lib+ vincristine (AL + VC $n = 14$) and FMD +VC ($n = 14$) (Figure 2a). Mice on FMD cycles exhibited acute weight loss each cycle, which was reversed during refeeding (Figure 2b). FMD alone did not improve median survival compared to the untreated standard chow group (25 vs. 23 days, p = n.s.) but did result in a significant improvement in overall survival (Figure 2c). Over half of the FMD + VC group survived more than 110 days without any leukemia symptoms, indicating a major improvement in cancer-free survival compared to the AL + VC group ($p < 0.05$ vs. AL + VC) (Figure 2c). In agreement with this finding, FMD + VC mice exhibited the smallest spleen size, weight, and GFP expression by qPCR (Figure 2d–f). Flow cytometry confirmed low numbers of GFP$^+$ tumor cells in the bone marrow and spleen. On the day of autopsy, a reduction of ALL cells in both organs and a complete absence of tumor cells in the FMD + VC mice with long survival was shown (Figure 2g,h; Supplementary Figure S1).

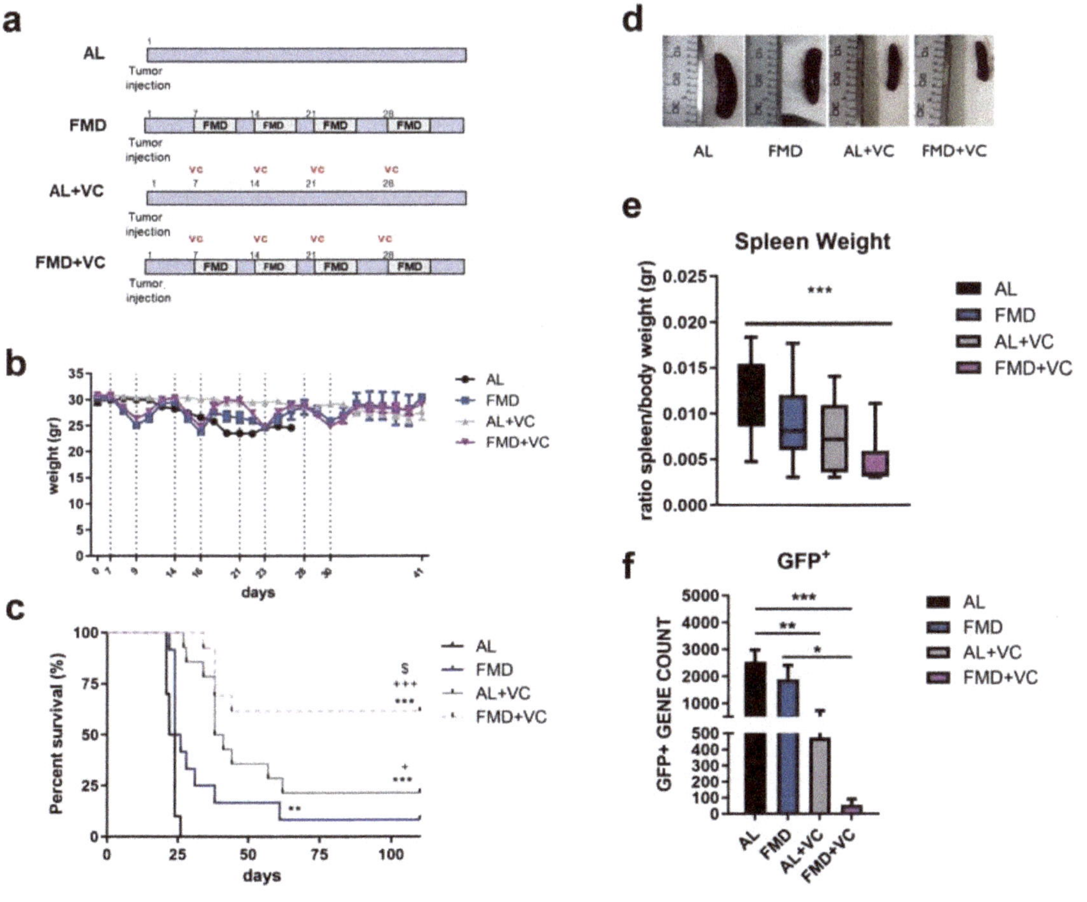

Figure 2. *Cont.*

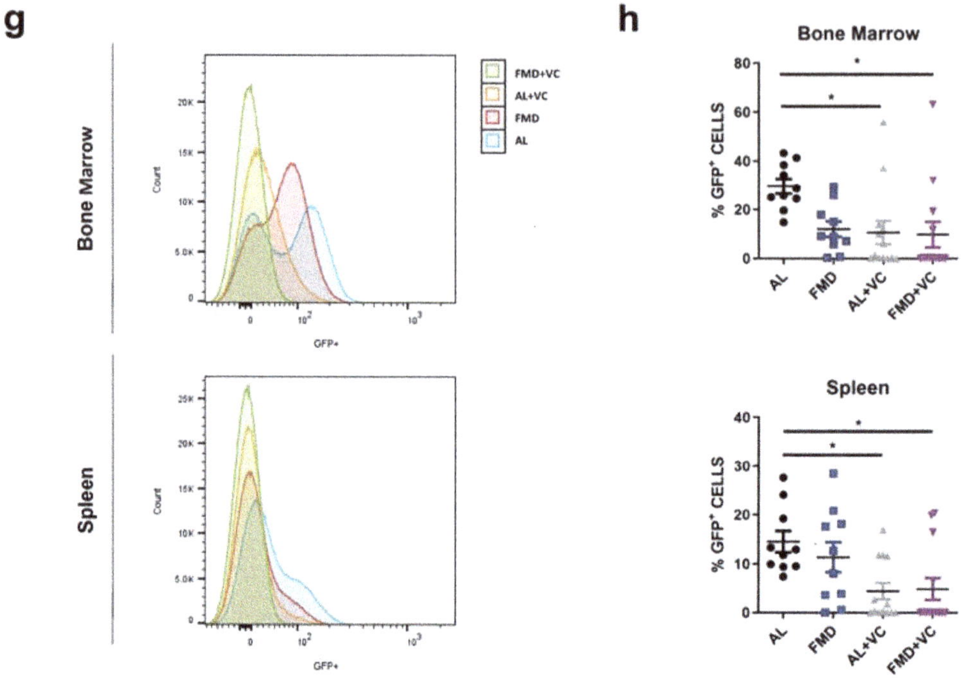

Figure 2. Effect of fasting-mimicking diet (FMD) in combination with vincristine in an in vivo model of BCR-ABL B-ALL. (**a**) Experimental scheme, (**b**) body weight (gr) and (**c**) survival curve of periodical FMD and vincristine in ALL in vivo model. Dashed vertical lines denote the FMD cycle. A total of 50 mice C57BL/6J (20 weeks old) were injected via retro-orbital injection with 1×10^4 GFP-expressing BCR-ABL syngeneic leukemia cancer cells/mice. One week later, the mice were divided into 4 groups: ad lib+ vehicle (AL $n = 10$), fasting-mimicking diet (FMD $n = 12$) 4 cycles of 3 days + vehicle, ad lib+ chemo drugs (vincristine (AL + VC $n = 14$)) I.P. 0.5 mg/kg once a week and FMD +VC ($n = 14$) ($p < 0.01$ **, $p < 0.001$ *** vs. AL: $p < 0.05$ + $p < 0.001$ +++ vs. FMD: $p < 0.05$ $ vs. AL + VC). (**d**) Representative spleen picture, (**e**) spleen weight (gr) and (**f**) GFP+ RNA expression data ($n = 6$/group). (**g**) Representative histograms and (**h**) FACS analyses quantification for GFP+ tumor cells in bone marrow and spleen. Data are expressed as mean ± s.e.m. * $p < 0.05$, ** $p < 0.01$, *** $p < 0.001$, one-way ANOVA.

3.3. *In Vitro Short-Term Starvation Reduces Cell Proliferation of Mouse and Human Leukemia Cells and Inhibits Autophagy*

We investigated the in vitro effects of short-term starvation (STS) on mouse (8093, M-ALL) and human (BV173, H-ALL) ALL cell lines grown under normal culture media (CTRL) or fasting-mimicking condition (STS medium with 0.5 gr/L glucose and 2% FBS) [17], with or without VC 5 nM. The STS medium reduced M-ALL and H-ALL proliferation (Figure 3a). STS in combination with VC decreased the cell number of both M-ALL and H-ALL cells by 70% by 48 h compared to the control medium. STS + VC also induced cell death as shown by a significant increase in lactate dehydrogenase (LDH) release (Figure 3b) and AnnexinV/eFluor780 viability dye staining. Compared to VC alone, 48 h of FMD + VC treatment increased apoptotic and necrotic M-ALL cells from 17.58 ± 2.18% to 42.32 ± 5.38% and H-ALL cells from 24.60 ± 4.44% to 60.66 ± 5.44% (Figure 3c,d). These analyses confirmed the effect of STS in enhancing VC-mediated cell death.

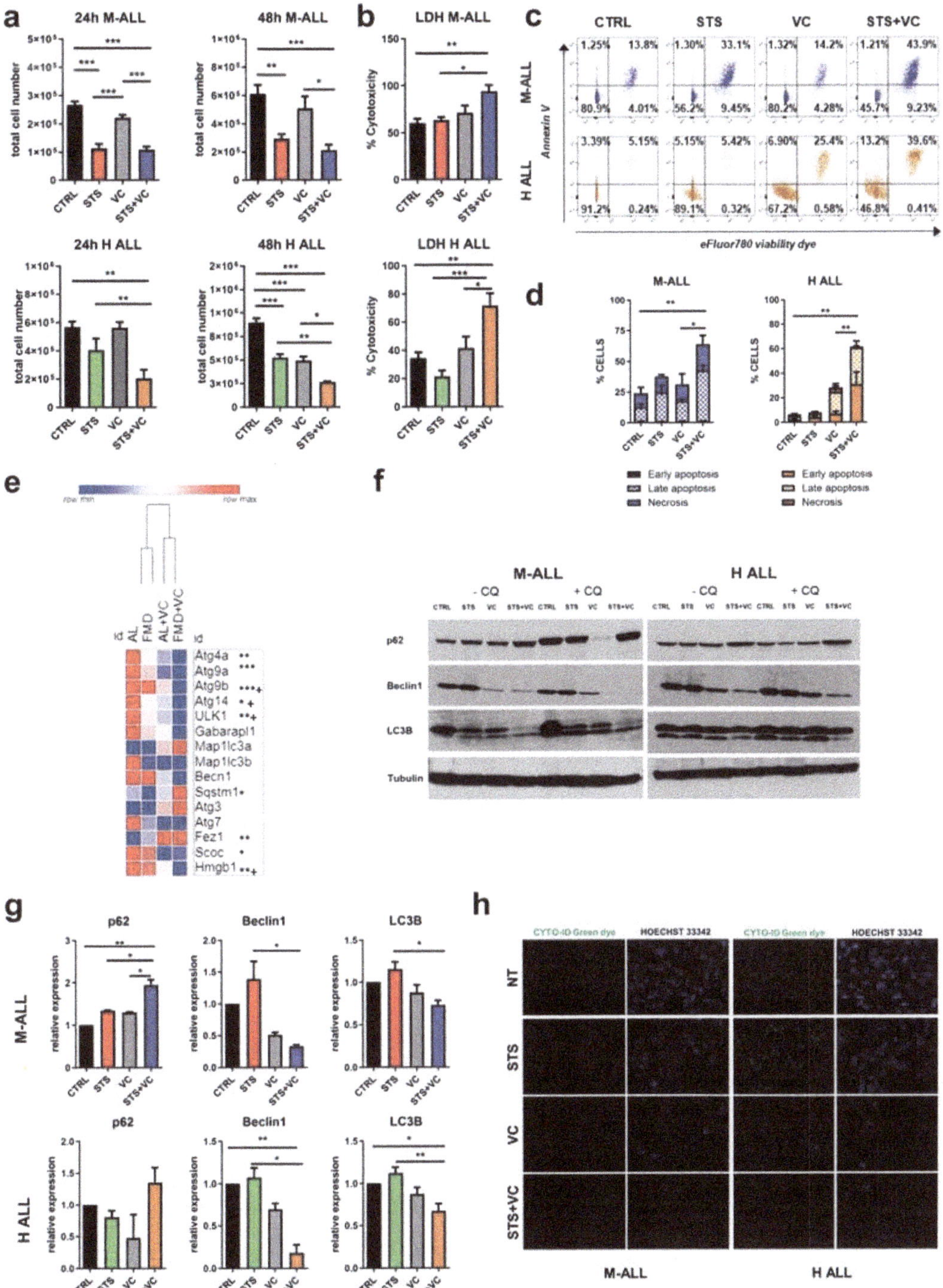

Figure 3. Effects of STS in combination with vincristine on mouse and human ALL cell lines. (**a**) Cell count and (**b**) LDH release of mouse and human ALL tumor cells were cultured in low-glucose

(0.5 g/L) and 2% serum (in vitro STS) or in a standard glucose and 10% serum (control CTRL) medium ± vincristine 5 nM for 24 and 48 h. (**c**) FACS analyses of AnnexinV and eFluor780 viability dye of M-ALL and H-ALL tumor cell line and (**d**) percentage data quantification of cells in early and late apoptosis and necrosis. (**e**) Heatmap displaying autophagy gene expression in spleen tissue from AL, FMD, AL + VC and FMD +VC mouse (n = 6/each group) * p < 0.05, ** p < 0.01, *** p < 0.001 vs. AL and + p < 0.05 vs. VC. (**f**) Protein analyses of LC3B, beclin1, p62 and tubulin with or without chloroquine (CQ) 100 μM in M-ALL and H-ALL cell line. (**g**) Relative protein quantification performed by densitometric analysis using ImageJ64 software. Data are expressed as mean ± s.e.m. * p < 0.05, ** p < 0.01, *** p < 0.001, one-way ANOVA ($n \geq 3$). (**h**) Cyto-ID green dye and Hoechst blue staining of M-ALL and H-ALL cells in normal or in STS media with or without VC 5 nM for 48 h.

To dissect the molecular mechanism responsible for the effect of FMD + VC on cancer-free survival, we performed RNA-seq analyses on spleen tissue isolated from BCR-ABL B-ALL mice at the end of the treatment (Figure 3e). Our data revealed in the FMD + VC group the downregulation of genes involved in the initiation of autophagy, including ULK1, Atg4, Atg9a and Atg9b, and the up-regulation of autophagy genes such as Sqstm1 (p62) and FEZ1 in comparison to AL and AL + VC mice. FEZ1 is a negative regulator of LC3 lipidation [18], and when overexpressed, inhibits p62 degradation possibly through sequestration of ULK1. However, because these gene expression levels might be confounded by a higher splenic ALL cell burden in the AL and FMD groups than those treated with VC (Figure 2f), comparison between the first two lanes of Figure 3e have limited value, whereas the third and fourth lane can be compared.

Consistent with this result, protein expression analyses of autophagic markers indicate downregulation of LC3B and beclin1 in spleen tissue from the FMD + VC group (Supplementary Figure S2a,b).

Autophagy plays a dual role in cancer pathogenesis: it can either promote or suppress tumor growth and survival [19–21]. To better understand the effect of STS and VC on autophagy, we analyzed key autophagic markers by Western blot in mouse and human ALL cells (Figure 3f,g). In agreement with the RNAseq analysis of spleen cells, STS + VC reduced expression of LC3B and beclin1 and increased the expression of the autophagy inhibitor p62. To quantify autophagy flux and LC3B turnover, chloroquine (CQ, 100 μM) was added to some samples 3–6 h before preparing the lysate. LC3B levels were not increased by chloroquine under FMD + VC conditions, indicating that autophagic flux is decreased.

These results were confirmed by Cyto-ID green, which showed strong staining in ALL cells that was reduced by VC and reduced further by STS + VC (Figure 3h). These results indicate that autophagy is highly active in ALL cells, but it is inhibited by FMD + VC.

3.4. *In Vivo Synergistic Effect of Autophagy Inhibition and FMD Treatment*

To investigate whether autophagy inhibition may enhance or mediate the effects of fasting/FMD and VC against ALL progression in vivo, a syngeneic mouse model of BCR-ABL was treated with the autophagy inhibitor chloroquine in combination with an ad lib diet or FMD ± VC (n = 8/groups) (Figure 4a). Body weight was monitored during the treatment (Figure 4b). Weekly co-treatment with CQ and vincristine resulted in a dramatic increase of long-term survival in ad lib mice, which was not increased further by FMD treatment (Figure 4c). The combined VC + CQ treatment reduced the size of the spleen (Figure 4d), as well as detectible GFP+ tumor cells, irrespective of FMD treatment (Figure 4e,f). Protein expression of LC3B, beclin1 and p62 in terminal sacrificed mouse spleen tissue confirmed the effect of either CQ + VC or FMD + VC in inhibiting autophagy (Figure 4g,h). These findings indicate that ALL cells are protected from VC treatment by an autophagy-dependent mechanism, which is reversed/blocked by co-treatment with either FMD or the autophagy inhibitor CQ.

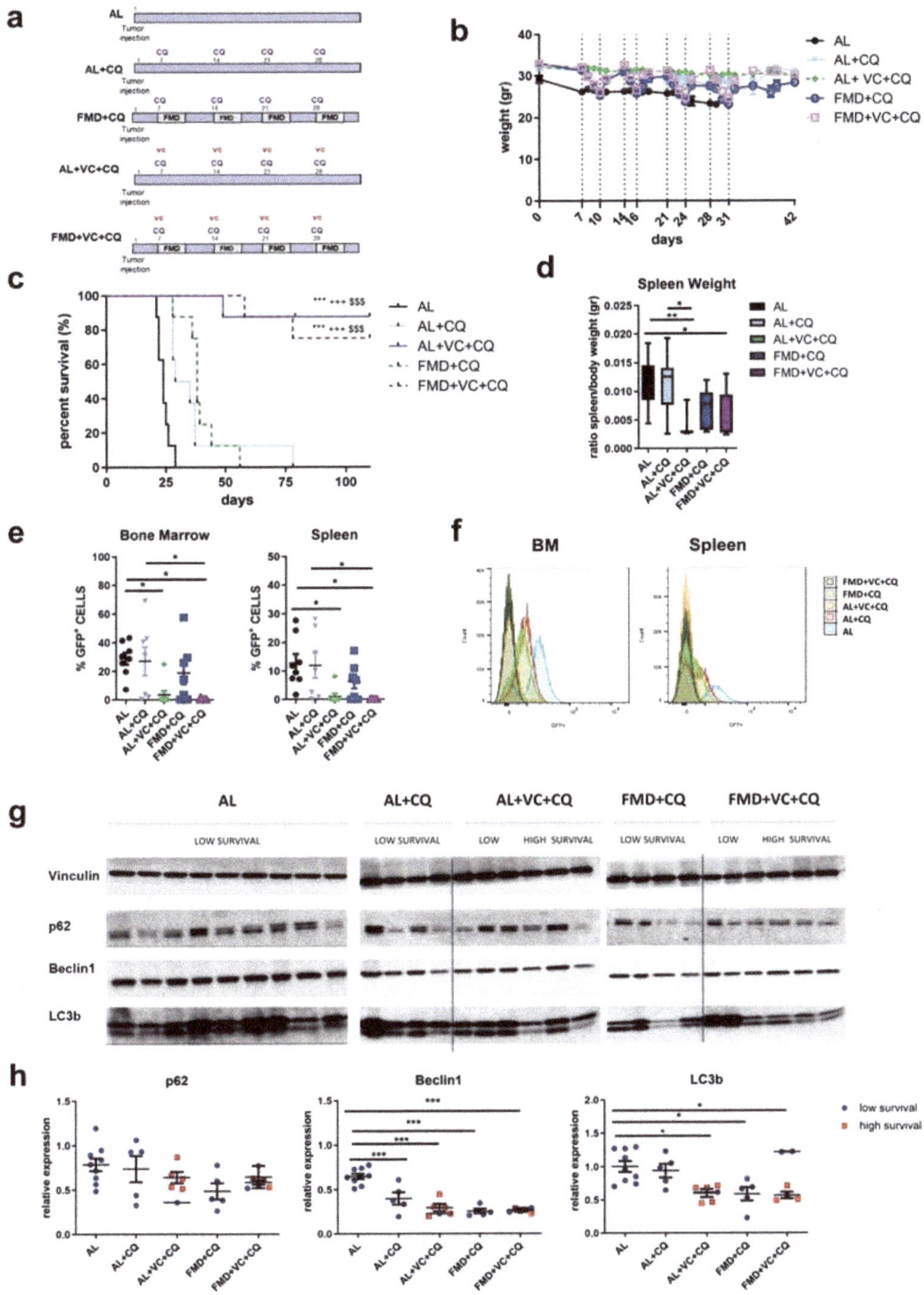

Figure 4. Effect of FMD in combination with vincristine and chloroquine in an in vivo model of BCR-ABL B-ALL. (**a**) Experimental scheme, (**b**) body weight (gr) and (**c**) survival curve. Dashed vertical lines denote the FMD cycle. A total of 40 mice C57BL/6J were injected via retro-orbital

injection with 1×10^4 GFP-expressing BCR-ABL syngeneic leukemia cancer cells/mice. One week later, the mice were divided into 5 groups: ad lib+ vehicle (AL), ad lib+ chloroquine once a week I.P. 50 mg/kg/day (AL + CQ $n = 8$), fasting-mimicking diet (FMD + CQ $n = 8$) 4 cycles (3–4 days) + CQ, ad lib+ VC + CQ once a week (AL + VC + CQ $n = 8$) and FMD +VC + CQ ($n = 8$) ($p < 0.001$ *** vs. AL, +++ vs. AL + CQ, $$$ vs. FMD + CQ). (**d**) spleen weight (gr). (**e**) FACS analyses representative histograms and (f) quantification for GFP+ tumor cells in bone marrow and spleen. (**g**) Protein analyses of LC3B, beclin1, p62 and vinculin in mouse spleen extract and (**h**) relative protein quantification performed by densitometric analysis using ImageJ64 software. Data are expressed as mean ± s.e.m. * $p < 0.05$, ** $p < 0.01$, *** $p < 0.001$, one-way ANOVA.

3.5. Reduced Expression of Autophagy Genes Enhances VC-Dependent Apoptosis in ALL Cells

To investigate the mechanisms of STS/VC-dependent ALL death, the viability of M-ALL and H-ALL cells grown in a normal or STS medium with CQ was measured. CQ treatment caused cytotoxicity of M-ALL and H-ALL cells cultured in either a normal or STS medium. Rapamycin, an autophagy activator, did not significantly affect cytotoxicity, though it caused a trend to reduce VC cytotoxicity in H-ALL (Figure 5a). CQ alone induced maximal toxicity in H-ALL cells, with or without VC, and this was potentiated in STS.

Similar patterns were observed when we inhibited autophagy by gene knockdown. Autophagy genes ULK1 and ATG9 were knocked down in M-ALL and H-ALL cell lines with two specific siRNA (ULK1-siRNA and ATG9-siRNA). We confirmed that the expression of ULK1 and ATG9 proteins was inhibited by siRNA transfection (Figure 5b). Knockdown of ULK1 and ATG9a inhibited cell proliferation (Figure 5c) and promoted cell death, resulting in increased LDH release (Figure 5d). These results confirm that the inhibition of autophagy inhibits the growth and survival of leukemia cells.

To validate the role of STS + VC in activating apoptosis, we measured the effect of the combined treatment on the phosphorylated form of p53 (Ser15) and the activation of the pro-apoptotic enzyme caspase-3. Activation of p53 by phosphorylation can lead to either cell cycle arrest or apoptosis. Protein analyses of phospho-p53 and cleaved caspase-3 showed a potentiating effect of STS on the toxicity of VC, particularly in H-ALL (Figure 5e,f). Taken together, these data suggest that fasting/FMD plus VC induce apoptosis in murine and human B-ALL cells in part by an autophagy- and p53-dependent mechanism.

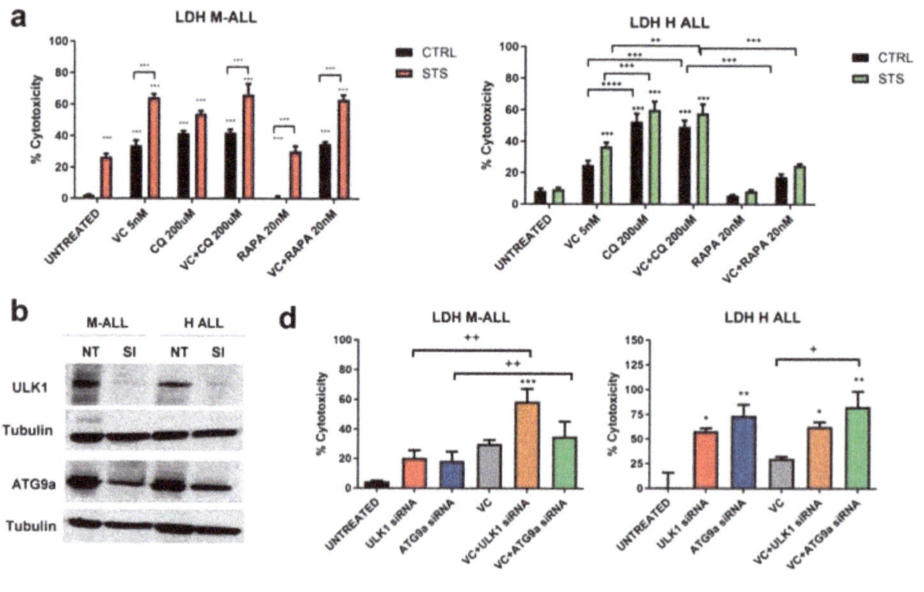

Figure 5. *Cont.*

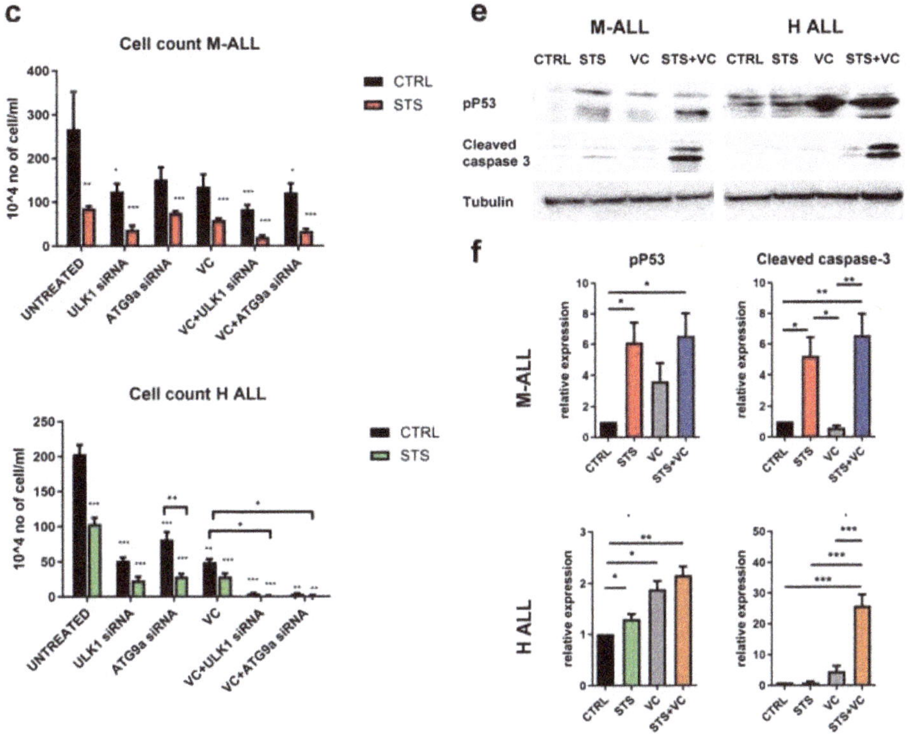

Figure 5. In vitro effects of STS in combination with vincristine on autophagy, cell death, and apoptosis. (**a**) LDH release of M-ALL and H-ALL cells cultured in STS or CTRL medium ± VC 5 nM for 48 h with or without chloroquine CQ 200 mM or rapamycin (RAPA) 20 nM. (**b**) Protein analyses of ULK1, ATG9a and tubulin after 48 h of transfection with the specific siRNA (Si) in normal medium. (**c**) Cell count and (**d**) LDH release of M-ALL and H-ALL after transfection with ULK1 siRNA (30 pM) or ATG9a (30 pM) siRNA with or without VC. (**e**) Western blot analyses and (**f**) relative protein quantification of phosphorylated p53, cleaved caspase 3 and tubulin in M-ALL and H-ALL cells ($n \geq 3$). Data are expressed as mean ± s.e.m. * $p < 0.05$, ** $p < 0.01$, *** $p < 0.001$, + $p < 0.05$, ++ $p < 0.01$, +++ $p < 0.001$, ++++ $p < 0.0001$ one-way ANOVA.

3.6. *Effect of FMD + VC on Immune Response against BCR-ABL B-ALL Cancer*

T cells can play an important role in the anti-tumor immune response. We previously showed that T cells mediate part of the effects of FMD against melanoma and breast cancer mouse models [22]. To determine whether FMD + VC may promote immune-dependent killing of B-ALL cells, we treated ALL mice with a single FMD cycle (n = 24 mice) and collected bone marrow, spleen and blood for flow cytometry analyses of the following T cell markers: CD3+CD4+, CD3+CD8+, CD4+CD25+, CD4+PD-1+, CD8+CD25+ and CD8+PD-1+ (Figure 6a,b). One cycle of FMD + VC shifted T cell populations toward anti-cancer immunity, increasing the pool of CD3+CD8+ in all three compartments, while reducing the pool of CD3+CD4+ cells in the bone marrow and spleen. Moreover FMD+VC significantly reduced the pool of both CD4+PD-1+ and CD8+PD-1+ cells (Figure 6b).

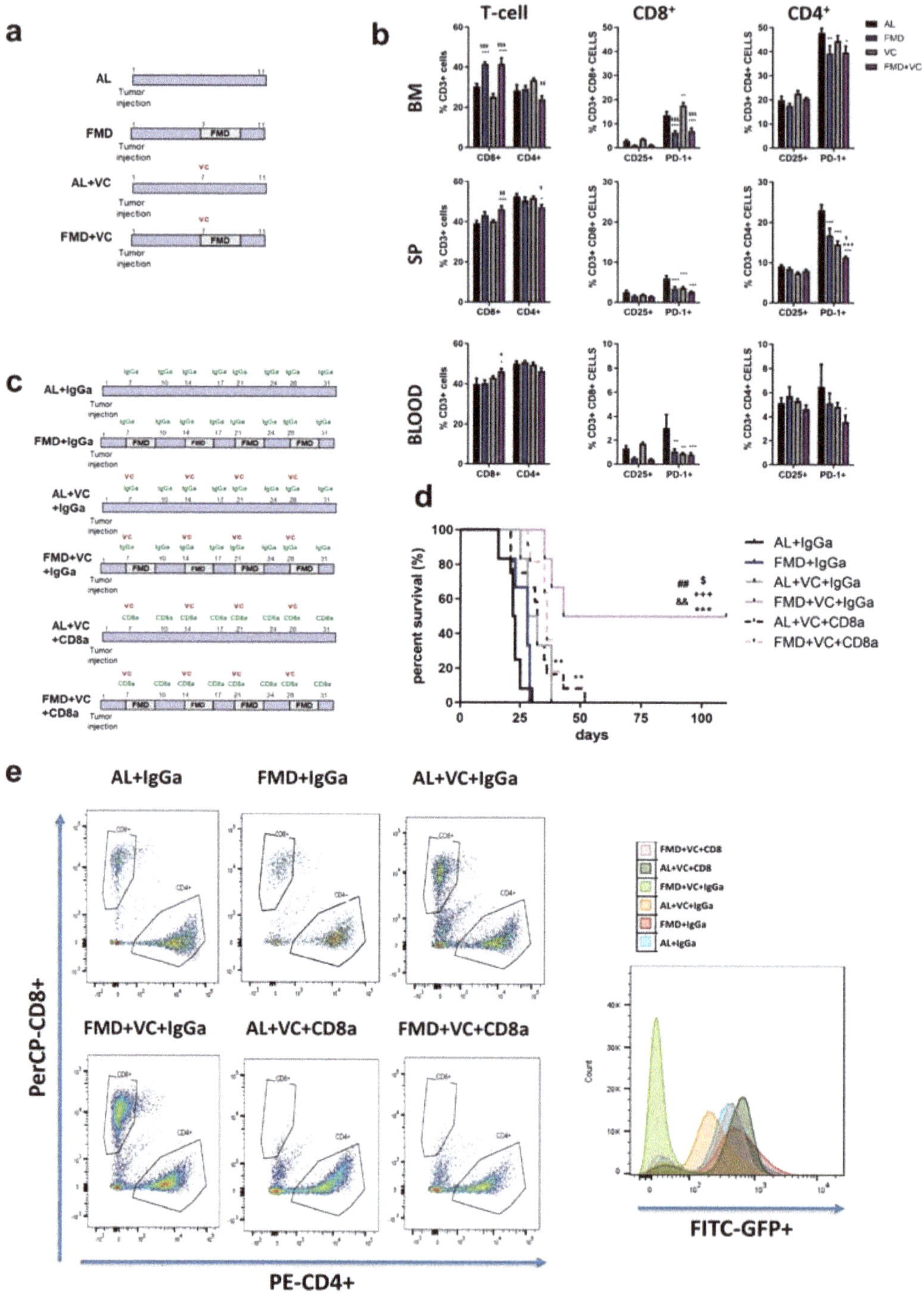

Figure 6. Effect of FMD in combination with vincristine on the immune response. (**a**) Experimental scheme of one cycle of FMD 3 days and vincristine in ALL in vivo model (n = 24). (**b**) FACS analyses

quantification for CD3+CD4+, CD3+CD8+, CD3+CD4+PD-1+, CD3+CD4+CD25+, CD3+CD8+PD-1+ and CD3+CD8+CD25+ in bone marrow, spleen and blood. Data are expressed as mean ± s.e.m. * $p < 0.05$, ** $p < 0.01$, *** $p < 0.001$ vs. CTRL, + $p < 0.05$, ++ $p < 0.01$, ++ $p < 0.001$ vs. FMD, \$ $p < 0.05$ vs. VC one-way ANOVA. (**c**) Experimental scheme and (**d**) survival curve of periodical FMD and VC in ALL in CD8+ in vivo depletion model. A total of 54 mice C57BL/6J (20 weeks old) were injected via retro-orbital injection with 1×10^4 GFP-expressing BCR-ABL syngeneic leukemia cancer cells/mice. One week later, the mice were divided into 6 groups: ad lib+ IgGa (AL + IgGa $n = 12$), ad lib+ VC once a week (AL + VC + IgGA $n = 6$), ad lib+ VC+ CD8a (AL + VC + CD8a $n = 12$), FMD (3 days) + IgGa ($n = 6$), FMD +VC + IgGa ($n = 6$) and FMD + VC + CD8a ($n = 12$) ($p < 0.001$ *** vs. AL + IgGa, +++ vs. FMD + IgGa, \$ $p < 0.05$ vs. VC + IgGa, $p < 0.01$ ## vs. VC + CD8a, && vs. FMD + VC + CD8a). (**e**) Representative FACS plots for CD3+CD4+ and CD3+CD8+ and histograms for GFP+ cancer cells in the bone marrow.

To test whether CD3+CD8+ T cells were required for FMD-dependent sensitization of leukemia cells to VC, we selectively depleted CD8+ T lymphocytes using a neutralizing monoclonal antibody (αCD8) or an IgGa (control) (Figure 6c,e). FMD + VC + IgGa control mice displayed the best outcome, with half of the mice showing long-term survival. However, CD8a treatment reversed this survival benefit (Figure 6d), indicating that the effects of FMD + VC are mediated at least in part by T cells.

4. Discussion

FMDs have the potential to enhance the efficacy of a wide variety of cancer treatments, weakening cancer cells by a process we termed differential stress sensitization (DSS) while strengthening normal cells by a response termed differential stress resistance (DSR) [11,17,23–25]. The effects of fasting/FMD in inducing DSS in both in vitro and in vivo models were previously shown to be mediated, in part, by the reduction of circulating IGF-1 and glucose levels [11,17,23,24]. In a mouse leukemia model, fasting alone reversed the progression of both B and T cell ALL but did not affect acute myeloid leukemia (AML) [26].

Here we show that cycles of FMD induce significant anti-leukemia efficacy and cancer-free survival when combined with vincristine, in part by activating T-cell-dependent anti-cancer effects. Fasting/FMD alone causes a trend for increasing autophagy, but when fasting/FMD is combined with vincristine, a significant and consistent downregulation of autophagy markers is observed. This role of autophagy in the FMD/VC-dependent toxicity to ALL cells is confirmed by the effect of the combination of vincristine with the autophagy inhibitor chloroquine, which also promotes p53 modulation, apoptosis and cancer-free survival in agreement with the established role of p53 in mediating cell death in AML and in solid malignancies [27,28].

Emerging data indicate that autophagy is a major contributor to chemotherapy resistance in AML, CLL, multiple myeloma, lymphoma and in nonhematologic cancers. Inactivation of autophagy by deletion of *Atg5* or *Atg7* prolonged survival in an AML mouse model. Furthermore, *Atg7*-deficient mice showed less chemoresistance to cytarabine treatment [29]. Pharmacological inhibition of ULK-1 when combined with TKI treatment reduces growth of a CML cell line and patient-derived xenografted CML cells in mouse models [30]. Thus, autophagy inhibition may represent an important therapeutic strategy against many malignancies [19,20,31–35]. The only inhibitors approved by the FDA are the antimalaria drug chloroquine and its derivative hydroxychloroquine (HCQ), which suppress autophagy by blocking autophagosome fusion and degradation in the final steps of autophagy. Both CQ and HCQ have been investigated in preclinical studies and in clinical trials in combination with chemotherapy, radiation therapy or other targeted therapies, and which have shown evidence of enhanced antitumor activity caused by these combination [19].

Notably, CQ alone was able to delay ALL growth but caused only a minor effect on the survival of mice (Figures 4 and 5). CQ disrupts the lysosomal functions, leading to cell death, and enhances VC efficacy compared with single-drug administration.

This role of autophagy in acting as an escape pathway in ALL cells treated with vincristine, which results in cell death when blocked by FMD, is in agreement with our recent study in which ER+ breast cancer cells treated with estrogen signaling and cdk4/6 inhibitors were killed by the effect of fasting/FMD cycles in blocking an escape based on insulin, leptin and IGF-1 signaling [36]. It is also in agreement with the effect of fasting/FMD in forcing the activation of PI3K-AKT and mTOR-dependent escape pathways, which resulted in the death of triple negative breast cancer cells when blocked [37]. Similarly, here, the toxicity of vincristine was limited by an autophagy-dependent escape mechanism, which was blocked by either FMD or chloroquine. However, it is important to point out that autophagy-inhibiting drugs are likely to have a more specific toxic effect on a limited range of cancer cell line, while fasting/FMD cycles instead appear to render many therapies more effective not only against a range of blood cancers but also against most solid tumors tested. Thus, the mechanism of action of FMD involves forcing many cancer cell types to reduce the number of redundant pathways that prevent cell death by generating energy or providing building blocks for the synthesis of DNA, proteins etc. These FMD-activated escape pathways can then be targeted by drugs, which in a number of mouse models, lead to tumor regression or cancer-free survival. The combination of FMD and these targeted drugs is likely to recruit immune cells by causing damage, including oxidative damage and DNA mutation in cancer cells. This may explain why FMD cycles are often effective against both immunocompetent and immunodeficient mouse cancer models, although blocking T cell function often reduces their efficacy.

In a previous study on a murine model of breast and melanoma cancer, we demonstrated that FMD in combination with chemotherapy was able to stimulate T-cell-dependent cytotoxicity against tumor cells, promoting the infiltration of CD3+CD8+ TILs and decrease of Tregs at the tumor site; an effect partially mediated by the downregulation of heme oxygenase-1 [22]. In this study, FACS analyses showed that mice treated with FMD + VC had a higher presence of CD3+CD8+ cells and a reduction of CD3+CD4+ cells, together with a lower pool of PD-1+ T cells in the bone marrow, spleen and blood. Thus, the combination of FMD + VC generated changes commonly associated with immunotherapy and consistent with the role of the immune responses in cancer-free survival. In fact, the in vivo depletion of CD8+ cells reversed the effect of FMD + VC on cancer-free survival (Figure 6). Thus, this study, together with our previous work [22], supports the role of fasting/FMDs in inhibiting autophagy and boosting immune response in combination with chemotherapy, which result in the killing of leukemia cells (Figure 7).

Different studies have shown a relationship between the immune system and autophagy [38]. In a murine metastatic liver tumor model, a combination of IL-2 with CQ increased long-term survival and enhanced immune cell proliferation and infiltration in the liver and spleen [39]. A clinical trial (NCT01550367) is currently testing the combination of HCQ and IL-2 as a treatment in patients with metastatic renal cell carcinoma.

In mouse melanoma and mouse and human ovarian cancer, it has been shown that blocking the PD-L1/PD-1 axis via anti-PD1 or anti-PD-L1 antibodies can trigger autophagy in tumor cells, and when coupled with autophagy inhibitors, enhance the response [40]. Moreover, different papers have shown that blocking autophagy recruits different immune cells in the tumor environment, promoting cancer regression [41–43].

Fasting/FMD and other dietary restrictions have also been tested clinically in a number of clinical trials. In a prospective, nonrandomized, controlled trial of 40 patients, the potential benefits of caloric restriction were shown (The Improving Diet and Exercise in ALL (IDEAL)) in the efficacy of chemotherapy in patients newly diagnosed with B-ALL. This intervention resulted in a low minimal residual disease risk, high-circulating adiponectin and low insulin resistance [44]. In a randomized controlled study of 131 patients with HER2-negative early-stage breast cancer, FMD cycles significantly enhanced the effects of neoadjuvant chemotherapy on the radiological and pathological tumor response [45]. A short-term fasting-mimicking diet was also well tolerated during chemotherapy in patients with ovarian cancers and appeared to improve quality of life and fatigue [46–48]. In conclusion, FMD cycles have high potential to be effective in increasing the toxicity of a range of therapies against ALL and other blood cancers and should be tested in randomized clinical trials, especially in combination with immunotherapy and low toxicity cancer therapies.

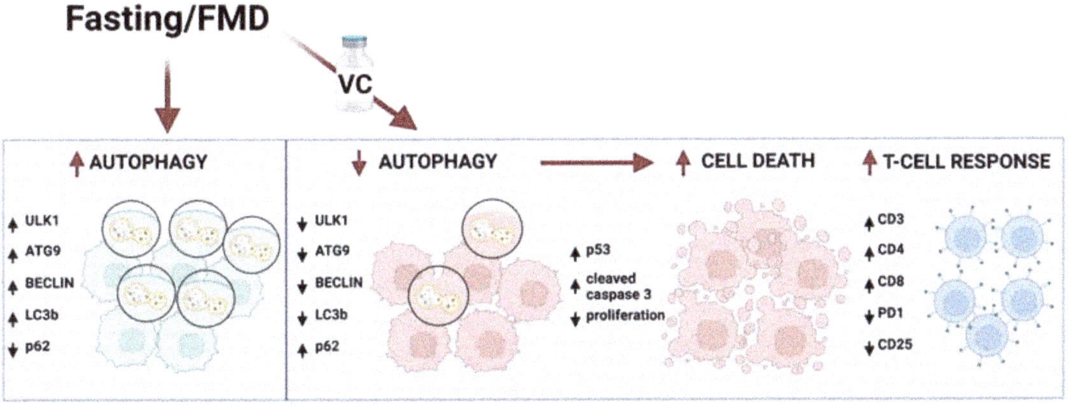

Figure 7. Schematic representation of FMD's effect in combination with vincristine in leukemia cells. The effect of fasting or FMD alone on autophagy markers are based on the results from this study or the literature, but in the presence of VC, the autophagy pathway is inhibited by downregulation of Ulk1, ATG9, Beclin1 and LC3b and upregulation of p62. Consequently, apoptosis is activated by induction of p53 and cleaved caspase 3, leading to leukemic cancer cell death and increased T cell immune response against leukemia cells by activation of CD3, CD4 and CD8 expression and inhibition of PD-1 and CD25.

In summary, we present a new strategy for improving leukemia treatment by combining FMD with chemotherapy to promote the killing of ALL cells in part by an immune-dependent mechanism. Fasting/FMD has been shown to reduce chemotherapy-associated toxicity in pre-clinical and clinical studies [47,49,50] and thus represents a safe and potentially effective treatment adjunct for leukemia patients which should be tested clinically.

5. Conclusions

FMD acts synergistically with vincristine to deplete B-ALL cells in mice. The beneficial effects of FMD on B-ALL progression and survival depend on autophagy inhibition and activation of anti-cancer immunity. The use of the autophagy inhibitor CQ can potentiate the effect of vincristine with or without fasting conditions. However, the use of autophagy inhibitors in cancer therapy could be challenging since, unlike FMD cycles, they may be effective only against a portion of cancers while also causing more side effects. These preclinical data provide proof of principle evidence for clinical trials testing the efficacy of the combination fasting/FMD with various therapies, including chemotherapy and immunotherapy.

Supplementary Materials: The following supporting information can be downloaded at: https://www.mdpi.com/article/10.3390/cancers15245870/s1, Table S1. Mouse FMD ingredients; Figure S1. Flow cytometry gating strategy to identify GFP+ BCR/ABL leukemia cells in the bone marrow and spleen; Figure S2. protein expression analyses of autophagic markers in-dicate downregulation of LC3B and beclin1 in spleen tissue from the FMD + VC group [51–54].

Author Contributions: R.B. and V.D.L. designed the study and wrote the manuscript. R.B. performed experiments and analyzed the data. J.T., R.C., N.G., S.M. and S.D.M. participated in performing experiments, provided intellectual expertise, and helped to interpret experimental results. F.R., M.P. and S.D.M. provided reagents, intellectual expertise and helped to interpret experimental results. All authors have read and agreed to the published version of the manuscript.

Funding: The research leading to these results has received funding from the P01 grant AG034906 to V.D.L. and from the AIRC IG 2018 ID 21820 project to V.D.L.

Institutional Review Board Statement: The animal study protocol was approved by the Institutional Animal Care and Use Committee (IACUC) of the University of Southern California (1540 Alcazar St, CHP 234 Los Angeles, CA 90033). (Protocol code: 20338, date of approval 25 March 2015).

Informed Consent Statement: Not applicable.

Data Availability Statement: All the data supporting the findings of this study are available within the article and its supplementary information files and from the corresponding author upon reasonable request.

Acknowledgments: We thank Gerardo Navarrete for preparing the diet and Marco Morselli for the RNA library preparation.

Conflicts of Interest: V.D.L. has equity interest in L-Nutra, a company that develops medical food. All other authors declare no competing interests.

References

1. Malard, F.; Mohty, M. Acute lymphoblastic leukaemia. *Lancet* **2020**, *395*, 1146–1162. [CrossRef] [PubMed]
2. Paul, S.; Kantarjian, H.; Jabbour, E.J. Adult Acute Lymphoblastic Leukemia. *Mayo Clin. Proc.* **2016**, *91*, 1645–1666. [CrossRef] [PubMed]
3. Friend, B.D.; Schiller, G.J. Closing the gap: Novel therapies in treating acute lymphoblastic leukemia in adolescents and young adults. *Blood Rev.* **2018**, *32*, 122–129. [CrossRef] [PubMed]
4. Sheng, X.; Mittelman, S.D. The role of adipose tissue and obesity in causing treatment resistance of acute lymphoblastic leukemia. *Front. Pediatr.* **2014**, *2*, 53. [CrossRef]
5. Orgel, E.; Sea, J.L.; Mittelman, S.D. Mechanisms by Which Obesity Impacts Survival from Acute Lymphoblastic Leukemia. *J. Natl. Cancer Inst. Monogr.* **2019**, *2019*, 152–156. [CrossRef] [PubMed]
6. Dushnicky, M.J.; Nazarali, S.; Mir, A.; Portwine, C.; Samaan, M.C. Is There a Causal Relationship between Childhood Obesity and Acute Lymphoblastic Leukemia? A Review. *Cancers* **2020**, *12*, 3082. [CrossRef]
7. Ehsanipour, E.A.; Sheng, X.; Behan, J.W.; Wang, X.; Butturini, A.; Avramis, V.I.; Mittelman, S.D. Adipocytes cause leukemia cell resistance to L-asparaginase via release of glutamine. *Cancer Res.* **2013**, *73*, 2998–3006. [CrossRef]
8. Sheng, X.; Tucci, J.; Parmentier, J.-H.; Ji, L.; Behan, J.W.; Heisterkamp, N.; Mittelman, S.D. Adipocytes cause leukemia cell resistance to daunorubicin via oxidative stress response. *Oncotarget* **2016**, *7*, 73147–73159. [CrossRef]
9. Tucci, J.; Alhushki, W.; Chen, T.; Sheng, X.; Kim, Y.M.; Mittelman, S.D. Switch to low-fat diet improves outcome of acute lymphoblastic leukemia in obese mice. *Cancer Metab.* **2018**, *6*, 15. [CrossRef]
10. Castillo, J.J.; Reagan, J.L.; Ingham, R.R.; Furman, M.; Dalia, S.; Merhi, B.; Nemr, S.; Zarrabi, A.; Mitri, J. Obesity but not overweight increases the incidence and mortality of leukemia in adults: A meta-analysis of prospective cohort studies. *Leuk. Res.* **2012**, *36*, 868–875. [CrossRef]
11. Raffaghello, L.; Lee, C.; Safdie, F.M.; Wei, M.; Madia, F.; Bianchi, G.; Longo, V.D. Starvation-dependent differential stress resistance protects normal but not cancer cells against high-dose chemotherapy. *Proc. Natl. Acad. Sci. USA* **2008**, *105*, 8215–8220. [CrossRef] [PubMed]
12. Safdie, F.; Brandhorst, S.; Wei, M.; Wang, W.; Lee, C.; Hwang, S.; Conti, P.S.; Chen, T.C.; Longo, V.D. Fasting enhances the response of glioma to chemo- and radiotherapy. *PLoS ONE* **2012**, *7*, e44603. [CrossRef] [PubMed]
13. Brandhorst, S.; Wei, M.; Hwang, S.; Morgan, T.E.; Longo, V.D. Short-term calorie and protein restriction provide partial protection from chemotoxicity but do not delay glioma progression. *Exp. Gerontol.* **2013**, *48*, 1120–1128. [CrossRef] [PubMed]
14. Brandhorst, S.; Choi, I.Y.; Wei, M.; Cheng, C.W.; Sedrakyan, S.; Navarrete, G.; Dubeau, L.; Yap, L.P.; Park, R.; Vinciguerra, M.; et al. A Periodic Diet that Mimics Fasting Promotes Multi-System Regeneration, Enhanced Cognitive Performance, and Healthspan. *Cell Metab.* **2015**, *22*, 86–99. [CrossRef] [PubMed]
15. Cheng, C.W.; Villani, V.; Buono, R.; Wei, M.; Kumar, S.; Yilmaz, O.H.; Cohen, P.; Sneddon, J.B.; Perin, L.; Longo, V.D. Fasting-Mimicking Diet Promotes Ngn3-Driven beta-Cell Regeneration to Reverse Diabetes. *Cell* **2017**, *168*, 775–788.e12. [CrossRef] [PubMed]
16. Heisterkamp, N.; Jenster, G.; ten Hoeve, J.; Zovich, D.; Pattengale, P.K.; Groffen, J. Acute leukaemia in bcr/abl transgenic mice. *Nature* **1990**, *344*, 251–253. [CrossRef] [PubMed]

17. Lee, C.; Raffaghello, L.; Brandhorst, S.; Safdie, F.M.; Bianchi, G.; Martin-Montalvo, A.; Pistoia, V.; Wei, M.; Hwang, S.; Merlino, A.; et al. Fasting cycles retard growth of tumors and sensitize a range of cancer cell types to chemotherapy. *Sci. Transl. Med.* **2012**, *4*, 124ra27. [CrossRef]
18. McKnight, N.C.; Jefferies, H.B.J.; Alemu, E.A.; Saunders, R.E.; Howell, M.; Johansen, T.; Tooze, S.A. Genome-wide siRNA screen reveals amino acid starvation-induced autophagy requires SCOC and WAC. *EMBO J.* **2012**, *31*, 1931–1946. [CrossRef]
19. Sui, X.; Chen, R.; Wang, Z.; Huang, Z.; Kong, N.; Zhang, M.; Han, W.; Lou, F.; Yang, J.; Zhang, Q.; et al. Autophagy and chemotherapy resistance: A promising therapeutic target for cancer treatment. *Cell Death Dis.* **2013**, *4*, e838. [CrossRef]
20. Sehgal, A.R.; Konig, H.; Johnson, D.E.; Tang, D.; Amaravadi, R.K.; Boyiadzis, M.; Lotze, M.T. You eat what you are: Autophagy inhibition as a therapeutic strategy in leukemia. *Leukemia* **2015**, *29*, 517–525. [CrossRef]
21. Rybstein, M.D.; Bravo-San Pedro, J.M.; Kroemer, G.; Galluzzi, L. The autophagic network and cancer. *Nat. Cell Biol.* **2018**, *20*, 243–251. [CrossRef] [PubMed]
22. Di Biase, S.; Lee, C.; Brandhorst, S.; Manes, B.; Buono, R.; Cheng, C.-W.; Cacciottolo, M.; Martin-Montalvo, A.; de Cabo, R.; Wei, M.; et al. Fasting-Mimicking Diet Reduces HO-1 to Promote T Cell-Mediated Tumor Cytotoxicity. *Cancer Cell* **2016**, *30*, 136–146. [CrossRef] [PubMed]
23. Buono, R.; Longo, V.D. Starvation, Stress Resistance, and Cancer. *Trends Endocrinol. Metab.* **2018**, *29*, 271–280. [CrossRef] [PubMed]
24. Lee, C.; Safdie, F.M.; Raffaghello, L.; Wei, M.; Madia, F.; Parrella, E.; Hwang, D.; Cohen, P.; Bianchi, G.; Longo, V.D. Reduced levels of IGF-I mediate differential protection of normal and cancer cells in response to fasting and improve chemotherapeutic index. *Cancer Res.* **2010**, *70*, 1564–1572. [CrossRef] [PubMed]
25. D'aronzo, M.; Vinciguerra, M.; Mazza, T.; Panebianco, C.; Saracino, C.; Pereira, S.P.; Graziano, P.; Pazienza, V. Fasting cycles potentiate the efficacy of gemcitabine treatment in in vitro and in vivo pancreatic cancer models. *Oncotarget* **2015**, *6*, 18545. [CrossRef]
26. Lu, Z.; Xie, J.; Wu, G.; Shen, J.; Collins, R.; Chen, W.; Kang, X.; Luo, M.; Zou, Y.; Huang, L.J.; et al. Fasting selectively blocks development of acute lymphoblastic leukemia via leptin-receptor upregulation. *Nat. Med.* **2017**, *23*, 79–90. [CrossRef]
27. Korolchuk, V.I.; Mansilla, A.; Menzies, F.M.; Rubinsztein, D.C. Autophagy inhibition compromises degradation of ubiquitin-proteasome pathway substrates. *Mol. Cell* **2009**, *33*, 517–527. [CrossRef]
28. Folkerts, H.; Hilgendorf, S.; Wierenga, A.T.J.; Jaques, J.; Mulder, A.B.; Coffer, P.J.; Schuringa, J.J.; Vellenga, E. Inhibition of autophagy as a treatment strategy for p53 wild-type acute myeloid leukemia. *Cell Death Dis.* **2017**, *8*, e2927. [CrossRef]
29. Sumitomo, Y.; Koya, J.; Nakazaki, K.; Kataoka, K.; Tsuruta-Kishino, T.; Morita, K.; Sato, T.; Kurokawa, M. Cytoprotective autophagy maintains leukemia-initiating cells in murine myeloid leukemia. *Blood* **2016**, *128*, 1614–1624. [CrossRef]
30. Ianniciello, A.; Zarou, M.M.; Rattigan, K.M.; Scott, M.; Dawson, A.; Dunn, K.; Brabcova, Z.; Kalkman, E.R.; Nixon, C.; Michie, A.M.; et al. ULK1 inhibition promotes oxidative stress-induced differentiation and sensitizes leukemic stem cells to targeted therapy. *Sci. Transl. Med.* **2021**, *13*, eabd5016. [CrossRef]
31. Maes, H.; Kuchnio, A.; Peric, A.; Moens, S.; Nys, K.; De Bock, K.; Quaegebeur, A.; Schoors, S.; Georgiadou, M.; Wouters, J.; et al. Tumor vessel normalization by chloroquine independent of autophagy. *Cancer Cell* **2014**, *26*, 190–206. [CrossRef] [PubMed]
32. Amaravadi, R.K.; Yu, D.; Lum, J.J.; Bui, T.; Christophorou, M.A.; Evan, G.I.; Thomas-Tikhonenko, A.; Thompson, C.B. Autophagy inhibition enhances therapy-induced apoptosis in a Myc-induced model of lymphoma. *J. Clin. Invest.* **2007**, *117*, 326–336. [CrossRef] [PubMed]
33. Cufí, S.; Vazquez-Martin, A.; Oliveras-Ferraros, C.; Corominas-Faja, B.; Cuyàs, E.; López-Bonet, E.; Martin-Castillo, B.; Joven, J.; Menendez, J.A. The anti-malarial chloroquine overcomes primary resistance and restores sensitivity to trastuzumab in HER2-positive breast cancer. *Sci. Rep.* **2013**, *3*, 2469. [CrossRef] [PubMed]
34. Ding, L.; Zhang, W.; Yang, L.; Pelicano, H.; Zhou, K.; Yin, R.; Huang, R.; Zeng, J. Targeting the autophagy in bone marrow stromal cells overcomes resistance to vorinostat in chronic lymphocytic leukemia. *OncoTargets Ther.* **2018**, *11*, 5151–5170. [CrossRef] [PubMed]
35. Galluzzi, L.; Bravo-San Pedro, J.M.; Demaria, S.; Formenti, S.C.; Kroemer, G. Activating autophagy to potentiate immunogenic chemotherapy and radiation therapy. *Nat. Rev. Clin. Oncol.* **2017**, *14*, 247–258. [CrossRef] [PubMed]
36. Caffa, I.; Spagnolo, V.; Vernieri, C.; Valdemarin, F.; Becherini, P.; Wei, M.; Brandhorst, S.; Zucal, C.; Driehuis, E.; Ferrando, L.; et al. Fasting-mimicking diet and hormone therapy induce breast cancer regression. *Nature* **2020**, *583*, 620–624. [CrossRef]
37. Salvadori, G.; Zanardi, F.; Iannelli, F.; Lobefaro, R.; Vernieri, C.; Longo, V.D. Fasting-mimicking diet blocks triple-negative breast cancer and cancer stem cell escape. *Cell Metab.* **2021**, *33*, 2247–2259.e6. [CrossRef]
38. Jiang, G.-M.; Tan, Y.; Wang, H.; Peng, L.; Chen, H.-T.; Meng, X.-J.; Li, L.-L.; Liu, Y.; Li, W.-F.; Shan, H. The relationship between autophagy and the immune system and its applications for tumor immunotherapy. *Mol. Cancer* **2019**, *18*, 17. [CrossRef]
39. Liang, X.; De Vera, M.E.; Buchser, W.J.; Chavez, A.R.d.V.; Loughran, P.; Stolz, D.B.; Basse, P.; Wang, T.; Van Houten, B.; Zeh, H.J.; et al. Inhibiting systemic autophagy during interleukin 2 immunotherapy promotes long-term tumor regression. *Cancer Res.* **2012**, *72*, 2791–2801. [CrossRef]
40. Clark, C.A.; Gupta, H.B.; Sareddy, G.; Pandeswara, S.; Lao, S.; Yuan, B.; Drerup, J.M.; Padron, A.; Conejo-Garcia, J.; Murthy, K.; et al. Tumor-Intrinsic PD-L1 Signals Regulate Cell Growth, Pathogenesis, and Autophagy in Ovarian Cancer and Melanoma. *Cancer Res.* **2016**, *76*, 6964–6974. [CrossRef]

41. Mgrditchian, T.; Arakelian, T.; Paggetti, J.; Noman, M.Z.; Viry, E.; Moussay, E.; Van Moer, K.; Kreis, S.; Guerin, C.; Buart, S.; et al. Targeting autophagy inhibits melanoma growth by enhancing NK cells infiltration in a CCL5-dependent manner. *Proc. Natl. Acad. Sci. USA* **2017**, *114*, E9271–E9279. [CrossRef] [PubMed]
42. Noman, M.Z.; Janji, B.; Kaminska, B.; Van Moer, K.; Pierson, S.; Przanowski, P.; Buart, S.; Berchem, G.; Romero, P.; Mami-Chouaib, F.; et al. Blocking hypoxia-induced autophagy in tumors restores cytotoxic T-cell activity and promotes regression. *Cancer Res.* **2011**, *71*, 5976–5986. [CrossRef] [PubMed]
43. Chen, D.; Xie, J.; Fiskesund, R.; Dong, W.; Liang, X.; Lv, J.; Jin, X.; Liu, J.; Mo, S.; Zhang, T.; et al. Chloroquine modulates antitumor immune response by resetting tumor-associated macrophages toward M1 phenotype. *Nat. Commun.* **2018**, *9*, 873. [CrossRef] [PubMed]
44. Orgel, E.; Framson, C.; Buxton, R.; Kim, J.; Li, G.; Tucci, J.; Freyer, D.R.; Sun, W.; Oberley, M.J.; Dieli-Conwright, C.; et al. Caloric and nutrient restriction to augment chemotherapy efficacy for acute lymphoblastic leukemia: The IDEAL trial. *Blood Adv.* **2021**, *5*, 1853–1861. [CrossRef] [PubMed]
45. de Groot, S.; Lugtenberg, R.T.; Cohen, D.; Welters, M.J.; Ehsan, I.; Vreeswijk, M.P.; Smit, V.T.; de Graaf, H.; Heijns, J.B.; Portielje, J.E.; et al. Fasting mimicking diet as an adjunct to neoadjuvant chemotherapy for breast cancer in the multicentre randomized phase 2 DIRECT trial. *Nat. Commun.* **2020**, *11*, 3083. [CrossRef] [PubMed]
46. Bauersfeld, S.P.; Kessler, C.S.; Wischnewsky, M.; Jaensch, A.; Steckhan, N.; Stange, R.; Kunz, B.; Brückner, B.; Sehouli, J.; Michalsen, A. The effects of short-term fasting on quality of life and tolerance to chemotherapy in patients with breast and ovarian cancer: A randomized cross-over pilot study. *BMC Cancer* **2018**, *18*, 476. [CrossRef] [PubMed]
47. Vernieri, C.; Fucà, G.; Ligorio, F.; Huber, V.; Vingiani, A.; Iannelli, F.; Raimondi, A.; Rinchai, D.; Frigè, G.; Belfiore, A.; et al. Fasting-Mimicking Diet Is Safe and Reshapes Metabolism and Antitumor Immunity in Patients with Cancer. *Cancer Discov.* **2022**, *12*, 90–107. [CrossRef]
48. Nencioni, A.; Caffa, I.; Cortellino, S.; Longo, V.D. Fasting and cancer: Molecular mechanisms and clinical application. *Nat. Rev. Cancer* **2018**, *18*, 707–719. [CrossRef]
49. Wei, M.; Brandhorst, S.; Shelehchi, M.; Mirzaei, H.; Cheng, C.W.; Budniak, J.; Groshen, S.; Mack, W.J.; Guen, E.; Di Biase, S.; et al. Fasting-mimicking diet and markers/risk factors for aging, diabetes, cancer, and cardiovascular disease. *Sci. Transl. Med.* **2017**, *9*, aai8700. [CrossRef]
50. Valdemarin, F.; Caffa, I.; Persia, A.; Cremonini, A.L.; Ferrando, L.; Tagliafico, L.; Tagliafico, A.; Guijarro, A.; Carbone, F.; Ministrini, S.; et al. Safety and Feasibility of Fasting-Mimicking Diet and Effects on Nutritional Status and Circulating Metabolic and Inflammatory Factors in Cancer Patients Undergoing Active Treatment. *Cancers* **2021**, *13*, 4013. [CrossRef]
51. Kim, D.; Langmead, B.; Salzberg, S.L. HISAT: A fast spliced aligner with low memory requirements. *Nat. Methods* **2015**, *12*, 357–360. [CrossRef] [PubMed]
52. Baruzzo, G.; Hayer, K.E.; Kim, E.J.; Di Camillo, B.; FitzGerald, G.A.; Grant, G.R. Simulation-based comprehensive benchmarking of RNA-seq aligners. *Nat. Methods* **2017**, *14*, 135–139. [CrossRef] [PubMed]
53. Lopez, D.; Montoya, D.; Ambrose, M.; Lam, L.; Briscoe, L.; Adams, C.; Modlin, R.L.; Pellegrini, M. SaVanT: A web-based tool for the sample-level visualization of molecular signatures in gene expression profiles. *BMC Genom.* **2017**, *18*, 824. [CrossRef] [PubMed]
54. Love, M.I.; Huber, W.; Anders, S. Moderated estimation of fold change and dispersion for RNA-seq data with DESeq2. *Genome Biol.* **2014**, *15*, 550. [CrossRef]

cancers MDPI

Review

Immune Response and Metastasis—Links between the Metastasis Driver MACC1 and Cancer Immune Escape Strategies

Sebastian Torke *, Wolfgang Walther and Ulrike Stein

Experimental and Clinical Research Center, Charité, Medical Centre Berlin and Max-Delbrück-Center for Molecular Medicine, 13125 Berlin, Germany; wowalt@mdc-berlin.de (W.W.); ustein@mdc-berlin.de (U.S.)
* Correspondence: sebastian.torke@mdc-berlin.de; Tel.: +49-30-9406-2389

Simple Summary: While both the fields of cancer metastasis research and cancer immunology have been heavily investigated, their combination—the immunology of metastasis—is underrepresented given the fact that metastasis is responsible for up to 90% of cancer deaths. Additionally, evading detection by the immune system is a key prerequisite for the spread of tumor cells to distant organs. In this review, we explore the connections between a master regulator of metastasis, MACC1, and both its direct and indirect links with immunological processes. Specifically, we highlight MACC1-induced alterations of signaling pathways and secreted factors and how this translates into changed immunological outcomes including effects on immune cell infiltration, activity, and their regulation through immunological checkpoints.

Abstract: Metastasis remains the most critical factor limiting patient survival and the most challenging part of cancer-targeted therapy. Identifying the causal drivers of metastasis and characterizing their properties in various key aspects of cancer biology is essential for the development of novel metastasis-targeting approaches. Metastasis-associated in colon cancer 1 (MACC1) is a prognostic and predictive biomarker that is now recognized in more than 20 cancer entities. Although MACC1 can already be linked with many hallmarks of cancer, one key process—the facilitation of immune evasion—remains poorly understood. In this review, we explore the direct and indirect links between MACC1 and the mechanisms of immune escape. Therein, we highlight the signaling pathways and secreted factors influenced by MACC1 as well as their effects on the infiltration and anti-tumor function of immune cells.

Keywords: metastasis; MACC1; immune evasion

1. Clinical Significance of MACC1 for Cancer Metastasis

The formation of metastasis remains the most lethal attribute of cancers, being responsible for the majority of cancer-related deaths, and—in some entities—making up the cause of about 90% of cancer deaths [1–3]. Although therapeutic strategies are successful in limiting cancer growth and progression, cancer cells will continue to evolve and later form metastases. This highlights the clinical need for the development and evaluation of biomarkers capable of predicting metastasis formation as well as therapy success. In the last ten years, metastasis-associated in colon cancer 1 (MACC1) has been recognized as such a biomarker in over 20 solid cancer entities [4]. It is a causal driver of metastasis, and its level in initially metastasis-free patients is a highly accurate predictor of metastasis formation and overall as well as metastasis-free patient survival. Importantly, multiple studies have so far confirmed that both intra-tumoral MACC1 expression as well as blood levels of MACC1 can serve as a marker predicting metastasis formation [5–8]. Despite its promising capabilities as a biomarker, MACC1 is so far not widely used in the clinic. Although the role of MACC1 for the initiation and propagation of metastasis has been well-studied, its physiological role in normal cell function remains largely unknown. One study, however, reported that MACC1 defects during embryonal development can lead to malformations [9]. Additionally, MACC1 has been linked to metabolic pathways that mainly regulate glucose and glutamine metabolism [10]. Functionally, MACC1 acts as a transcription factor within the hepatocyte growth factor (HGF)/c-MET signaling pathway, therein relaying extracellular signals [4]. Furthermore, MACC1 is involved in the regulation of cellular functions through many other pathways including mitogen-activated protein kinase kinase (MEK)/extracellular signal-regulated kinases (ERK), phosphoinositide 3-kinases (PI3K)/protein kinase B (AKT)/β-catenin, signal transducers and activators of transcription (STAT)1/3, and twist-related protein (TWIST)/vascular endothelial growth factor (VEGF). Therefore, functionally, MACC1 is involved in mediating cellular proliferation, metabolic activity, cancer cell stemness properties, and angiogenesis, which all lead to the promotion of a metastasis-associated phenotype [4]. Conclusively, it is not surprising that MACC1 is therefore being evaluated as a target of metastasis-specific therapeutic intervention.

While various molecules have been identified that are capable of reducing the expression of MACC1, the most promising approach to date is to use agents from the class of statins. These drugs were initially developed to lower plasma cholesterol levels but have recently shown great potential in inhibiting a MACC1-induced metastatic phenotype in vitro and metastasis in animal models. Moreover, most strikingly, an evaluation of the clinical data from over 300,000 patients including more than 50,000 cancer patients revealed strong evidence of a cancer-preventive effect by statins [11,12]. While all of these highlight the importance and potential that MACC1 possesses as a biomarker for metastasis, additional research is needed to fully elucidate MACC1's specific physiological functions as well as further understand its role in mediating key cancer processes.

One such process that to date remains only vaguely comprehended is in what manner MACC1 is involved in regulating how tumor cells interact with their environment, and specifically with close-by immune cells. This is of special importance as one key prerequisite for the formation of metastasis is the development of distinct attributes that allow cancerous cells to evade their detection and attack by surveilling immune cells [13]. These immune-evasive features will need to already develop early within the primary tumor, but they become increasingly relevant within the circulation where circulating tumor cells (CTCs) come in close contact with immune cells, and furthermore in the secondary organs in which metastatic cells aim to take root. In this review, we provide a comprehensive overview over the known links between the metastasis-driver MACC1 and the mechanisms of immune evasion.

2. MACC1 Correlates with Immune Cell Infiltration

To date, the experimental evidence of MACC1 expression possessing immunological consequences has largely been centered around correlations of MACC1 levels with intratumoral immune cell infiltration. The first study, which used bioinformatic analysis of colon adenocarcinoma (COAD) patients based on The Cancer Genome Atlas (TCGA), identified MACC1 as a positive regulator of the infiltration of natural killer (NK) cells, macrophages, and neutrophils. However, this study could not ascertain any effects on the levels of infiltrating dendritic cells, B cells and T helper, cytotoxic, or regulatory T cells [14]. A second study in breast cancer confirmed the positive effect of MACC1 on the infiltration of macrophages, in this case, specifically of anti-inflammatory, tumor-associated macrophages (TAMs). Here, however, the effect on NK cell infiltration was negatively associated with MACC1, as was the infiltration of cluster of differentiation (CD)8+ cytotoxic T lymphocytes (CTLs) [15]. Of note, one study reported that the levels of MACC1 were closely related with the expression of the immunological checkpoint programmed cell death 1 ligand 1 (PD-L1). Here, the ectopic manipulation (up- or downregulation) of MACC1 translated into corresponding changes of PD-L1, and ultimately into alterations of the anti-tumor effects and immune cell-mediated killing capacity of peripheral blood mononuclear cells (PBMCs) in a co-culture setting [16]. Although this provides the first experimental evidence of the correlation between MACC1 and the immune system, further links can be drawn between MACC1 target genes and immune cell infiltration and function.

3. MACC1 Influences Immune Cell Infiltration and Tumor-Immunity through Positive Feedback Loop and Vascularization

One of the main functions of MACC1 is to act as a transcription factor. In this regard, a number of its target genes have been described to possess the potential for inducing effects of immunological relevance. First and foremost, in a positive feedback loop following the activation of MACC1 downstream of HGF/c-Met, MACC1 induces the expression of c-Met [17,18]. c-Met itself possesses various properties related to the immune system. Most notably, HGF stimulation can induce the expression of PD-L1 through c-MET and thereby contribute to cancer cell immune evasion [19,20], a process likely dependent on MACC1 as a core regulator/effector of this signaling pathway. Moreover, HGF/c-Met regulates the recruitment of immune cells into the tumor microenvironment (TME) and can create an immunosuppressive milieu (e.g., by inducing anti-inflammatory T helper cells and macrophages as well as shifting the cytokine milieu toward the secretion of immune system-regulating cytokines). Specifically, HGF-stimulation can diminish the production of IFN-γ, whilst interleukin (IL)-4 and IL-10 secretion increases [19,21,22].

Additionally, one of the functional outcomes of HGF/c-Met stimulation is the MACC1-mediated induction of angiogenesis through increasing the production and secretion of VEGF [23,24]. VEGF, in turn, can directly influence the recruitment, differentiation, and activity of cells belonging to the innate as well as adaptive immune system. Specifically, the secretion of VEGF leads to the specific recruitment of a variety of regulatory immune cells such as regulatory T cells (Treg) and myeloid-derived suppressor cells (MDSCs). Furthermore, VEGF pushes macrophage differentiation into an anti-inflammatory M2-subset. Additionally, VEGF influences the maturation of dendritic cells (DCs) and increases their expression of immuno-regulatory molecules such as PD-L1, which act as immune checkpoints and can limit the activation and function of other immune cells in close proximity to those DCs. All of these mechanisms promote an anti-inflammatory milieu and a pro-tumor environment. Additionally, VEGF can directly influence the proliferation, recruitment, and cytotoxicity of killer cells such as CTLs and NK cells, decreasing their recognition and effector functions against cancer cells [25,26]. Notably, MACC1 increases the vessel density and vascularization through VEGF/Twist. However, VEGF also influences the expression of adhesion molecules on endothelial cells, specifically blocking the upregulation of molecules such as intercellular adhesion molecules (ICAMs), vascular cell adhesion protein (VCAM), or selectins that are induced by inflammatory mediators. This decreases the extravasation and tumor infiltration of various immune cells normally mediated by those adhesion molecules. Additionally, VEGF can also specifically increase other endothelial adhesion molecules such as CLEVER-1 and Fas-ligand (FasL). CLEVER-1 is a scavenging receptor and correlates with the selectively increased infiltration of anti-inflammatory and pro-tumor immune cells such as Tregs and M2 macrophages. Endothelial expression of FasL initiates apoptosis pathways specifically in activated T cells, with Tregs being resistant to FasL-mediated killing. Ultimately, this is associated with decreased numbers of intra-tumoral CTLs [27]. Interestingly, a direct link has already been described between the expression of MACC1 and the Fas/FasL apoptosis pathway.

4. MACC1 Mediates Immune Evasion through STAT1/3 and Fas

The tumor-intrinsic expression of MACC1 is relevant for the sensitization to death receptor-mediated apoptosis via Fas/FasL. In this context, the presence of MACC1 increases the activation and phosphorylation of STAT proteins that induce the expression of myeloid cell leukemia-1 (MCL1), which in turn decreases the expression of Fas [4]. This effect can protect cancer cells from Fas-agonist induced apoptosis induction and likely, in a more physiological context, from immune cell-induced destruction. Interestingly, STAT proteins can also mediate other mechanisms of immune evasion. Of note, the STAT signaling pathway—when induced by interferon (IFN)-γ—can drive the overexpression of PD-L1, thereby inhibiting the anti-cancer functions of tumor-targeting immune cells [28]. Interestingly, there is already a reported connection between MACC1 and PD-L1. In gastric cancer tissues, the levels of MACC1 were positively correlated with the expression of PD-L1. Furthermore, alterations in the expression of MACC1 through genetic manipulation such as silencing or induced overexpression translated into corresponding changes of PD-L1 expression. However, it is unclear which cells within the TME are expressing PD-L1 in this study. Therefore, additional investigations are needed to elucidate whether there is a direct link of MACC1 and the expression of immunological checkpoints [16]. In a more general context, STAT1 expression not only correlates with the expression of PD-L1 and its interaction partner programmed cell death protein 1 (PD-1), but can also be associated with disease stage and tumor grade [29]. Moreover, one transcriptional target of STAT proteins is c-Myc, which can be induced by STAT3 in early tumorigenesis and in response to T cell immunosurveillance. The induction of c-Myc can then drive immunoediting, leading to decreased T cell recognition and generally an immune-suppressive TME [30]. However, STAT proteins are also responsible for the regulation of the expression of major histocompatibility complex (MHC) molecules induced by IFN-γ. Therein, they are centrally involved in how cancer cells present antigens and how they are recognized and targeted by immune cells. Importantly, many cancers dysregulate pathways of antigen presentation, however, the used strategies for achieving this are plentiful. Overall, the outcome of STAT protein expression and function can vary, even within a specific cellular subset. However, if pro- or anti-tumor functions predominate, depends on the individual context as well as additional contributing factors [31,32]. Finally, as STAT proteins are major immune regulators, one key aspect is that they can induce the production of cytokines by cancer cells [33,34].

5. The Role of Cytokines and Stemness Factors for MACC1 and the Immune System

In this context, cancer cells can use inflammatory mediators such as cytokines to promote their own growth. Moreover, chronic inflammatory conditions can initiate the aberrant changes leading to cancer formation, and these conditions can also continuously drive cancer progression. MACC1, in particular, has been identified as one target initiated through the tumor-promoting effects of two major regulators: the cytokines tumor necrosis factor (TNF)-α and IFN-γ [35]. Here, the signaling cascades activated by the cytokines lead to the activation of nuclear factor 'kappa-light-chain-enhancer' of activated B-cells (NFkB), resulting in the induction of transcription factors binding to the MACC1 promoter and thereby to a shift toward a metastasis-associated phenotype mediated by MACC1 itself. Although these cytokines naturally influence many cells within the TME, this is an interesting example of how the inflammatory properties surrounding cancerous cells can contribute to metastasis. Interestingly, large amounts of various cytokines can be produced by cancer stem cells [36], which have already been linked to MACC1.

While cancer stem cells (CSCs) only make up a small fraction of the total tumor cells, they likely provide the basis for the majority of malignant human tumors and are critically important in initiating metastasis [37,38]. In this context, MACC1 has been identified to regulate the cancer cell stemness properties through the induction of transcription factors known as core regulators of multipotency such as octamer-binding transcription factor 4 (Oct4) and Nanog [39]. Importantly, CSCs possess a variety of strategies to avoid immune recognition. Most notably, they employ immune checkpoint molecules to evade cell contact-induced killing by immune cells and can secrete extracellular vesicles (EVs), growth factors, metabolites, and cytokines to modulate the TME in a wider scope [40]. Specifically, CSCs exhibit high levels of expression for the inhibitory molecules PD-L1, cytotoxic T-lymphocyte-associated protein 4 (CTLA-4), T-cell immunoglobulin, and mucin-domain containing-3 (TIM-3) as well as lymphocyte-activation gene 3 (LAG3). Additionally, CSCs often downregulate the expression of MHC class I molecules in comparison to more differentiated cancer cells, a feature that conveys their protection from T cell recognition and destruction. Interestingly, CSCs are also capable of reducing the expression of tumor-associated antigens, thereby additionally limiting immune cell surveillance [41]. Furthermore, CSCs increase the expression of CD47, a transmembrane protein that transmits inhibitory signals toward macrophages and hinders phagocytosis, thereby acting as an immune checkpoint for the innate immune system [42]. Through the secretion of soluble factors, CSCs can further manipulate their surroundings. EVs, for instance, are capable of suppressing T cell functions, inhibiting dendritic cell maturation, and pushing macrophages into the anti-inflammatory M2 subtype [41]. In addition, CSCs secrete a variety of cytokines including IL-4, IL-8, transforming growth factor (TGF)-β, macrophage colony-stimulating factor (M-CSF), and granulocyte-macrophage colony-stimulating factor (GM-CSF), all possessing immunosuppressive capabilities and collaborating to produce a pro-tumor environment through the manipulation of the anti-tumor activity of immune cells [42]. Importantly, these functional outcomes can be directly linked to core multipotency factors such as Oct4 [43], which in turn can be induced by active MACC1 [39]. Along the same lines, Nanog conveys this immune-resistant phenotype and can also be upregulated by MACC1 [39,44]. Finally, MACC1 can induce stemness properties by stimulating LGR5 expression [45].

6. MACC1 Manipulates the TME via PI3K/Akt and Wnt Pathways

Another link of MACC1 with mechanisms of immune evasion is mediated through the regulation of the PI3K/AKT pathway. Various studies have identified MACC1 as a positive regulator of PI3K/AKT signaling [46,47], at least in part through a negative regulation of the phosphatase and tensin homolog (PTEN) [4,48,49]. Activation of this pathway can create an inhibitory myeloid TME and reduce the overall numbers of infiltrating CTLs [50]. Furthermore, through the loss of activity from PTEN, specific changes in the cytokine milieu such as a high production of chemokine (C-X-C motif) ligand (CXCL)17 can induce the recruitment of MDSCs, Tregs, and anti-inflammatory M2 macrophages. Moreover, the loss of PTEN or its secreted variant PTEN-L can diminish the production and release of pro-inflammatory cytokines such as TNF-α and IL-6. Importantly, reduced PTEN-activity is also linked to an increased expression of immunological checkpoints such as PD-1 and PD-L1 [51]. All of these described alterations collaborate to create an immunosuppressive environment as well as overall favorable conditions for continuous tumor cell growth.

One of the outcomes of increased AKT kinase activity is the stabilization of β-catenin and the promotion of its nuclear translocation, thereby connecting the PI3K/Akt pathway closely with the Wnt/β-catenin pathway. Various studies have already linked MACC1 with the Wnt pathway [52]. In this context, MACC1 has been established as a transcriptional target of Wnt/β-catenin [53], and more importantly, MACC1 can regulate the expression of Wnt target genes such as VEGF, c-Myc, cyclin D1/E, and matrix metalloproteinases [4,39]. The activation of Wnt signaling in tumor cells can then, in turn, promote immune evasion through the direct induction of PD-L1 and CD47 expression, and furthermore via an alteration of the infiltration and function of immune cells within the TME. Specifically, increased levels of Wnt-induced secreted protein 1 (WISP1) facilitate a pro-tumor microenvironment by promoting TAMs whereas the downregulation of chemokines such as chemokine (C-C motif) ligand (CCL)4 and CCL5 reduce the numbers of DCs [54,55]. Additionally, an inverse correlation has been described between Wnt/β-catenin signaling activity and the infiltration of T cells into cancerous tissue [56]. One key target gene of this signaling pathway is S100A4, also referred to as *metastasin*. This master regulator of metastatic functions including epithelial–mesenchymal transition (EMT) [57] has been shown to be a transcriptional target of MACC1 [5]. S100A4 can stimulate tumor cells to secrete numerous inflammatory cytokines, most notably IL-8 and CCL2, which then shape the TME toward a pro-tumor state and favor metastatic growth [58]. Furthermore, S100A4 is also positively linked with the expression of PD-L1 and the inhibition of anti-tumor T cell activity. Of note, this is not only a direct effect, but can also act on the surrounding cells, as S100A4 can be transmitted (e.g., via extracellular vesicles) [59]. Interestingly, MACC1 and S100A4 have been independently established as key drivers of metastasis and can individually be used to stratify patients according to their specific metastatic risk. However, combining both markers yields the greatest potential for patient stratification. Additionally, therapeutic strategies have been developed to target both MACC1 as well as S100A4. Here, again, combining the interventional therapies works synergistically and has the highest therapeutic value in vitro and in limiting metastasis formation in animal models [5].

7. Additional Mechanisms by Which MACC1 Might Influence Immune Cell Infiltration and Function

Interestingly, MACC1 can also influence metastasis and immune evasion via a rather unexpected pathway—by modulating the mechanisms of endocytosis. MACC1 has been described to promote receptor internalization and recycling as part of clathrin-mediated endocytosis (CME) [60]. In the context of metastasis, the stimulation or continuation of CME can facilitate stronger or longer lasting signals from growth factor receptors such as the EGFR, thereby stabilizing oncogenic signaling [61]. Interestingly, endocytic pathways can also influence tumor immunity and tumor-induced immune suppression through the regulation of tumor surface antigens. Here, especially the degradation of molecules involved in antigen presentation, namely MHC class I, is of considerable importance [62,63]. Endocytosis, as a pathway that is based around membrane-vesicles deemed for the degradation and recycling of intra- and extracellular components, is furthermore mechanistically closely related to another pathway, autophagy [64]. Importantly, MACC1 has been described as a positive regulator of 5′ adenosine monophosphate-activated protein kinase (AMPK)/Unc-51-like autophagy-activating kinases (ULK1) induced autophagy. Tumor cells employing this tool can reroute MHC class I molecules toward lysosomes, leading to their destruction and an overall diminished T cell recognition. Through autophagy, tumor cells can also alter the secretion of cytokines into the TME, affecting, for instance, CXCL1, CXCL2, CXCL5, and CXCL12 production. This ultimately leads to an attraction of MDSCs and immunosuppressive macrophages as well as an overall pro-tumor environment. Furthermore, autophagy diminishes the release of the chemokine CCL5, which facilitates the infiltration of NK cells into the TME. Additionally, through autophagic processes, tumor cells become capable of degrading granzyme B, a factor secreted by activated CTLs and NK cells with the aim to induce apoptosis in target cells, thereby effectively blocking immune cell induced tumor cell destruction [65]. Autophagy is also a process providing additional energy when the metabolism of a cell is challenged [66]. Interestingly, there is a more direct link described between MACC1 and the metabolic system.

First and foremost, MACC1 is capable of altering important metabolic pathways, leading to elevations in glucose and glutamine uptake. In this way, MACC1 functions through the upregulation of specific transporter proteins such as glucose transporter (GLUT)1 and GLUT4 as well as via the regulation of glycolytic enzymes including hexokinase, pyruvate dehydrogenase kinase, and lactate dehydrogenase [10,67]. Through the induction of these metabolic changes—namely by promoting the so-called Warburg-effect, leading to an increased use of glucose—cancer cells promote nutrient depletion, oxygen-deprivation, increased acidity, and the release of potentially toxic metabolites into their surrounding area [68]. These changes promote a TME that is highly immunosuppressive, hinders immune cell infiltration (immune exclusion), and can even mediate the loss of efficacy of adoptive cell therapies and immune checkpoint inhibitors [69]. Additionally, it has been described that hypoxia can induce the expression of PD-L1 on various cell types of the TME including tumor cells. Furthermore, the combination of glucose and oxygen deprivation can diminish the overall antigen presentation on cancer cells via the MHC class I system [70,71].

Although also linked to metabolism but independent from cancer formation, a relationship between MACC1 and obesity has been described. In this study, higher levels of cell-free MACC1 in plasma were observed in obese adults in comparison to normal weight control individuals. Reducing the total body fat resulted in lowered MACC1 levels. Additionally, rats with high-fat diet-induced obesity demonstrated higher MACC1 levels and a more severe colon tumor outcome [72]. Importantly, obesity can promote chronic inflammation and induce systemic changes in T cell and macrophage populations, exacerbating various diseases including cancer [73].

8. Outlook and Conclusions

Since its first description in 2009, MACC1 has been recognized as a prognostic and predictive biomarker as well as a driver of metastasis formation in over 20 cancer entities. MACC1 levels analyzed from tumor tissue or liquid biopsies can be used to determine the individual metastatic risk and predict patient survival and therapy outcomes. However, there is only limited data of MACC1 as a biomarker of immune system status and anti-tumor immunity. Several studies have identified correlations between MACC1 expression levels and the infiltration of immune cells into the tumor surrounding areas [14,15]. Furthermore, a bioinformatic analysis of COAD using multiple databases identified MACC1 as a potential predictor for immune responses as well as a novel target for immunotherapy-based intervention strategies [74]. Further evidence is provided by the general use of statins, potent transcriptional inhibitors of MACC1 expression, as they are widely used and are often also prescribed to patients treated for cancer. Here, multiple studies have reported that statins are not only capable of preventing cancer development [12], but also augment already established immune-targeting treatment regimens. In this case, statins can enhance therapy responses when targeting immunological checkpoints such as PD-L1, a factor that can be regulated in its expression by MACC1 [16,75–77]. However, direct experimental links are still rare and future studies need to provide additional and more extensive connections between MACC1 as a metastatic switch and its impact on the immune system. Here, the description of a direct link between intra-tumoral MACC1 expression and the state/function of nearby immune cells would especially help further the understanding of the immunological consequences of MACC1. Additionally, it would be interesting to study whether there is any difference in the MACC1 expression levels between immune checkpoint therapy responders or non-responders.

In this review, we have demonstrated that the intra-tumoral expression of MACC1 facilitates cancer immune evasion through multiple processes (Figure 1, Table 1), all leading toward the altered expression of immune system-controlling surface molecules or cytokines, ultimately inducing changes in the composition and function of the TME. While MACC1 is already established as a marker of metastasis, we highlight here that it possesses additional value as a predictor for cancer immune evasion by manipulating immune cell infiltration and function. Furthermore, the points made in this review indicate that there is a close relationship between metastasis and immune evasion that has been largely underestimated in the past. Here, additional work should be performed to analyze the links between molecules associated with or causally driving metastasis and their potential immunological consequences. Taken together, we have demonstrated that MACC1 shapes the environment for tumor cells to promote the formation of metastasis by mediating various strategies relevant to immune escape mechanisms of cancer cells.

Table 1. Immunological consequences of MACC1 expression. Upwards-pointing arrows indicate an upregulation or increase, downwards-pointing arrows a downregulation or decrease.

MACC1 Effect	Direct/Indirect Consequences	Effect on Immune System	Investigated Entity	References
MACC1 expression itself	• PD-L1 ↑	• Reduced anti-tumor immune function	• Gastric cancer	[16]
Positive feedback to HGF/c-Met	• PD-L1 ↑ • Anti-inflammatory TME (Th cells, macrophages) • Immune system-regulating cytokines ↑	• Altered immune cell infiltration • Reduced anti-tumor immune function	• Renal cancer • Autoimmune diseases	[19–22]

Table 1. *Cont.*

MACC1 Effect	Direct/Indirect Consequences	Effect on Immune System	Investigated Entity	References
Induction of VEGF	• ICAM, VCAM, Selectins ↓ • CLEVER-1, FasL ↑	• Specific recruitment of Tregs and MDSCs • M2 macrophage polarization • Recruitment/anti-tumor effect of CTLs/NK cells ↓	• Endothelial cells • Breast cancer	[25–27]
Activation of STAT1/3	• Fas expression ↓ • PD-L1 ↑ • C-Myc ↑	• Protection against receptor-induced apoptosis • Reduced anti-tumor immune function • Immunoediting ↑ • Immuno-suppressive TME	• Renal cancer • Oral/gastric cancer • Skin cancer • Lung cancer • Breast cancer	[28,30–34]
Facilitation of cancer stemness via Oct4 and Nanog	• Immune checkpoints ↑ • MHC class I ↓ • CD47 ↑ • Cyto-/Chemokines and EVs to modulate TME	• Immuno-suppressive TME • Reduced anti-tumor immune function • Evasion of T cell recognition	• Colorectal cancer • Brain cancer • Breast cancer • Colon cancer • Head and neck cancer	[36,39–44]
PI3K/Akt signaling	• PTEN ↓ • Changed cyto-/chemokine milieu • PD-1/PD-L1 ↑	• Inhibitory myeloid TME • Infiltration of CTLs ↓ • Recruitment of MDSCs, Tregs and TAM	• Skin cancer • Lung cancer • Breast cancer • Colorectal cancer • Brain cancer	[50,51]
Wnt signaling	• Wnt target genes (VEGF, c-Myc, cyclin D1/E, MMPs) • PD-L1/CD47 ↑ • Changed cyto-/chemokine milieu • Induction of S100A4	• Reduced infiltration of DCs and CTLs • Promoting TAMs	• Brain cancer • Skin cancer • Lung cancer	[54–59]
Endocytosis and autophagy	• Stronger/longer signals from growth factor receptors • Tumor surface antigen ↓ • MHC class I ↓ • Changed cyto-/chemokine milieu	• Evasion of T cell recognition • Pro-tumor environment characterized by MDSCs and TAMs • Infiltration of NK cells ↓ • Degradation of granzyme B → protection from immune cell destruction	• Pancreatic cancer • Lung cancer	[62,64,65]
Metabolism	• Glucose transporters/glycolytic enzymes ↑ • PD-L1 ↑ • MHC class I ↓	• Nutrient depletion, oxygen-deprivation, increased acidity, and accumulation of metabolites • Highly immunosuppressive TME • Immune exclusion	• Breast cancer • Lung cancer • Fibrosarcoma • Skin cancer	[68–71]

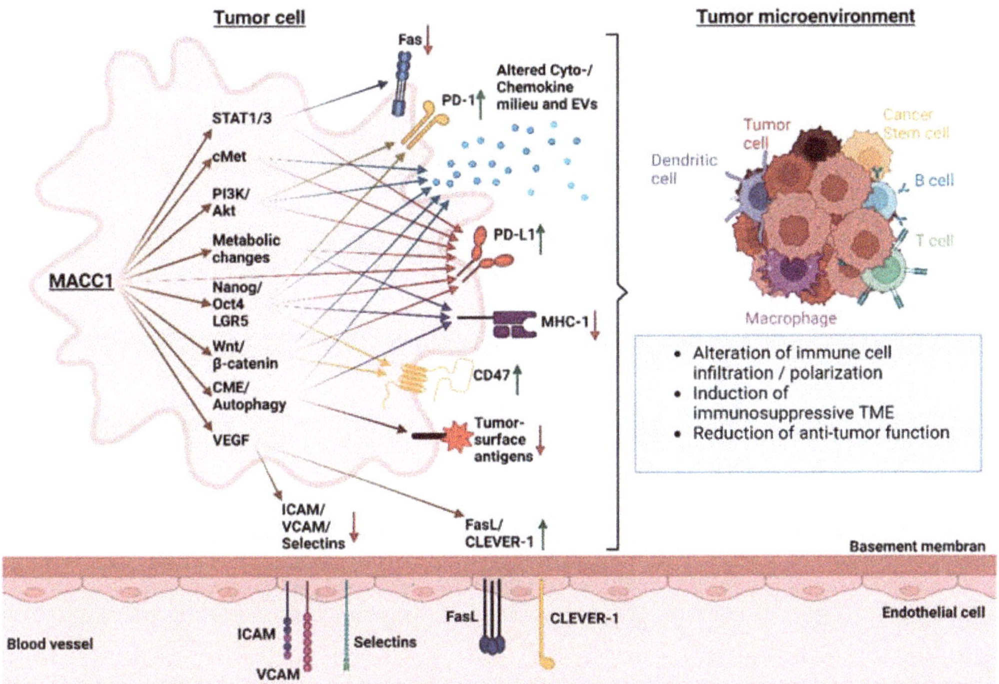

Figure 1. Links between MACC1 and mechanisms of immune escape. MACC1 directly affects the expression or function of the indicated factors, which in turn leads to immunological consequences marked by the arrows. Up- (green arrow pointing upward) or downregulation (red arrow pointing downward) are next to the immunologically relevant markers. Consequently, this affects the tumor microenvironment and creates a tumor-favorable milieu.

Author Contributions: S.T. performed the literature search as well as the drafting and writing of the manuscript. S.T. prepared the table and figure and performed all revision modifications. U.S. and W.W. were involved in reviewing and editing the manuscript. All authors have read and agreed to the published version of the manuscript.

Funding: This research received no external funding.

Acknowledgments: The authors are thankful to Nina Heisterkamp for proof-reading the manuscript.

Conflicts of Interest: All authors declare no conflict of interests.

References

1. Dillekås, H.; Rogers, M.S.; Straume, O. Are 90% of deaths from cancer caused by metastases? *Cancer Med.* **2019**, *8*, 5574–5576. [CrossRef] [PubMed]
2. Seyfried, T.N.; Huysentruyt, L.C. On the Origin of Cancer Metastasis. *Crit. Rev. Oncog.* **2013**, *18*, 43–73. [CrossRef] [PubMed]
3. Fares, J.; Fares, M.Y.; Khachfe, H.H.; Salhab, H.A.; Fares, Y. Molecular principles of metastasis: A hallmark of cancer revisited. *Signal Transduct. Target. Ther.* **2020**, *5*, 1–17. [CrossRef] [PubMed]
4. Radhakrishnan, H.; Walther, W.; Zincke, F.; Kobelt, D.; Imbastari, F.; Erdem, M.; Kortüm, B.; Dahlmann, M.; Stein, U. MACC1-the first decade of a key metastasis molecule from gene discovery to clinical translation. *Cancer Metastasis Rev.* **2018**, *37*, 805–820. [CrossRef] [PubMed]
5. Kortüm, B.; Radhakrishnan, H.; Zincke, F.; Sachse, C.; Burock, S.; Keilholz, U.; Dahlmann, M.; Walther, W.; Dittmar, G.; Kobelt, D.; et al. Combinatorial treatment with statins and niclosamide prevents CRC dissemination by unhinging the MACC1-β-catenin-S100A4 axis of metastasis. *Oncogene* **2022**, *41*, 4446–4458. [CrossRef] [PubMed]
6. Treese, C.; Werchan, J.; von Winterfeld, M.; Berg, E.; Hummel, M.; Timm, L.; Rau, B.; Daberkow, O.; Walther, W.; Daum, S.; et al. Inhibition of MACC1-Induced Metastasis in Esophageal and Gastric Adenocarcinomas. *Cancers* **2022**, *14*, 1773. [CrossRef]

7. Link, T.; Kuhlmann, J.D.; Kobelt, D.; Herrmann, P.; Vassileva, Y.D.; Kramer, M.; Frank, K.; Göckenjan, M.; Wimberger, P.; Stein, U. Clinical relevance of circulating MACC1 and S100A4 transcripts for ovarian cancer. *Mol. Oncol.* **2019**, *13*, 1268–1279. [CrossRef] [PubMed]
8. Hagemann, C.; Neuhaus, N.; Dahlmann, M.; Kessler, A.F.; Kobelt, D.; Herrmann, P.; Eyrich, M.; Freitag, B.; Linsenmann, T.; Monoranu, C.M.; et al. Circulating MACC1 Transcripts in Glioblastoma Patients Predict Prognosis and Treatment Response. *Cancers* **2019**, *11*, 825. [CrossRef]
9. Melvin, V.S.; Feng, W.; Hernandez-Lagunas, L.; Artinger, K.B.; Williams, T. A Morpholino-based screen to identify novel genes involved in craniofacial morphogenesis. *Dev. Dyn. Off. Publ. Am. Assoc. Anat.* **2013**, *242*, 817–831. [CrossRef]
10. Lisec, J.; Kobelt, D.; Walther, W.; Mokrizkij, M.; Grötzinger, C.; Jaeger, C.; Baum, K.; Simon, M.; Wolf, J.; Beindorff, N.; et al. Systematic Identification of MACC1-Driven Metabolic Networks in Colorectal Cancer. *Cancers* **2021**, *13*, 978. [CrossRef]
11. Juneja, M.; Kobelt, D.; Walther, W.; Voss, C.; Smith, J.; Specker, E.; Neuenschwander, M.; Gohlke, B.-O.; Dahlmann, M.; Radetzki, S.; et al. Statin and rottlerin small-molecule inhibitors restrict colon cancer progression and metastasis via MACC1. *PLoS Biol.* **2017**, *15*, e2000784. [CrossRef] [PubMed]
12. Gohlke, B.-O.; Zincke, F.; Eckert, A.; Kobelt, D.; Preissner, S.; Liebeskind, J.M.; Gunkel, N.; Putzker, K.; Lewis, J.; Preissner, S.; et al. Real-world evidence for preventive effects of statins on cancer incidence: A trans-Atlantic analysis. *Clin. Transl. Med.* **2022**, *12*, e726. [CrossRef] [PubMed]
13. Nguyen, P.H.D.; Wasser, M.; Tan, C.T.; Lim, C.J.; Lai, H.L.H.; Seow, J.J.W.; DasGupta, R.; Phua, C.Z.J.; Ma, S.; Yang, J.; et al. Trajectory of immune evasion and cancer progression in hepatocellular carcinoma. *Nat. Commun.* **2022**, *13*, 1441. [CrossRef] [PubMed]
14. Liu, B.-X.; Yang, J.; Zeng, C.; Chen, Y. MACC1 Correlates with Tumor Progression and Immune Cell Infiltration of Colon Adenocarcinoma and is Regulated by the lncRNA ZFAS1/miR-642a-5p Axis. *J. Oncol.* **2022**, *2022*, 8179208. [CrossRef] [PubMed]
15. Ali, D.A.; El-Guindy, D.M.; Elrashidy, M.A.; Sabry, N.M.; Kabel, A.M.; Gaber, R.A.; Ibrahim, R.R.; Samy, S.M.; Shalaby, M.M.; Salama, S.A.; et al. The Prognostic Significance of MACC1 Expression in Breast Cancer and Its Relationship to Immune Cells in the Tumor Microenvironment and Patient Survival. *Med. Kaunas Lith.* **2021**, *57*, 934. [CrossRef] [PubMed]
16. Tong, G.; Cheng, B.; Li, J.; Wu, X.; Nong, Q.; He, L.; Li, X.; Li, L.; Wang, S. MACC1 regulates PDL1 expression and tumor immunity through the c-Met/AKT/mTOR pathway in gastric cancer cells. *Cancer Med.* **2019**, *8*, 7044–7054. [CrossRef]
17. Wen, J.; Xie, Y.; Zhang, Y.; Li, J.; Li, J.; Zhang, Y.; Lu, X.; Zhang, Y.; Liu, Y.; Liu, T.; et al. MACC1 Contributes to the Development of Osteosarcoma Through Regulation of the HGF/c-Met Pathway and Microtubule Stability. *Front. Cell Dev. Biol.* **2020**, *8*, 00825. [CrossRef] [PubMed]
18. Sheng, X.-J.; Li, Z.; Sun, M.; Wang, Z.-H.; Zhou, D.-M.; Li, J.-Q.; Zhao, Q.; Sun, X.-F.; Liu, Q.-C. MACC1 induces metastasis in ovarian carcinoma by upregulating hepatocyte growth factor receptor c-MET. *Oncol. Lett.* **2014**, *8*, 891–897. [CrossRef] [PubMed]
19. Papaccio, F.; Della Corte, C.M.; Viscardi, G.; Di Liello, R.; Esposito, G.; Sparano, F.; Ciardiello, F.; Morgillo, F. HGF/MET and the Immune System: Relevance for Cancer Immunotherapy. *Int. J. Mol. Sci.* **2018**, *19*, 3595. [CrossRef]
20. Balan, M.; Mier y Teran, E.; Waaga-Gasser, A.M.; Gasser, M.; Choueiri, T.K.; Freeman, G.; Pal, S. Novel roles of c-Met in the survival of renal cancer cells through the regulation of HO-1 and PD-L1 expression. *J. Biol. Chem.* **2015**, *290*, 8110–8120. [CrossRef]
21. Futamatsu, H.; Suzuki, J.; Mizuno, S.; Koga, N.; Adachi, S.; Kosuge, H.; Maejima, Y.; Hirao, K.; Nakamura, T.; Isobe, M. Hepatocyte growth factor ameliorates the progression of experimental autoimmune myocarditis: A potential role for induction of T helper 2 cytokines. *Circ. Res.* **2005**, *96*, 823–830. [CrossRef] [PubMed]
22. Zambelli, A.; Biamonti, G.; Amato, A. HGF/c-Met Signalling in the Tumor Microenvironment. In *Tumor Microenvironment: Signaling Pathways—Part B*; Birbrair, A., Ed.; Advances in Experimental Medicine and Biology; Springer International Publishing: Cham, Switzerland, 2021; pp. 31–44. [CrossRef]
23. Wang, L.; Zhou, R.; Zhao, Y.; Dong, S.; Zhang, J.; Luo, Y.; Huang, N.; Shi, M.; Bin, J.; Liao, Y.; et al. MACC-1 Promotes Endothelium-Dependent Angiogenesis in Gastric Cancer by Activating TWIST1/VEGF-A Signal Pathway. *PLoS ONE* **2016**, *11*, e0157137. [CrossRef] [PubMed]
24. Lv, M.; Jiao, Y.; Yang, B.; Ye, M.; Di, W.; Su, W.; Zhong, J. MACC1 as a Potential Target for the Treatment and Prevention of Breast Cancer. *Biology* **2023**, *12*, 455. [CrossRef] [PubMed]
25. Geindreau, M.; Ghiringhelli, F.; Bruchard, M. Vascular Endothelial Growth Factor, a Key Modulator of the Anti-Tumor Immune Response. *Int. J. Mol. Sci.* **2021**, *22*, 4871. [CrossRef] [PubMed]
26. Li, Y.-L.; Zhao, H.; Ren, X.-B. Relationship of VEGF/VEGFR with immune and cancer cells: Staggering or forward? *Cancer Biol. Med.* **2016**, *13*, 206–214. [CrossRef]
27. Hendry, S.A.; Farnsworth, R.H.; Solomon, B.; Achen, M.G.; Stacker, S.A.; Fox, S.B. The Role of the Tumor Vasculature in the Host Immune Response: Implications for Therapeutic Strategies Targeting the Tumor Microenvironment. *Front. Immunol.* **2016**, *7*, 621. [CrossRef]
28. Radhakrishnan, H.; Ilm, K.; Walther, W.; Shirasawa, S.; Sasazuki, T.; Daniel, P.T.; Gillissen, B.; Stein, U. MACC1 regulates Fas mediated apoptosis through STAT1/3—Mcl-1 signaling in solid cancers. *Cancer Lett.* **2017**, *403*, 231–245. [CrossRef] [PubMed]
29. Liu, F.; Liu, J.; Zhang, J.; Shi, J.; Gui, L.; Xu, G. Expression of STAT1 is positively correlated with PD-L1 in human ovarian cancer. *Cancer Biol. Ther.* **2020**, *21*, 963–971. [CrossRef] [PubMed]

30. Tsai, C.-H.; Chuang, Y.-M.; Li, X.; Yu, Y.-R.; Tzeng, S.-F.; Teoh, S.T.; Lindblad, K.E.; Matteo, M.D.; Cheng, W.-C.; Hsueh, P.-C.; et al. Immunoediting instructs tumor metabolic reprogramming to support immune evasion. *Cell Metab.* **2023**, *35*, 118–133.e7. [CrossRef]
31. Owen, K.L.; Brockwell, N.K.; Parker, B.S. JAK-STAT Signaling: A Double-Edged Sword of Immune Regulation and Cancer Progression. *Cancers* **2019**, *11*, 2002. [CrossRef]
32. Zhang, Y.; Liu, Z. STAT1 in Cancer: Friend or Foe? *Discov. Med.* **2017**, *24*, 19–29. [PubMed]
33. Ni, Y.; Low, J.T.; Silke, J.; O'Reilly, L.A. Digesting the Role of JAK-STAT and Cytokine Signaling in Oral and Gastric Cancers. *Front. Immunol.* **2022**, *13*, 835997. [CrossRef]
34. Yu, H.; Pardoll, D.; Jove, R. STATs in cancer inflammation and immunity: A leading role for STAT3. *Nat. Rev. Cancer* **2009**, *9*, 798–809. [CrossRef]
35. Kobelt, D.; Zhang, C.; Glauben, R.; Siegmund, B.; Stein, U. Pro-inflammatory TNF-α and IFN-γ promote tumor growth and metastasis via induction of MACC1. *Front. Immunol.* **2020**, *11*, 525727. [CrossRef]
36. Lei, M.M.L.; Lee, T.K.W. Cancer Stem Cells: Emerging Key Players in Immune Evasion of Cancers. *Front. Cell Dev. Biol.* **2021**, *9*, 692940. [CrossRef]
37. Shiozawa, Y.; Nie, B.; Pienta, K.J.; Morgan, T.M.; Taichman, R.S. Cancer Stem Cells and their Role in Metastasis. *Pharmacol. Ther.* **2013**, *138*, 285–293. [CrossRef] [PubMed]
38. Steinbichler, T.B.; Savic, D.; Dudás, J.; Kvitsaridze, I.; Skvortsov, S.; Riechelmann, H.; Skvortsova, I.-I. Cancer stem cells and their unique role in metastatic spread. *Semin. Cancer Biol.* **2020**, *60*, 148–156. [CrossRef]
39. Lemos, C.; Hardt, M.S.; Juneja, M.; Voss, C.; Förster, S.; Jerchow, B.; Haider, W.; Bläker, H.; Stein, U. MACC1 Induces Tumor Progression in Transgenic Mice and Colorectal Cancer Patients via Increased Pluripotency Markers Nanog and Oct4. *Clin. Cancer Res.* **2016**, *22*, 2812–2824. [CrossRef] [PubMed]
40. Wu, B.; Shi, X.; Jiang, M.; Liu, H. Cross-talk between cancer stem cells and immune cells: Potential therapeutic targets in the tumor immune microenvironment. *Mol. Cancer* **2023**, *22*, 38. [CrossRef]
41. Tsuchiya, H.; Shiota, G. Immune evasion by cancer stem cells. *Regen. Ther.* **2021**, *17*, 20–33. [CrossRef]
42. Galassi, C.; Musella, M.; Manduca, N.; Maccafeo, E.; Sistigu, A. The Immune Privilege of Cancer Stem Cells: A Key to Understanding Tumor Immune Escape and Therapy Failure. *Cells* **2021**, *10*, 2361. [CrossRef] [PubMed]
43. Ma, T.; Hu, C.; Lal, B.; Zhou, W.; Ma, Y.; Ying, M.; Prinos, P.; Quiñones-Hinojosa, A.; Lim, M.; Laterra, J.; et al. Reprogramming Transcription Factors Oct4 and Sox2 Induce a BRD-Dependent Immunosuppressive Transcriptome in GBM-Propagating Cells. *Cancer Res.* **2021**, *81*, 2457–2469. [CrossRef] [PubMed]
44. Noh, K.H.; Kim, B.W.; Song, K.-H.; Cho, H.; Lee, Y.-H.; Kim, J.H.; Chung, J.-Y.; Kim, J.-H.; Hewitt, S.M.; Seong, S.-Y.; et al. Nanog signaling in cancer promotes stem-like phenotype and immune evasion. *J. Clin. Investig.* **2012**, *122*, 4077–4093. [CrossRef] [PubMed]
45. Erdem, M.; Lee, K.H.; Hardt, M.; Regan, J.L.; Kobelt, D.; Walther, W.; Mokrizkij, M.; Regenbrecht, C.; Stein, U. MACC1 Regulates LGR5 to Promote Cancer Stem Cell Properties in Colorectal Cancer. *Cancers* **2024**, *16*, 604. [CrossRef] [PubMed]
46. Liu, J.; Pan, C.; Guo, L.; Wu, M.; Guo, J.; Peng, S.; Wu, Q.; Zuo, Q. A new mechanism of trastuzumab resistance in gastric cancer: MACC1 promotes the Warburg effect via activation of the PI3K/AKT signaling pathway. *J. Hematol. Oncol. J Hematol Oncol.* **2016**, *9*, 76. [CrossRef]
47. Wang, J.; Wang, W.; Cai, H.; Du, B.; Zhang, L.; Ma, W.; Hu, Y.; Feng, S.; Miao, G. MACC1 facilitates chemoresistance and cancer stem cell-like properties of colon cancer cells through the PI3K/AKT signaling pathway. *Mol. Med. Rep.* **2017**, *16*, 8747–8754. [CrossRef] [PubMed]
48. Hohmann, T.; Hohmann, U.; Dehghani, F. MACC1-induced migration in tumors: Current state and perspective. *Front. Oncol.* **2023**, *13*, 1165676. [CrossRef] [PubMed]
49. Qian, L.-Q.; Li, X.-Q.; Ye, P.-H.; Su, H.-Y.; Wang, G.; Liu, Y.; Shen, G.-H.; Gao, Q.-G. Downregulation of MACC1 inhibits the viability, invasion and migration and induces apoptosis in esophageal carcinoma cells through the phosphatase and tensin homolog/phosphoinositide 3-kinase/protein kinase B signaling pathway. *Oncol. Lett.* **2017**, *14*, 4897–4905. [CrossRef] [PubMed]
50. Collins, N.B.; Al Abosy, R.; Miller, B.C.; Bi, K.; Zhao, Q.; Quigley, M.; Ishizuka, J.J.; Yates, K.B.; Pope, H.W.; Manguso, R.T.; et al. PI3K activation allows immune evasion by promoting an inhibitory myeloid tumor microenvironment. *J. Immunother. Cancer* **2022**, *10*, e003402. [CrossRef]
51. Vidotto, T.; Melo, C.M.; Castelli, E.; Koti, M.; dos Reis, R.B.; Squire, J.A. Emerging role of PTEN loss in evasion of the immune response to tumours. *Br. J. Cancer* **2020**, *122*, 1732–1743. [CrossRef]
52. Meng, F.; Li, H.; Shi, H.; Yang, Q.; Zhang, F.; Yang, Y.; Kang, L.; Zhen, T.; Dai, S.; Dong, Y.; et al. MACC1 Down-Regulation Inhibits Proliferation and Tumourigenicity of Nasopharyngeal Carcinoma Cells through Akt/β-Catenin Signaling Pathway. *PLoS ONE* **2013**, *8*, e60821. [CrossRef] [PubMed]
53. Kim, H.J.; Moon, S.J.; Kim, S.-H.; Heo, K.; Kim, J.H. DBC1 regulates Wnt/β-catenin-mediated expression of MACC1, a key regulator of cancer progression, in colon cancer. *Cell Death Dis.* **2018**, *9*, 831. [CrossRef] [PubMed]
54. Katoh, M.; Katoh, M. WNT signaling and cancer stemness. *Essays Biochem.* **2022**, *66*, 319–331. [CrossRef] [PubMed]
55. Tao, W.; Chu, C.; Zhou, W.; Huang, Z.; Zhai, K.; Fang, X.; Huang, Q.; Zhang, A.; Wang, X.; Yu, X.; et al. Dual Role of WISP1 in maintaining glioma stem cells and tumor-supportive macrophages in glioblastoma. *Nat. Commun.* **2020**, *11*, 3015. [CrossRef] [PubMed]

56. Martin-Orozco, E.; Sanchez-Fernandez, A.; Ortiz-Parra, I.; Ayala-San Nicolas, M. WNT Signaling in Tumors: The Way to Evade Drugs and Immunity. *Front. Immunol.* **2019**, *10*, 2854. [CrossRef] [PubMed]
57. Nirala, B.; Baskin, D.; Yun, K. Cell-autonomous and non-autonomous functions of S100A4 in regulating stemness, mesenchymal transition, and metastasis. *Oncoscience* **2017**, *4*, 166–167. [CrossRef] [PubMed]
58. Bettum, I.J.; Vasiliauskaite, K.; Nygaard, V.; Clancy, T.; Pettersen, S.J.; Tenstad, E.; Mælandsmo, G.M.; Prasmickaite, L. Metastasis-associated protein S100A4 induces a network of inflammatory cytokines that activate stromal cells to acquire pro-tumorigenic properties. *Cancer Lett.* **2014**, *344*, 28–39. [CrossRef] [PubMed]
59. Wu, X.; Zhang, H.; Jiang, G.; Peng, M.; Li, C.; Lu, J.; Jiang, S.; Yang, X.; Jiang, Y. Exosome-transmitted S100A4 induces immunosuppression and non-small cell lung cancer development by activating STAT3. *Clin. Exp. Immunol.* **2022**, *210*, 309–320. [CrossRef]
60. Imbastari, F.; Dahlmann, M.; Sporbert, A.; Mattioli, C.C.; Mari, T.; Scholz, F.; Timm, L.; Twamley, S.; Migotti, R.; Walther, W.; et al. MACC1 regulates clathrin-mediated endocytosis and receptor recycling of transferrin receptor and EGFR in colorectal cancer. *Cell. Mol. Life Sci. CMLS* **2021**, *78*, 3525–3542. [CrossRef]
61. Khan, I.; Steeg, P.S. Endocytosis: A pivotal pathway for regulating metastasis. *Br. J. Cancer* **2021**, *124*, 66–75. [CrossRef]
62. Dersh, D.; Yewdell, J.W. Immune MAL2-practice: Breast cancer immunoevasion via MHC class I degradation. *J. Clin. Investig.* **2021**, *131*, 144344. [CrossRef] [PubMed]
63. Wu, B.; Wang, Q.; Shi, X.; Jiang, M. Targeting Endocytosis and Cell Communications in the Tumor Immune Microenvironment. *Cell Commun. Signal.* **2022**, *20*, 161. [CrossRef] [PubMed]
64. Birgisdottir, Å.B.; Johansen, T. Autophagy and endocytosis—Interconnections and interdependencies. *J. Cell Sci.* **2020**, *133*, jcs228114. [CrossRef] [PubMed]
65. Xia, H.; Green, D.R.; Zou, W. Autophagy in tumour immunity and therapy. *Nat. Rev. Cancer* **2021**, *21*, 281–297. [CrossRef] [PubMed]
66. Kim, K.H.; Lee, M.-S. Autophagy—A key player in cellular and body metabolism. *Nat. Rev. Endocrinol.* **2014**, *10*, 322–337. [CrossRef] [PubMed]
67. Lin, L.; Huang, H.; Liao, W.; Ma, H.; Liu, J.; Wang, L.; Huang, N.; Liao, Y.; Liao, W. MACC1 supports human gastric cancer growth under metabolic stress by enhancing the Warburg effect. *Oncogene* **2015**, *34*, 2700–2710. [CrossRef] [PubMed]
68. Leone, R.D.; Powell, J.D. Metabolism of immune cells in cancer. *Nat. Rev. Cancer* **2020**, *20*, 516–531. [CrossRef]
69. Heuser, C.; Renner, K.; Kreutz, M.; Gattinoni, L. Targeting lactate metabolism for cancer immunotherapy—A matter of precision. *Semin. Cancer Biol.* **2023**, *88*, 32–45. [CrossRef]
70. Cruz-Bermúdez, A.; Laza-Briviesca, R.; Casarrubios, M.; Sierra-Rodero, B.; Provencio, M. The Role of Metabolism in Tumor Immune Evasion: Novel Approaches to Improve Immunotherapy. *Biomedicines* **2021**, *9*, 361. [CrossRef]
71. Patterson, L.F.S.; Vardhana, S.A. Metabolic regulation of the cancer-immunity cycle. *Trends Immunol.* **2021**, *42*, 975–993. [CrossRef]
72. Bähr, I.; Jaeschke, L.; Nimptsch, K.; Janke, J.; Herrmann, P.; Kobelt, D.; Kielstein, H.; Pischon, T.; Stein, U. Obesity, colorectal cancer and MACC1 expression: A possible novel molecular association. *Int. J. Oncol.* **2022**, *60*, 17. [CrossRef] [PubMed]
73. Rathmell, J.C. Obesity, Immunity, and Cancer. *N. Engl. J. Med.* **2021**, *384*, 1160–1162. [CrossRef] [PubMed]
74. Hu, Y.; Wang, M.; Wang, K.; Gao, J.; Tong, J.; Zhao, Z.; Li, M. A potential role for metastasis-associated in colon cancer 1 (MACC1) as a pan-cancer prognostic and immunological biomarker. *Math. Biosci. Eng. MBE* **2021**, *18*, 8331–8353. [CrossRef]
75. Kansal, V.; Burnham, A.J.; Kinney, B.L.C.; Saba, N.F.; Paulos, C.; Lesinski, G.B.; Buchwald, Z.S.; Schmitt, N.C. Statin drugs enhance responses to immune checkpoint blockade in head and neck cancer models. *J. Immunother. Cancer* **2023**, *11*, e005940. [CrossRef] [PubMed]
76. Choe, E.-J.; Lee, C.-H.; Bae, J.-H.; Park, J.-M.; Park, S.-S.; Baek, M.-C. Atorvastatin Enhances the Efficacy of Immune Checkpoint Therapy and Suppresses the Cellular and Extracellular Vesicle PD-L1. *Pharmaceutics* **2022**, *14*, 1660. [CrossRef]
77. Vos, W.G.; Lutgens, E.; Seijkens, T.T.P. Statins and immune checkpoint inhibitors: A strategy to improve the efficacy of immunotherapy for cancer? *J. Immunother. Cancer* **2022**, *10*, e005611. [CrossRef]

cancers

MDPI

Review

Exploring TSGA10 Function: A Crosstalk or Controlling Mechanism in the Signaling Pathway of Carcinogenesis?

Farzad Taghizadeh-Hesary [1], Mobina Ghadyani [2], Fatah Kashanchi [3] and Babak Behnam [4,*]

1 ENT and Head and Neck Research Center and Department, The Five Senses Health Institute, School of Medicine, Iran University of Medical Sciences, Tehran 14496-14535, Iran
2 Chester Medical School, University of Chester, Chester CH2 1BR, UK
3 Laboratory of Molecular Virology, George Mason University, Manassas, VA 20110, USA
4 Avicenna Biotech Research, Germantown, MD 20871, USA
* Correspondence: babakb@avicennabio.com or babak.behnam@gmail.com

Simple Summary: This research aims to explore the role of the TSGA10 protein in cancer development, specifically in how it might influence the growth and spread of cancer cells. Scientists are particularly interested in TSGA10 because it is found in both normal reproductive tissues and cancer cells, yet seems to slow down cancer progression. The key question is why cancer cells would produce a protein that could hinder their own survival. To investigate this, the authors propose several hypotheses about how TSGA10 might be involved in carcinogenesis. They will analyze both published and unpublished studies and data to understand how TSGA10 functions at different stages of cancer. By uncovering these mechanisms, this research could lead to new targeted therapies that use TSGA10 to combat cancer more effectively, offering fresh insights and potential breakthroughs in cancer treatment.

Abstract: Cancer-specific antigens have been a significant area of focus in cancer treatment since their discovery in the mid-twentieth century. Cancer germline antigens are a class of antigens specifically overexpressed in germline tissues and cancer cells. Among these, TSGA10 (testis-specific gene antigen 10) is of great interest because of its crucial impact on cancer progression. Early studies explored *TSGA10* expression in a variety of cancer types. More recent studies revealed that TSGA10 can suppress tumor progression by blocking cancer cell metabolism, angiogenesis, and metastasis. An open question regarding the TSGA10 is why cancer cells must express a protein that prevents their progression. To answer this question, we conducted a comprehensive review to engage the TSGA10 in the context of the current understanding of "malignant transformation". This review demonstrated that TSGA10 expression level in cancer cells depends on the cancer stage across malignant transformation. In addition, we evaluated how TSGA10 expression can prevent the "cancer hallmarks". Given this information, TSGA10 can be of great interest in developing effective targeted anti-cancer therapies.

Keywords: cancer; cancer germline antigen; TSGA10; tumor suppressor; tumor microenvironment

1. Introduction

Cancer germline antigens (CGAs) have emerged as intriguing players in normal development and cancer progression. They are predominantly expressed in the testes, ovaries, and placenta, contributing to vital processes like spermatogenesis, yet they also make unexpected appearances in different types of cancer cells [1]. Among these, TSGA10 (testis-specific gene antigen 10) is considered due to its unique impacts on cancer phenotypes. In a nutshell, TSGA10 is an 82-kilodalton protein encoded by the *TSGA10* gene located on chromosome 2q11.2, containing at least 22 exons [2,3]. *TSGA10* coding gene was discovered by Modarressi et al. (2001) based on mRNA extraction from human testis tissue [4]. In spermatids, TSGA10 is cleaved into two fragments upon translation: a 27 kDa N-terminus fragment located in the fibrous sheath of the sperm tail and a 55 kDa C-terminus fragment located in the mid-piece of sperm [5].

Studies have explored the physiologic functions of TSGA10 in spermatogenesis [6], embryogenesis [7], and neural development [7]. However, its function in carcinogenesis is still a matter of debate. CGAs usually contribute to cancer cell proliferation, invasion, and migration [1]. However, the information regarding TSGA10 is contradictory. Some studies have introduced it as a CGA [7–12]; however, more recent studies have explored its tumor-suppressive effects. Mansouri et al. (2016) realized that TSGA10 induction could inhibit the angiogenesis and invasion of HeLa cells in vitro [13]. In line with this, Jahani et al. (2020) found that TSGA10 overexpression in a breast cancer cell line (MCF-7) can reduce their metabolic and metastatic activities [14].

In summary, some studies introduced TSGA10 as a CGA, while others found it a tumor suppressor. This duality has sparked curiosity among researchers worldwide to explore the role of TSGA10 in cancer progression. An open question regarding the TSGA10 is why cancer cells need to express a protein that prevents their progression. To answer this question, we reviewed the TSGA10 literature from the scope of "cancer hallmarks" and "malignant transformation". The available experimental studies are primarily based on in vitro studies on cancer cell lines [13–15]. This issue might affect the results due to the ignorance of cancer cells as dynamic entities in living organisms, their heterogeneity, and the impacts of the surrounding tumor microenvironment (TME) [16]. This conceptual review was, therefore, conducted to re-evaluate the available literature regarding the TSGA10 in cancer progression by considering malignant transformation and cancer hallmarks.

The following two sections summarize the literature regarding the TSGA10's role in physiologic development and tumorigenesis. Section 4 provides a synopsis of the literature pertaining to malignant transformation and cancer hallmarks. Sections 5 and 6 provide an interpretation of the TSGA10 literature based on malignant transformation and cancer hallmarks, respectively. The last two sections present the clinical implications and conclusions of this conceptual review.

2. A State-of-the-Art Literature Review of TSGA10 Role in the Physiologic Development

This section addresses the highlights of TSGA10 in the physiologic development. In 2001, Modarressi et al. isolated the *TSGA10* gene in the human testis and introduced its structure [4]. Upon expression, TSGA10 is spliced into two ends with distinct roles. In a mouse model, Behnam et al. (2006) demonstrated that the C-terminus of TSGA10 was implicated in the differentiation of the tail bud, small intestine, vertebrae, and the brain cortex, while the N-terminus was expressed during the development of digits [7]. Simultaneously, Aarabi et al. (2006) noted that TSGA10 expression in human testis was limited to germ cells, and lack of TSGA10 expression might negatively affect spermatogenesis and male fertility [17]. In mid-2006, Hägele et al. unveiled a crucial effect of TSGA10 by showing that TSGA10 could prevent the nuclear translocation of hypoxia-inducible factor (HIF)-1 in spermatozoa [18]. In 2010, Roghanian et al. explored that TSGA10 was expressed in dendritic cells and macrophages and interacted with vimentin through its leucine zipper motif [5]. Bioinformatic analyses have identified different proteins interacting with TSGA10 (Figure 1).

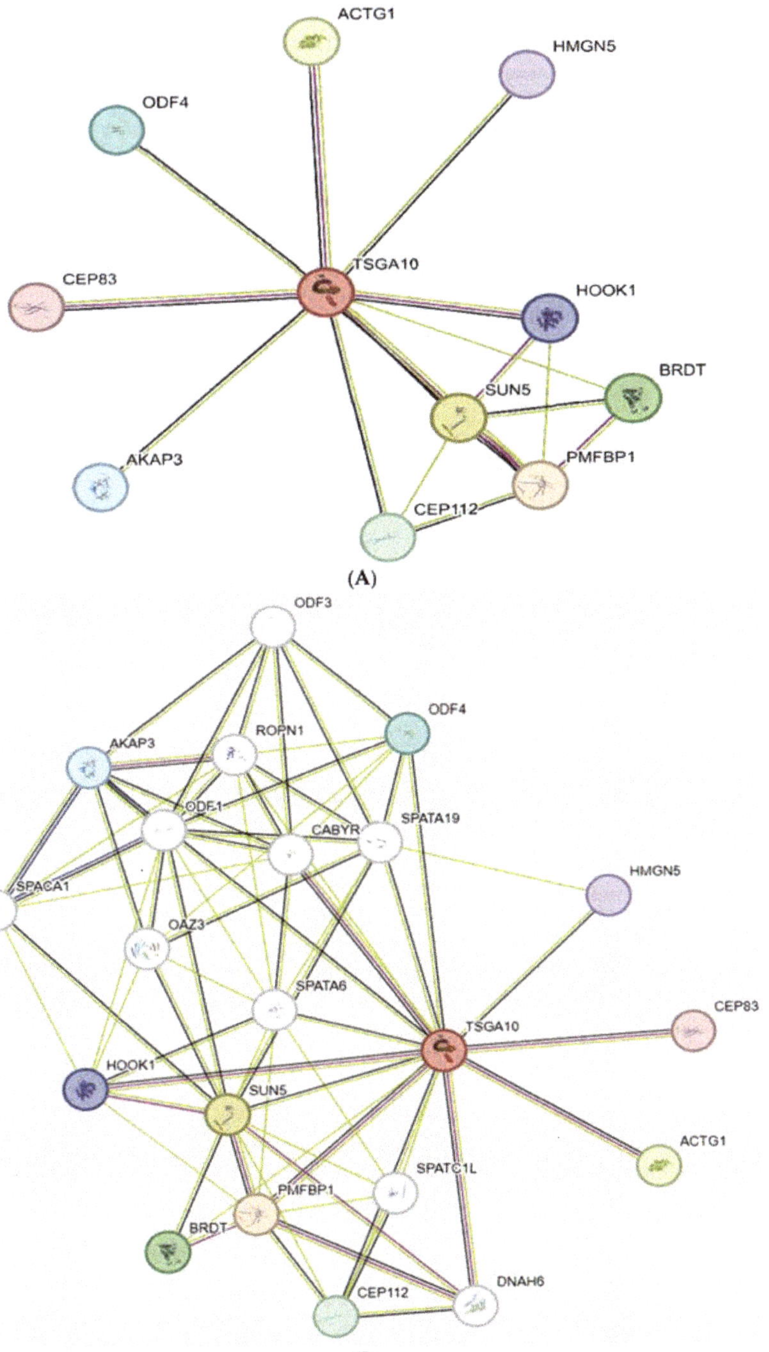

Figure 1. (**A**,**B**) Molecular models of TSGA10 interacting proteins and partners with other signaling pathways. Retrieved from STRING interaction network source: https://cn.string-db.org/ (accessed on 27 August 2024).

3. A State-of-the-Art Literature Review of TSGA10 Role in Human Malignancies

Upon introduction in 2001, Tanaka et al. (2004) realized that TSGA10 was overexpressed in a subset of melanoma (5%), colon cancer (5%), hepatocellular carcinoma (20%), ovarian cancer (35%), and prostate cancer (15%) cells [2]. Two years later, Mobasheri et al. (2006) realized that TSGA10 was expressed in 84% of patients with acute lymphoblastic leukemia (ALL) [10]. Another study demonstrated that TSGA10 was downregulated in 93% of patients with acute myeloid leukemia (AML) compared with healthy controls [19]. In an experiment on anaplastic astrocytoma cells, Behnam et al. (2009) demonstrated that C-terminus TSGA10 was located in the perinuclear region, and the N-terminus end was located in the nucleus [12]. In an in vivo study on a mouse model of esophageal squamous cell carcinoma (ESCC), Yuan et al. (2013) indicated that TSGA10 could serve as a tumor suppressor by activating the p53 or Rb signaling pathways. The investigators also demonstrated that this effect could be reversed by microRNA-577 (miR-577) binding to the 3′ untranslated regions (3′ UTRs) of TSGA10 mRNA [20]. In 2016, Mansouri et al. found that TSGA10 induction could effectively reduce the rate of angiogenesis and invasion in HeLa cells. The investigators realized that these inhibitory effects were caused by the disruption of the HIF-1 axis [13]. Asgharzadeh et al. demonstrated that the C-terminus end of TSGA10 interacted with HIF-1 with high affinity [21]. Salehipour et al. (2017) found different transcription patterns of TSGA10 in breast cancer compared with testis. This study demonstrates that TSGA10 transcripts in breast cancer cells tend to have shorter 5′ UTRs with fewer upstream open reading frames [3]. Kazerani et al. found similar findings in high-grade brain tumors compared with low-grade tumors. The authors concluded that shorter 5′ UTRs in high-grade tumors might reduce the translation efficiency of TSGA10, providing proper conditions for angiogenesis and metastasis [22]. In 2018, Bao et al. demonstrated that miR-23a-containing exosomes secreted from nasopharyngeal carcinoma cells induced angiogenesis by directly targeting the TSGA10 [23]. In line with this finding, Zhang et al. (2019) found that HIF-1 enhanced the proliferation, invasion, and migration of ESCC cells by targeting TSGA10 in a miR-10b-3p-dependent manner [24]. Hoseinkhani et al. (2019) demonstrated the negative correlation of TSGA10 with HIF-1 and VEGF (vascular endothelial growth factor) expression in patients with AML [19]. In early 2020, Jahani et al. found that TSGA10 overexpression in breast cancer cells could reduce cell proliferation and induce the G_2/M cell cycle arrest. In addition, TSGA10 induction could reduce the cancer cells' metabolism and metastatic ability [14]. Table 1 summarizes the key studies exploring the role of TSGA10 in different malignancies.

Table 1. Comparative TSGA10 alteration in different cancers.

Cancer Types	Discussed Mechanisms	TSGA10 over Expression	TSGA10 Downregulation
Esophageal Squamous Cell Carcinoma [20,24]	TSGA10 acts as a tumor suppressor as it inhibits tumor growth by regulating the cell cycle and inducing apoptosis. Typically, downregulated in more advanced stages, larger and poorly differentiated ESCC, which leads to increased cell proliferation and malignancy.	Can it help regulate tumorigenesis?	MiR-577 functions as an oncomir as it promotes cancer progression by targeting and downregulating TSGA10. Under hypoxic conditions, the expression of miR-10b-3p would be enhanced, therefore targeting TSGA10 and reducing its expression.

Table 1. *Cont.*

Cancer Types	Discussed Mechanisms	TSGA10 over Expression	TSGA10 Downregulation
Primary cutaneous T-cell lymphoma (CTCL) [25]	TSGA10 acts as a tumor-associated antigen and a candidate for targeted immunotherapy in primary CTCL and suggests a role in the immune response against tumor cells.	TSGA10 is overexpressed as a potential tumor-associated antigen in primary CTCL.	Likely reduce the immune system's ability to recognize and target the cancer cells; hence, less effective immune surveillance and, potentially, cancer progression.
Breast Cancer [3,14,21,26,27]	A paradoxical relationship is observed between TSGA10 expression and cellular migration. The high-affinity interaction of TSGA10 C-terminal domain with HIF-1α affects 8 key proteins (VEGFA, HSP90AA1, AKT1, ARNT, TP53, VHL, JUN, and EGFR) in cancer progression.	TSGA10 overexpression is associated with reduced metastasis. TSGA10 overexpression decreases metastatic and metabolic activities, thereby reducing cell proliferation and metastasis.	TSGA10 is typically downregulated in breast cancer, which leads to cancer progression and metastasis.
Brain Tumor [12,22]	Unknown. TSGA10 is specifically expressed in astrocytes.	TSGA10 is overexpressed in brain tumors.	TSGA10 Downregulation may disrupt normal cell cycle control, which could lead to decreased cell proliferation.
Nasopharyngeal Carcinoma [23]	miR-23a regulated angiogenesis by directly targeting TSGA10. Metastasis-associated miR-23a from NPC-derived exosomes plays an important role in mediating angiogenesis by targeting TSGA10.	Overexpression of TSGA10 can counteract the effects of miR-23a and result in inhibiting proliferation, angiogenesis, and cell migration and invasion.	Suppression of TSGA10 is associated with tumorigenesis via enhancing the migration of endothelial cells, suggesting that angiogenesis is regulated by miR-23a as it directly targets TSGA10 and represses its antiangiogenic functions.
Hepatocellular Carcinoma (HCC) [28]	TSGA10 acts as an immunogenic protein that can elicit an immune response; hence, TSGA10 plays a significant role in the progression and prognosis of hepatocellular carcinoma.	TSGA10's overexpression is linked to tumor aggressiveness, poor patient outcomes, and serves as a potential immunogenic target.	Downregulation of TSGA10 is associated with increased cell proliferation and reduced apoptosis.
AML/ALL [10,19]	*TSGA10* acts as a tumor suppressor gene in AML, as it negatively regulates the expression of VEGF by interacting with HIF-1α. TSGA10 may be involved in the proliferation of leukemic cells.	TSGA10 Overexpression leads to VEGF and HIF-1α downregulation, consequently inhibiting tumor growth and angiogenesis. TSGA10 is overexpressed in ALL, leading to proliferation of leukemic cells.	Decreased expression of TSGA10 in AML leads to increased VEGF and HIF-1α levels, promoting tumor growth and angiogenesis.
Pan-cancer studies - Melanoma - HCC - Colon - Ovarian - Prostate [2,11]	Irregular expression of TSGA10 in various cancers can affect the proliferation of cancer cells, suggesting its role in tumorigenesis.	TSGA10 is overexpressed in a subset of melanoma (5%), colon cancer (5%), HCC (20%), ovarian cancer (35%), and prostate cancer (15%), leading to increased cell division and growth, altered apoptosis, enhanced cell migration and invasion, and activation of oncogenic pathways.	Downregulation of TSGA10 can lead to a potential tumor suppression.

The following section provides a synopsis of malignant transformation and cancer hallmarks.

4. Stepwise Cancer Progression and Cancer Hallmarks

Extensive research on cancer biology revealed that cancer cells are not static but are dynamic, acquiring new phenotypes and capabilities during progression following sustained randomized (but programmed) changes in their genotype [29]. This process is called "malignant transformation" that makes the tumor mass "heterogenous", containing cancer cells at different phases of malignancy with different phenotypes as well as different levels of resistance to anti-cancer treatments [30,31]. It has been demonstrated that cancer cells, after development, traverse a multistep journey, obtaining different characteristics named "cancer hallmarks" [29]. With advances in our understanding of cancer biology, the cancer hallmarks have evolved from six items in 2000 (including sustained proliferative signals, evading growth suppressors, resisting cell death, active tissue invasion and metastasis, sustained angiogenesis, and enabling replicative immortality) to ten hallmarks in 2011 (plus genome instability, tumor-promoting inflammation, deregulating cellular metabolism, and immune escape) [32,33]. In 2021, Prof. Douglas Hanahan put forward another four hallmarks, including non-mutational epigenetic reprogramming, unlocking phenotypic plasticity, senescence, and the influence of polymorphic microbes, to better illustrate the cancer phenotypes [29]. Figure 2 illustrates a summary of changes in cancer phenotype across the malignant transformation pathway.

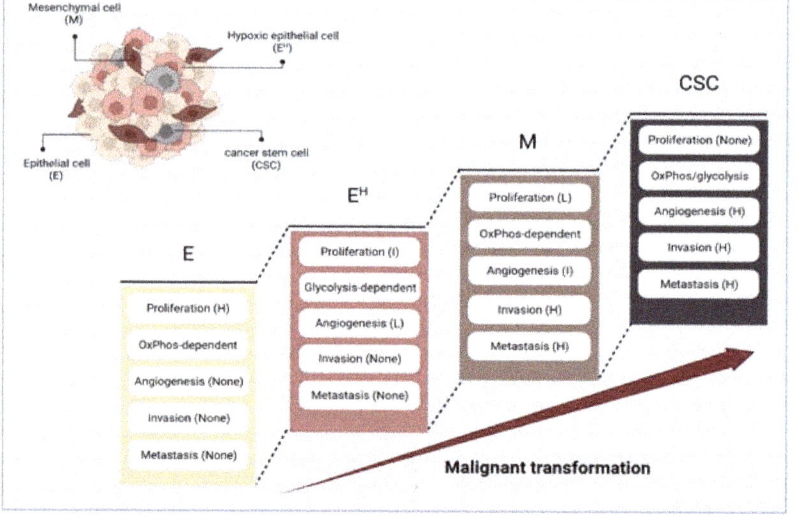

Figure 2. Phenotypes and capabilities of cancer cells across malignant transformation, including early-epithelial phase (E), epithelial phase at hypoxic conditions (E^H), mesenchymal transition (M), and obtaining stemness (CSC). CSC indicates cancer stem cells; E, epithelial cancer cells at normoxia; E^H, epithelial cancer cells at hypoxia; H, high-level; I, intermediate-level; L, low-level; M, mesenchymal cancer cells.

In the initial stage of malignant transformation (proliferative phase), cancer cells try to overcome growth suppressors to secure a sustained replication. In this phase, cancer cells benefit from normoxia and sufficient micronutrients for continuous proliferation. In the proliferative phase, cancer cells' metabolism primarily relies on oxidative phosphorylation (OxPhos) to provide the building blocks essential for replication, including amino acids, fatty acids, and nucleosides [34].

Following the increase in number of cancer cells and disturbance of the supply-demand balance, cancer cells obtain new features enabling them to survive and progress in the hypoxic, hypoglycemic, and acidic TME. In this stage, the cancer cells' metabolism primarily depends on glycolysis. Even in this hard-to-survive condition, cancer cells can continue proliferating using glycolysis intermediates to generate the macromolecules essential for cell division [33]. The phenotypic transformation of cancer cells in this phase is conducted by several transcription factors, including HIF-1 [35]. Under normoxia, HIF-1 is ubiquitinated, following hydroxylation by prolyl hydroxylase domain (PHD) proteins and activation of von Hippel Lindau protein (pVHL) [36]. This process does not happen in hypoxic conditions, and the intact HIF-1 can conduct the transcription of numerous mediators supporting cancer cells to survive and progress in the harsh TME [37].

With the advances in tumor growth, certain cancer cells enter a dedicated transition pathway to obtain mesenchymal phenotype, a process called epithelial–mesenchymal transition (EMT). During EMT, epithelial cancer cells lose their intercellular connections and proliferative ability and obtain mesenchymal phenotypes with enhanced invasive and migratory abilities [38]. In this stage, the cancer cells' metabolism depends on OxPhos [38]. EMT is primarily regulated by several dedicated transcription factors, including Snail, Twist, and ZEB. It has been demonstrated that ZEB and Twist were transcribed by HIF-1 [39,40], and Snail required HIF-1 for its stability [41]; therefore, EMT is an HIF-dependent process.

Following progression, a set of cancer cells obtain "stemness" phenotypes. These cells are called cancer stem cells (CSCs). The source of CSCs is the topic of debate and expanding research. Some believe they develop from normal stem cells or their progenitors (a.k.a. transit-amplifying cells) following successive oncogenic mutations [42], while others know their origins in differentiated cells [43]. An evolving idea considers EMT as the driving force of stemness. Proponents of this concept believe that EMT converts cancer cells in a terminally differentiated state to a metastable state, providing an opportunity to express new genes and, thereby, obtain new phenotypes [44]. In support, it has been demonstrated that the induction of EMT transcription factors (Snail, Twist, and ZEB) has enhanced the expression of stem cell markers (e.g., CD44) and their tumor sphere-forming ability (reviewed in [44]). CSCs' metabolism relies on both OxPhos and glycolysis. It has been demonstrated that CSCs could switch their metabolism to glycolysis in hypoxia and OxPhos in normoxia [45].

This section illustrated the synopsis of the literature on the multistep progression of cancer. In each step, the cancer cell has a specific genetic signature replying to its metabolism and objectives. The next section discusses the role of TSGA10 in cancer progression based on the current understanding of malignant transformation.

5. Interpretation of TSGA10 Studies Based on the Malignant Transformation

Based on the current knowledge of the malignant transformation, our interpretation of TSGA10 in cancer progression is as follows:

In vivo studies demonstrated wide-range expression levels of TSGA10 in cancer cells (Figure 2 of [2] and Figure 1 of [19]). This finding might be due to the "heterogeneity" of cancer cells in the tumor mass with different phases of malignant transformation. In support, Kazerani et al. found a higher expression of TSGA10 in low-grade brain tumors compared with high-grade tumors [22]. Another study demonstrated a similar pattern in ESCC cells. Yuan et al. realized that TSGA10 expression significantly decreased with tumor grade, primary tumor size, and clinical stage [20]. Figure 3 illustrates our assumption on trends in TSGA10 expression during distinct steps of malignant transformation and its association with HIF-1 expression. Given the stepwise progression of cancers [46], we propose the following scenario to explain the changes in the TSGA10 level.

At the early phase of cancer initiation (proliferative phase), cancer cells primarily tend to proliferate to generate the tumor mass. Considering the importance of TSGA10 in centrosome assembly [47], it can be concluded that cancer cells upregulate the TSGA10 expression to respond to this endpoint (the increasing trend in Figure 3). However, in the following, cancer cells need to downregulate TSGA10 to respond to their progression. Cancer cells require a number of tools to progress, even in the hypoxic TME. One of these advanced tools is HIF-1, a transcription factor that improves the cancer cells' survival, progression, and resistance in multiple ways [48]. It has been demonstrated that TSGA10 has counter-regulatory effects with HIF-1. In human sperms, the C-terminal domain of TSGA10 prevents the nuclear localization of HIF-1 during spermatogenesis [18]. This counteraction has also been demonstrated in human cancers. Jahani et al. (2020) found that TSGA10 overexpression in breast cancer cells could reduce the expression of HIF-1 target genes, including MMP7 (matrix metalloproteinase-7), GLUT1 (glucose transporter 1), CXCR4 (C-X-C chemokine receptor type 4), CXCL12 (C-X-C motif chemokine 12), LOXL2 (lysyl oxidase-like 2), and vimentin [14]. These inhibitory effects would disrupt the cancer cells' metabolism (by reducing GLUT1) and their invasion and migratory abilities (by reducing MMP7, CXCR4, LOXL2, CXCL12, and vimentin). LOXL2 is a catalytic enzyme that cleaves the collagen cross-linking. It has been demonstrated that LOXL2 was an essential mediator of angiogenesis [49]. Therefore, the inhibitory effect of TSGA10 on LOXL2 expression can potentially inhibit angiogenesis, a point not mentioned in Jahani et al.'s article. Meanwhile, Amoorahim et al. (2020) demonstrated that TSGA10 overexpression in human umbilical vein endothelial cells (HUVECs) could inhibit endothelial cell proliferation and migration, thereby, angiogenesis, by disrupting the HIF-2α axis [50]. Jahani's and Amoorahim's studies demonstrate that TSGA10 overexpression can inhibit cancer cell metastasis by reducing its migratory ability and disrupting angiogenesis.

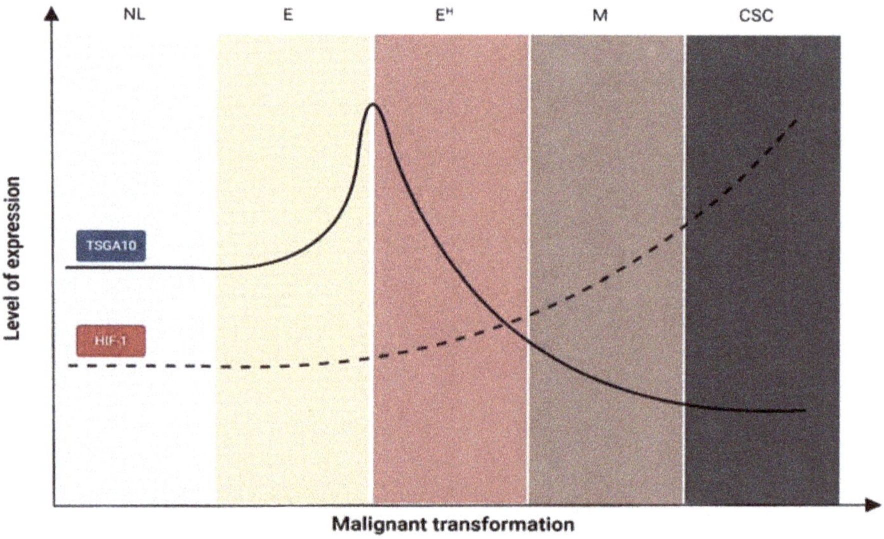

Figure 3. Changes in the TSGA10 and HIF-1 expression levels across the malignant transformation. CSC indicates cancer stem cells; E, epithelial cancer cells at normoxia; E^H, epithelial cancer cells at hypoxia; M, mesenchymal cancer cells; NL, normal (noncancerous) cells.

It has been demonstrated that HIF-1 has inhibitory effects on TSGA10 in cancer cells. An in vitro study on ESCC cells demonstrated that HIF-1 overexpression could downregulate TSGA10 expression by inducing miR-10b-3p expression. This study indicated that miR-10b-3p overexpression could improve cancer cell invasion and metastasis in a mouse xenograft model [24]. Another study demonstrated that miR-577 could directly regulate TSGA10 expression in ESCC cells by binding to the 3′UTR of the *TSGA10* gene. In this study, Yuan et al. found that miR-577 overexpression effectively boosted cancer cell proliferation and enhanced the transition from the G_1 to S phase by downregulating TSGA10 expression [20]. These pieces of information illustrate that cancer cells need to decrease the TSGA10 expression in the advanced stages of malignant transformation to increase their ability to metastasize (the decreasing trend in Figure 3).

6. Interpretation of TSGA10 Studies Based on the Cancer Hallmarks

This section outlines how TSGA10 can prevent distinct cancer hallmarks:

Enabling replicative immortality and **resisting cell death:** When cells undergo malignant transformation, they employ a unique mechanism to evade replicative senescence and subsequent cell death. This involves elongating their telomeres using telomerase, allowing them to continue proliferation and avoid cell death [51]. It has been demonstrated that telomerase activity in cancer cells is HIF1-dependent [52]. Recent evidence in breast cancer stem cells shows that HIF-1 is essential for NANOG-mediated telomerase reverse transcriptase (TERT) gene expression [53]. As noted, TSGA10 prevents the HIF-1 axis [13,14]. Therefore, cancer cells must downregulate TSGA10 to have a continuous replication;

Genomic instability: Under physiologic conditions, it is imperative to synchronize centriole duplication with DNA replication to guarantee that each daughter cell obtains only one pair of centrioles. However, cancer cells develop "centrosome amplification", which is the aberration in centrosome shape, size, position, and number. This condition increases the chance of aneuploidy and genomic instability across the cell divisions [54]. The physiologic assembly of centromeres relies on a set of regulators, including centrosomal protein 135 (CEP135), which is a centriole assembly protein [55]. It has been demonstrated that CEP135 was dysregulated in some breast cancers [56]. Carvalho-Santos et al. demonstrated that TSGA10 interacted with CEP135 and contributed to the physiologic assembly of centriole and basal body [57]. Therefore, downregulation of TSGA10 during cancer progression can increase the chance of genomic instability and provide less differentiated cancer cells, favoring cancer progression;

Deregulating cellular metabolism: The metabolism of cancer cells varies from normal cells. Cancer cells can run glycolysis even in the presence of enough oxygen pressure, a process known as "aerobic glycolysis". This characteristic enables cancer cells to survive even in the harsh conditions of TME, like hypoxia, acidosis, and low-glucose levels [58]. It has been demonstrated that mitochondria are the main regulators of cancer cell metabolism [59]. The available literature regarding the association between TSGA10 and mitochondria is limited. Luo et al. (2020) demonstrated that TSGA10 expression is essential to organize the mitochondria along the midpiece of sperm [60]. Similar to this effect can also be speculated in cancer cells, considering the following two assumptions:

(a) Mitochondria trafficking is an essential component in malignant transformation. It has been demonstrated that cancer cells with high affinity to metastasis had fragmented mitochondria in their periphery, likely to provide enough energy for invasion. However, mitochondria in cancer cells with less metastatic affinity are mainly located in the perinuclear region in the fused form [61];

(b) Jahani et al. demonstrated that TSGA10 induction in MCF-7 breast cancer cells decreased ROS production [14]. Given that mitochondria are the main source of ROS in cancer cells [62], Jahani et al.'s finding can reflect the decrease in mitochondrial metabolism following TSGA10 activation.

This finding, in addition to the perinuclear localization of C-terminus TSGA10 [12], put forward a concept that TSGA10 overexpression can translocate the mitochondria to the perinuclear region, facilitating their fusion and reducing their metabolic activity. This concept needs to be examined in future experimental studies;

Inducing or accessing vasculature: Cancer cells require access to oxygen and micronutrients to have a sustained proliferation. This access is achieved by releasing angiogenic factors (e.g., VEGF) and the breakdown of extracellular matrix using MMPs [63]. As noted, TSGA10 has a negative correlation with VEGF [19]. Furthermore, Asgharzadeh et al. demonstrated that the interaction between TSGA10 and HIF-1 can modulate the epidermal growth factor receptor (EGFR) [21]. Given the importance of EGFR in the angiogenesis process [64], TSGA10 may influence angiogenesis through its interaction with EGFR. In addition, Jahani et al. demonstrated that TSGA10 could reduce the expression of MMP-7 in cancer cells [14]. Therefore, a decrease in cellular TSGA10 level is required for angiogenesis and vascular access;

Activating invasion and metastasis: Cancer metastasis is in two forms: single-cell and collective. Each type of metastasis requires epithelial cells to be transformed into the mesenchymal counterpart through EMT [63]. It has been well established that EMT is a HIF-dependent process [65]. Therefore, cancer cells must downregulate their TSGA10 level to obtain mesenchymal phenotype;

Avoiding immune destruction: Cancer cells exploit different mechanisms to shield from immunosurveillance, such as reducing antigen presentation, expressing immune checkpoints, and converting the condition of surrounding TME too harsh for immune cell recruitment (e.g., acidity) [59]. Among these, the expression of CGAs has been put forward as an immune escape mechanism. Kortleve et al. (2022) evaluated the association between the expression level of fifteen CGAs (including TSGA10) and immune escape in a pan-cancer model. Their study demonstrated that TSGA10 expression negatively correlated with the tumor-infiltrating lymphocytes and MHC molecules [66]. This study also indicated the negative correlation between TSGA10 and cancer-associated fibroblast (CAF) infiltration [66]. With reduction in TSGA10 level, the rates of CAFs in TME is elevated, which provides several benefits for cancer cells promotion by providing substrates for OxPhos of cancer cells (including pyruvate, lactate, and glutamate) [67], transferring mitochondria to cancer cells via nanotubes to support their metabolism [68], inducing tumor-promoting autophagy by releasing β-HB, IGF1/2, and CXCL12 [69], and inhibiting anti-tumor immune response [70]. The causative association between TSGA10 and the immune microenvironment needs to be addressed in immune-competent models and assessment of immunophenotypes and immunokinetics;

Tumor-promoting inflammation: TME consists of noncancerous cells modified to support the cancer cells' survival, progression, and treatment resistance. The major cellular components of TME are cancer-associated fibroblasts (CAFs), myeloid-derived suppressor cells (MDSCs), tumor-associated macrophages (TAMs), tumor-associated neutrophils (TANs), and regulatory T cells (Tregs) [71]. As noted earlier, a decrease in TSGA10 expression (in advanced phases) is correlated with CAF infiltration into TME [66]. It has been demonstrated that CAFs can induce the recruitment of tumor-promoting immune cells (TAMs, MDSCs, TANs, and Tregs) toward TME in several mechanisms [70]. Therefore, a decrease in TSGA10 expression in advanced phases of malignant transformation can contribute to developing tumor-promoting inflammation. Co-culture and organoid models could help dissect the TME-specific effects of TSGA10.

7. The Potential Clinical Implications of TSGA10 Upregulation in Cancer Cells

Mesenchymal–epithelial transition: Given the close association between TSGA10 and centrosomes [47], the involvement of centrosomes in cellular polarity (as a feature of epithelial cells versus mesenchymal cells), and a reduction in invasion capabilities of cancer cells with TSGA10 overexpression, one may conclude that TSGA10 may serve as a running factor of the mesenchymal–epithelial transition (MET). In support, it has been demonstrated that TSGA10 overexpression in MCF-7 breast cancer cells upregulated the expression of E-cadherin [14], a classical biomarker of epithelial cells [72]. In addition, Jahani et al. found that TSGA10 upregulation led to the downregulation of vimentin, a classic biomarker of mesenchymal cells [14]. It has been established that vimentin was one of the main drivers of EMT, by which cancer cells obtain special phenotypes to invade the extracellular matrix and withstand the external sheer forces during metastasis [63].

Supporting evidence is the interaction between centrosomal CEP135 and TSGA10. CEP135 plays a crucial role in centrosome organization, a process essential for maintaining cellular polarity [73]. Carvalho-Santos et al. demonstrated that TSGA10 interacted with CEP135 and likely contributed to the physiologic assembly of centriole and basal body [57]. With this information, one may conclude that TSGA10 upregulation can contribute to the cancer cells to retrieve their polarity. It has also been demonstrated that more organized centrosomes could prevent aneuploidy and genomic instability [74]. Therefore, upregulation of TSGA10 might prevent the cancer cells from being promoted to the less differentiated phenotypes. Considering the importance of EMT in stemness [44] and the high resistance of CSCs to radiotherapy [75], chemotherapy [76], and immunotherapy [77], TSGA10 induction can serve as a potential modality to reduce treatment resistance.

Exosome secretion: Tumor progression is significantly influenced by the secretion of extracellular vesicles, which are produced in larger quantities by cancer cells compared to normal cells. These vesicles, including exosomes, carry biomolecules such as microRNAs (miRNAs) that inhibit multiple target genes and alter intercellular communication, promoting metastasis [78]. Based on the available evidence, TSGA10 can interact with several proteins involved in exosome formation and secretion, including actin [5], Rab27 [47], HIF-1 [14], and ODF2 [47]. TSGA10 interacts with actin-rich structures, influencing exosome release and connecting to Rab27 proteins that regulate exosome secretion. Its overexpression may induce apoptosis and autophagy by affecting exosome dynamics. Additionally, TSGA10 interacts with HIF-1a and p53 [13,20], potentially promoting exosome release in cancer cells. These interactions suggest that TSGA10 plays a significant role in cancer progression through exosome regulation, although further research is needed to fully understand its mechanisms.

Radiotherapy: This modality is the mainstay of cancer treatments and is extensively applied to different malignancies [79,80]. The main cytotoxic effect of radiotherapy is targeting the vital macromolecules of cancer cells, especially DNA [81]. It has been demonstrated that the G_1 and S phases of the cell cycle are more radioresistant, and the G_2 and M phases are more radiosensitive [82]. As alluded to above, TSGA10 induction can lead to G_2/M arrest in cancer cells [14]. This effect can serve as a potential approach to improve the radiosensitivity of cancer cells. In addition, it has been demonstrated that TSGA10 overexpression in endothelial cells could induce cell cycle arrest [50]. Ironically, tumor neovascularization enhances hypoxia by forming immature, leaky vessels undergoing collapse in the extracellular matrix with high interstitial fluid pressure [83]. Therefore, TSGA10 overexpression in tumor endothelial cells can enhance the tumor oxygen pressure that inherently improves radiosensitivity [84]. The potential effects of TSGA10 induction on radiosensitivity are speculative and need testing in appropriate preclinical models.

Immunotherapy: It has been demonstrated that HIF-1 expression improved the cancer cells' ability to evade the immune system [85]. The HIF-1 axis suppresses the innate and adaptive immune response by inducing the secretion of immunosuppressive factors (prostaglandin E_2 and transforming growth factor-β) [86], expression of programmed death protein-ligand 1 (PD-L1) on cancer cells [87], and reducing tumor-associated antigen presentation via major histocompatibility complex class I (MHC-1) [88]. In addition, HIF-1 signaling can induce MDSC accumulation in TME [89], which inhibits the immune response in several ways, including by attracting M2 macrophages and Tregs into TME, impairing lymphocyte adhesion to endothelial cells and expression of immune checkpoint molecules (PD-L1) [90]. Given the inhibitory effects of TSGA10 on the HIF-1 axis, TSGA10 overexpression can provide an opportunity to enhance the anti-tumor immune response and, thereby, response to immunotherapies. The potential impact of TSGA10 induction on immunotherapy requires evaluation in suitable experimental studies.

Anti-mitochondrial therapy: Emerging evidence on mitochondrial metabolism has been put forward as a determining factor in cancer biology and treatment resistance [59]. This concept has recently evolved into introducing mitochondrial metabolism as a new aspect of personalized cancer treatment [91]. A big hurdle in targeted anti-mitochondrial therapies is that mitochondria are present in all human cells, and anti-mitochondrial therapies can harm normal cells. This issue limits the application of broad-spectrum anti-mitochondrial agents. As noted earlier, TSGA10 might target the mitochondrial metabolism and trafficking in cancer cells. The specific expression of TSGA10 in cancer cells can provide an opportunity to limit the anti-mitochondrial effects on cancer cells. This concept can be engaged to design targeted therapies to limit mitochondrial metabolism, specifically in cancer cells.

Future works are suggested to address the following objectives: (a) to perform multi-omics and lineage tracing to better characterize TSGA10's context-specific functions; (b) to develop robust in vitro and in vivo models manipulating TSGA10 to validate proposed mechanisms; (c) to use unbiased screening to identify TSGA10 interactome and downstream effectors; (d) to evaluate combination therapies engaging TSGA10 along with standard care in clinically relevant models; and (e) to address key questions around therapeutic index, delivery challenges, biomarkers, and resistance for clinical translation.

8. Conclusions

The available TSGA10 literature poses a big question, "Why do cancer cells need to express a protein that prevents their progression?". This review outlines that TSGA10 expression level in cancer cells depends on the cancer stage across malignant transformation. It demonstrates that in early phases, cancer cells overexpress TSGA10 to respond to their endpoint to proliferate; however, as cancer cells progress to the advanced phases of malignancy, the level of TSGA10 is reduced to allow HIF-1 to take the wheel of cancer cells and conduct the EMT and metastasis. In other words, we assume that TSGA10 is a "basic tool" for the early progression of cancer. Once cancer cells plan to enter the advanced phases, they improve their equipment to more "advanced tools" to lose their polarity, develop genomic instability, improve their mitochondrial metabolism, undergo EMT, and recruit more CAFs into TME to support their progression. The various tumor-suppressive effects of TSGA10 on cancer biology can provide a potential opportunity to enhance the efficacy of different cancer therapies. The mechanisms proposed may not prove causation. Further experimental studies are needed to explore the role of TSGA10 in cancer progression. In addition, engaging TSGA10 induction as a therapeutic approach assumes it can be safely upregulated in cancer cells without adverse effects, which requires investigation.

Author Contributions: Conceptualization, B.B., F.T.-H. and F.K.; methodology, B.B. and F.T.-H.; software, F.T.-H. and M.G.; validation, B.B., F.T.-H. and F.K.; formal analysis, B.B. and F.T.-H.; investigation, B.B., F.T.-H. and M.G.; resources, B.B., F.T.-H. and F.K.; data curation, F.T.-H. and M.G.; writing—original draft preparation, F.T.-H. and M.G.; writing—review and editing, B.B., F.T.-H. and M.G.; visualization, B.B. and F.T.-H.; supervision, B.B.; project administration, B.B. All authors have read and agreed to the published version of the manuscript.

Funding: This research received no external funding.

Conflicts of Interest: B.B. was employed by the company NSF International and Avicenna Biotech Research. The remaining authors declare that the research was conducted in the absence of any commercial or financial relationships that could be construed as a potential conflict of interest. The authors declare no conflict of interest.

References

1. Li, X.F.; Ren, P.; Shen, W.Z.; Jin, X.; Zhang, J. The Expression, Modulation and Use of Cancer-Testis Antigens as Potential Biomarkers for Cancer Immunotherapy. *Am. J. Transl. Res.* **2020**, *12*, 7002–7019. [PubMed]
2. Tanaka, R.; Ono, T.; Sato, S.; Nakada, T.; Koizumi, F.; Hasegawa, K.; Nakagawa, K.; Okumura, H.; Yamashita, T.; Ohtsuka, M.; et al. Over-Expression of the Testis-Specific Gene TSGA10 in Cancers and Its Immunogenicity. *Microbiol. Immunol.* **2004**, *48*, 339–345. [CrossRef] [PubMed]
3. Salehipour, P.; Nematzadeh, M.; Mobasheri, M.B.; Afsharpad, M.; Mansouri, K.; Modarressi, M.H. Identification of New TSGA10 Transcript Variants in Human Testis with Conserved Regulatory RNA Elements in 5'untranslated Region and Distinct Expression in Breast Cancer. *Biochim. Biophys. Acta Gene Regul. Mech.* **2017**, *1860*, 973–982. [CrossRef] [PubMed]
4. Modarressi, M.H.; Cameron, J.; Taylor, K.E.; Wolfe, J. Identification and Characterisation of a Novel Gene, TSGA10, Expressed in Testis. *Gene* **2001**, *262*, 249–255. [CrossRef] [PubMed]
5. Roghanian, A.; Jones, D.C.; Pattisapu, J.V.; Wolfe, J.; Young, N.T.; Behnam, B. Filament-Associated TSGA10 Protein Is Expressed in Professional Antigen Presenting Cells and Interacts with Vimentin. *Cell. Immunol.* **2010**, *265*, 120–126. [CrossRef]

6. Asgari, R.; Bakhtiari, M.; Rezazadeh, D.; Yarani, R.; Esmaeili, F.; Mansouri, K. TSGA10 as a Potential Key Factor in the Process of Spermatid Differentiation/Maturation: Deciphering Its Association with Autophagy Pathway. *Reprod. Sci.* **2021**, *28*, 3228–3240. [CrossRef] [PubMed]
7. Behnam, B.; Modarressi, M.H.; Conti, V.; Taylor, K.E.; Puliti, A.; Wolfe, J. Expression of Tsga10 Sperm Tail Protein in Embryogenesis and Neural Development: From Cilium to Cell Division. *Biochem. Biophys. Res. Commun.* **2006**, *344*, 1102–1110. [CrossRef] [PubMed]
8. Reimand, K.; Perheentupa, J.; Link, M.; Krohn, K.; Peterson, P.; Uibo, R. Testis-expressed protein TSGA10 an auto-antigen in autoimmune polyendocrine syndrome type I. *Int. Immunol.* **2008**, *20*, 39–44. [CrossRef]
9. Smith, C.J.A.; Oscarson, M.; Rönnblom, L.; Alimohammadi, M.; Perheentupa, J.; Husebye, E.S.; Gustafsson, J.; Nordmark, G.; Meloni, A.; Crock, P.A.; et al. TSGA10—A Target for Autoantibodies in Autoimmune Polyendocrine Syndrome Type 1 and Systemic Lupus Erythematosus. *Scand. J. Immunol.* **2011**, *73*, 147–153. [CrossRef]
10. Mobasheri, M.B.; Modarressi, M.H.; Shabani, M.; Asgarian, H.; Sharifian, R.A.; Vossough, P.; Shokri, F. Expression of the Testis-Specific Gene, TSGA10, in Iranian Patients with Acute Lymphoblastic Leukemia (ALL). *Leuk. Res.* **2006**, *30*, 883–889. [CrossRef]
11. Mobasheri, M.B.; Jahanzad, I.; Mohagheghi, M.A.; Aarabi, M.; Farzan, S.; Modarressi, M.H. Expression of Two Testis-Specific Genes, TSGA10 and SYCP3, in Different Cancers Regarding to Their Pathological Features. *Cancer Detect. Prev.* **2007**, *31*, 296–302. [CrossRef] [PubMed]
12. Behnam, B.; Chahlavi, A.; Pattisapu, J.; Wolfe, J. TSGA10 Is Specifically Expressed in Astrocyte and Over-Expressed in Brain Tumors. *Avicenna J. Med. Biotechnol.* **2009**, *1*, 161–166.
13. Mansouri, K.; Mostafie, A.; Rezazadeh, D.; Shahlaei, M.; Modarressi, M.H. New Function of *TSGA10* Gene in Angiogenesis and Tumor Metastasis: A Response to a Challengeable Paradox. *Hum. Mol. Genet.* **2016**, *25*, 233–244. [CrossRef] [PubMed]
14. Jahani, M.; Shahlaei, M.; Norooznezhad, F.; Miraghaee, S.S.; Hosseinzadeh, L.; Moasefi, N.; Khodarahmi, R.; Farokhi, A.; Mahnam, A.; Mansouri, K. TSGA10 Over Expression Decreases Metastasic and Metabolic Activity by Inhibiting HIF-1 in Breast Cancer Cells. *Arch. Med. Res.* **2020**, *51*, 41–53. [CrossRef] [PubMed]
15. Valipour, E.; Nooshabadi, V.T.; Mahdipour, S.; Shabani, S.; Farhady-Tooli, L.; Majidian, S.; Noroozi, Z.; Mansouri, K.; Motevaseli, E.; Modarressi, M.H. Anti-Angiogenic Effects of Testis-Specific Gene Antigen 10 on Primary Endothelial Cells. *Gene* **2020**, *754*, 144856. [CrossRef]
16. Taghizadeh-Hesary, F. "Reinforcement" by Tumor Microenvironment: The Seventh "R" of Radiobiology. *Int. J. Radiat. Oncol. Biol. Phys.* **2023**, *119*, 727–733. [CrossRef]
17. Aarabi, M.; Soltanghoraee, H.; Amirjannati, N.; Ghaffari, M.; Sadeghi, M.R.; Akhondi, M.M.; Modarresi, M.H. Testis Specific Gene 10 Expression in the Testes of Patients with Non-Obstructive Azoospermia. *J. Reprod. Infertil.* **2006**, *7*, 179–186.
18. Hägele, S.; Behnam, B.; Borter, E.; Wolfe, J.; Paasch, U.; Lukashev, D.; Sitkovsky, M.; Wenger, R.H.; Katschinski, D.M. TSGA10 Prevents Nuclear Localization of the Hypoxia-Inducible Factor (HIF)-1α. *FEBS Lett.* **2006**, *580*, 3731–3738. [CrossRef]
19. Hoseinkhani, Z.; Rastegari-Pouyani, M.; Oubari, F.; Mozafari, H.; Rahimzadeh, A.B.; Maleki, A.; Amini, S.; Mansouri, K. Contribution and Prognostic Value of *TSGA10* Gene Expression in Patients with Acute Myeloid Leukemia (AML). *Pathol. Res. Pract.* **2019**, *215*, 506–511. [CrossRef]
20. Yuan, X.; He, J.; Sun, F.; Gu, J. Effects and Interactions of MiR-577 and TSGA10 in Regulating Esophageal Squamous Cell Carcinoma. *Int. J. Clin. Exp. Pathol.* **2013**, *6*, 2651–2667.
21. Asgharzadeh, M.R.; Pourseif, M.M.; Barar, J.; Eskandani, M.; Jafari Niya, M.; Mashayekhi, M.R.; Omidi, Y. Functional Expression and Impact of Testis-Specific Gene Antigen 10 in Breast Cancer: A Combined in Vitro and in Silico Analysis. *Bioimpacts* **2019**, *9*, 145–159. [CrossRef] [PubMed]
22. Kazerani, R.; Salehipour, P.; Shah Mohammadi, M.; Amanzadeh Jajin, E.; Modarressi, M.H. Identification of TSGA10 and GGNBP2 Splicing Variants in 5′ Untranslated Region with Distinct Expression Profiles in Brain Tumor Samples. *Front. Oncol.* **2023**, *13*, 1075638. [CrossRef]
23. Bao, L.; You, B.; Shi, S.; Shan, Y.; Zhang, Q.; Yue, H.; Zhang, J.; Zhang, W.; Shi, Y.; Liu, Y.; et al. Metastasis-associated miR-23a from nasopharyngeal carcinoma-derived exosomes mediates angiogenesis by repressing a novel target gene TSGA10. *Oncogene* **2018**, *37*, 2873–2889. [CrossRef] [PubMed]
24. Zhang, Q.; Zhang, J.; Fu, Z.; Dong, L.; Tang, Y.; Xu, C.; Wang, H.; Zhang, T.; Wu, Y.; Dong, C.; et al. Hypoxia-Induced microRNA-10b-3p Promotes Esophageal Squamous Cell Carcinoma Growth and Metastasis by Targeting TSGA10. *Aging* **2019**, *11*, 10374–10384. [CrossRef] [PubMed]
25. Theinert, S.M.; Pronest, M.M.; Peris, K.; Sterry, W.; Walden, P. Identification of the Testis-Specific Protein 10 (TSGA10) as Serologically Defined Tumour-Associated Antigen in Primary Cutaneous T-Cell Lymphoma: CTCL-Associated Antigen TSGA10. *Br. J. Dermatol.* **2005**, *153*, 639–641. [CrossRef] [PubMed]
26. Fan, Y.; Kao, C.; Yang, F.; Wang, F.; Yin, G.; Wang, Y.; He, Y.; Ji, J.; Liu, L. Integrated Multi-Omics Analysis Model to Identify Biomarkers Associated With Prognosis of Breast Cancer. *Front. Oncol.* **2022**, *12*, 899900. [CrossRef]
27. Dianatpour, M.; Mehdipour, P.; Nayernia, K.; Mobasheri, M.-B.; Ghafouri-Fard, S.; Savad, S.; Modarressi, M.H. Expression of Testis Specific Genes TSGA10, TEX101 and ODF3 in Breast Cancer. *Iran. Red Crescent Med. J.* **2012**, *14*, 730–734. [CrossRef] [PubMed]

28. Zhang, Z.; Chen, W.; Luo, C.; Zhang, W. Exploring a Four-Gene Risk Model Based on Doxorubicin Resistance-Associated lncRNAs in Hepatocellular Carcinoma. *Front. Pharmacol.* **2022**, *13*, 1015842. [CrossRef]
29. Hanahan, D. Hallmarks of Cancer: New Dimensions. *Cancer Discov.* **2022**, *12*, 31–46. [CrossRef]
30. Dagogo-Jack, I.; Shaw, A.T. Tumour Heterogeneity and Resistance to Cancer Therapies. *Nat. Rev. Clin. Oncol.* **2018**, *15*, 81–94. [CrossRef]
31. Taghizadeh-Hesary, F.; Houshyari, M.; Farhadi, M. Mitochondrial Metabolism: A Predictive Biomarker of Radiotherapy Efficacy and Toxicity. *J. Cancer Res. Clin. Oncol.* **2023**, *149*, 6719–6741. [CrossRef] [PubMed]
32. Hanahan, D.; Weinberg, R.A. The Hallmarks of Cancer. *Cell* **2000**, *100*, 57–70. [CrossRef] [PubMed]
33. Hanahan, D.; Weinberg, R.A. Hallmarks of Cancer: The next Generation. *Cell* **2011**, *144*, 646–674. [CrossRef]
34. Gregorio, J.; Petricca, S.; Iorio, R.; Toniato, E.; Flati, V. Mitochondrial and Metabolic Alterations in Cancer Cells. *Eur. J. Cell Biol.* **2022**, *101*, 151225. [CrossRef]
35. Shi, R.; Liao, C.; Zhang, Q. Hypoxia-Driven Effects in Cancer: Characterization, Mechanisms, and Therapeutic Implications. *Cells* **2021**, *10*, 678. [CrossRef] [PubMed]
36. Jun, J.C.; Rathore, A.; Younas, H.; Gilkes, D.; Polotsky, V.Y. Hypoxia-Inducible Factors and Cancer. *Curr. Sleep. Med. Rep.* **2017**, *3*, 1–10. [CrossRef]
37. Harrison, H.; Pegg, H.J.; Thompson, J.; Bates, C.; Shore, P. HIF1-Alpha Expressing Cells Induce a Hypoxic-like Response in Neighbouring Cancer Cells. *BMC Cancer* **2018**, *18*, 674. [CrossRef]
38. Huang, Z.; Zhang, Z.; Zhou, C.; Liu, L.; Huang, C. Epithelial-Mesenchymal Transition: The History, Regulatory Mechanism, and Cancer Therapeutic Opportunities. *MedComm* **2020**, *3*, 144. [CrossRef] [PubMed]
39. Joseph, J.V.; Conroy, S.; Pavlov, K.; Sontakke, P.; Tomar, T.; Eggens-Meijer, E.; Balasubramaniyan, V.; Wagemakers, M.; Dunnen, W.F.A.; Kruyt, F.A.E. Hypoxia Enhances Migration and Invasion in Glioblastoma by Promoting a Mesenchymal Shift Mediated by the HIF1α–ZEB1 Axis. *Cancer Lett.* **2015**, *359*, 107–116. [CrossRef] [PubMed]
40. Yang, M.H.; Wu, M.Z.; Chiou, S.H.; Chen, P.M.; Chang, S.Y.; Liu, C.J.; Teng, S.C.; Wu, K.J. Direct regulation of TWIST by HIF-1alpha promotes metastasis. *Nat. Cell Biol.* **2008**, *10*, 295–305. [CrossRef]
41. Lv, L.; Yuan, J.; Huang, T.; Zhang, C.; Zhu, Z.; Wang, L.; Jiang, G.; Zeng, F. Stabilization of Snail by HIF-1α and TNF-α Is Required for Hypoxia-Induced Invasion in Prostate Cancer PC3 Cells. *Mol. Biol. Rep.* **2014**, *41*, 4573–4582. [CrossRef] [PubMed]
42. Chaffer, C.L.; Weinberg, R.A. How Does Multistep Tumorigenesis Really Proceed? *Cancer Discov.* **2015**, *5*, 22–24. [CrossRef]
43. Bisht, S.; Nigam, M.; Kunjwal, S.S.; Sergey, P.; Mishra, A.P.; Sharifi-Rad, J. Cancer Stem Cells: From an Insight into the Basics to Recent Advances and Therapeutic Targeting. *Stem Cells Int.* **2022**, *2022*, 9653244. [CrossRef]
44. Wang, H.; Unternaehrer, J.J. Epithelial-Mesenchymal Transition and Cancer Stem Cells: At the Crossroads of Differentiation and Dedifferentiation. *Dev. Dyn.* **2019**, *248*, 10–20. [CrossRef]
45. Yasuda, T.; Ishimoto, T.; Baba, H. Conflicting Metabolic Alterations in Cancer Stem Cells and Regulation by the Stromal Niche. *Regen. Ther.* **2021**, *17*, 8–12. [CrossRef] [PubMed]
46. Akbari, H.; Taghizadeh-Hesary, F.; Heike, Y.; Bahadori, M. Cell Energy: A New Hypothesis in Decoding Cancer Evolution. *Arch. Iran. Med.* **2019**, *22*, 733–735. [PubMed]
47. Behnam, B.; Mobahat, M.; Fazilaty, H.; Wolfe, J.; Omran, H. TSGA10 Is a Centrosomal Protein, Interacts with ODF2 and Localizes to Basal Body. *J. Cell Sci. Ther.* **2015**, *6*, 217.
48. Ma, Z.; Xiang, X.; Li, S.; Xie, P.; Gong, Q.; Goh, B.-C.; Wang, L. Targeting Hypoxia-Inducible Factor-1, for Cancer Treatment: Recent Advances in Developing Small-Molecule Inhibitors from Natural Compounds. *Semin. Cancer Biol.* **2022**, *80*, 379–390. [CrossRef]
49. Zaffryar-Eilot, S.; Marshall, D.; Voloshin, T.; Bar-Zion, A.; Spangler, R.; Kessler, O.; Ghermazien, H.; Brekhman, V.; Suss-Toby, E.; Adam, D.; et al. Lysyl Oxidase-like-2 Promotes Tumour Angiogenesis and Is a Potential Therapeutic Target in Angiogenic Tumours. *Carcinogenesis* **2013**, *34*, 2370–2379. [CrossRef]
50. Amoorahim, M.; Valipour, E.; Hoseinkhani, Z.; Mahnam, A.; Rezazadeh, D.; Ansari, M.; Shahlaei, M.; Gamizgy, Y.H.; Moradi, S.; Mansouri, K. TSGA10 Overexpression Inhibits Angiogenesis of HUVECs: A HIF-2α Biased Perspective. *Microvasc. Res.* **2020**, *128*, 103952. [CrossRef]
51. Robinson, N.J.; Schiemann, W.P. Telomerase in Cancer: Function, Regulation, and Clinical Translation. *Cancers* **2022**, *14*, 808. [CrossRef] [PubMed]
52. Yatabe, N.; Kyo, S.; Maida, Y.; Nishi, H.; Nakamura, M.; Kanaya, T.; Tanaka, M.; Isaka, K.; Ogawa, S.; Inoue, M. HIF-1-mediated activation of telomerase in cervical cancer cells. *Oncogene* **2004**, *23*, 3708–3715. [CrossRef] [PubMed]
53. Lu, H.; Lyu, Y.; Tran, L.; Lan, J.; Xie, Y.; Yang, Y.; Murugan, N.L.; Wang, Y.J.; Semenza, G.L. HIF-1 Recruits NANOG as a Coactivator for TERT Gene Transcription in Hypoxic Breast Cancer Stem Cells. *Cell Rep.* **2021**, *36*, 109757. [CrossRef]
54. Mittal, K.; Kaur, J.; Jaczko, M.; Wei, G.; Toss, M.S.; Rakha, E.A.; Janssen, E.A.M.; Søiland, H.; Kucuk, O.; Reid, M.D.; et al. Centrosome Amplification: A Quantifiable Cancer Cell Trait with Prognostic Value in Solid Malignancies. *Cancer Metastasis Rev.* **2021**, *40*, 319–339. [CrossRef]
55. Lin, Y.C.; Chang, C.W.; Hsu, W.B.; Tang, C.J.; Lin, Y.N.; Chou, E.J.; Wu, C.T.; Tang, T.K. Human Microcephaly Protein CEP135 Binds to hSAS-6 and CPAP, and Is Required for Centriole Assembly. *EMBO J.* **2013**, *32*, 1141–1154. [CrossRef] [PubMed]
56. Ganapathi Sankaran, D.; Stemm-Wolf, A.J.; Pearson, C.G. CEP135 Isoform Dysregulation Promotes Centrosome Amplification in Breast Cancer Cells. *Mol. Biol. Cell* **2019**, *30*, 1230–1244. [CrossRef] [PubMed]

57. Carvalho-Santos, Z.; Machado, P.; Branco, P.; Tavares-Cadete, F.; Rodrigues-Martins, A.; Pereira-Leal, J.B.; Bettencourt-Dias, M. Stepwise Evolution of the Centriole-Assembly Pathway. *J. Cell Sci.* **2010**, *123*, 1414–1426. [CrossRef]
58. Al Tameemi, W.; Dale, T.P.; Al-Jumaily, R.M.K.; Forsyth, N.R. Hypoxia-Modified Cancer Cell Metabolism. *Front. Cell Dev. Biol.* **2019**, *7*, 4. [CrossRef]
59. Taghizadeh-Hesary, F.; Akbari, H.; Bahadori, M.; Behnam, B. Targeted Anti-Mitochondrial Therapy: The Future of Oncology. *Genes* **2022**, *13*, 1728. [CrossRef]
60. Luo, G.; Hou, M.; Wang, B.; Liu, Z.; Liu, W.; Han, T.; Zhang, D.; Zhou, X.; Jia, W.; Tan, Y.; et al. Tsga10 Is Essential for Arrangement of Mitochondrial Sheath and Male Fertility in Mice. *Andrology* **2021**, *9*, 368–375. [CrossRef]
61. Boulton, D.P.; Caino, M.C. Mitochondrial Fission and Fusion in Tumor Progression to Metastasis. *Front. Cell Dev. Biol.* **2022**, *10*, 849962. [CrossRef] [PubMed]
62. Sarmiento-Salinas, F.L.; Delgado-Magallón, A.; Montes-Alvarado, J.B.; Ramírez-Ramírez, D.; Flores-Alonso, J.C.; Cortés-Hernández, P.; Reyes-Leyva, J.; Herrera-Camacho, I.; Anaya-Ruiz, M.; Pelayo, R.; et al. Breast Cancer Subtypes Present a Differential Production of Reactive Oxygen Species (ROS) and Susceptibility to Antioxidant Treatment. *Front. Oncol.* **2019**, *9*, 480. [CrossRef]
63. Castaneda, M.; Hollander, P.; Kuburich, N.A.; Rosen, J.M.; Mani, S.A. Mechanisms of cancer metastasis. *Semin. Cancer Biol.* **2022**, *87*, 17–31. [CrossRef] [PubMed]
64. Rao, L.; Giannico, D.; Leone, P.; Solimando, A.G.; Maiorano, E.; Caporusso, C.; Duda, L.; Tamma, R.; Mallamaci, R.; Susca, N.; et al. HB-EGF–EGFR Signaling in Bone Marrow Endothelial Cells Mediates Angiogenesis Associated with Multiple Myeloma. *Cancers* **2020**, *12*, 173. [CrossRef] [PubMed]
65. Tam, S.Y.; Wu, V.W.C.; Law, H.K.W. Hypoxia-Induced Epithelial-Mesenchymal Transition in Cancers: HIF-1α and Beyond. *Front. Oncol.* **2020**, *10*, 486. [CrossRef] [PubMed]
66. Kortleve, D.; Coelho, R.M.L.; Hammerl, D.; Debets, R. Cancer Germline Antigens and Tumor-Agnostic CD8(+) T Cell Evasion. *Trends Immunol.* **2022**, *43*, 391–403. [CrossRef] [PubMed]
67. Nocquet, L.; Juin, P.P.; Souazé, F. Mitochondria at Center of Exchanges between Cancer Cells and Cancer-Associated Fibroblasts during Tumor Progression. *Cancers* **2020**, *12*, 3017. [CrossRef]
68. Ippolito, L.; Morandi, A.; Taddei, M.L.; Parri, M.; Comito, G.; Iscaro, A.; Raspollini, M.R.; Magherini, F.; Rapizzi, E.; Masquelier, J.; et al. Cancer-Associated Fibroblasts Promote Prostate Cancer Malignancy via Metabolic Rewiring and Mitochondrial Transfer. *Oncogene* **2019**, *38*, 5339–5355. [CrossRef]
69. Wang, Y.; Gan, G.; Wang, B.; Wu, J.; Cao, Y.; Zhu, D.; Xu, Y.; Wang, X.; Han, H.; Li, X.; et al. Cancer-Associated Fibroblasts Promote Irradiated Cancer Cell Recovery Through Autophagy. *EBioMedicine* **2017**, *17*, 45–56. [CrossRef]
70. Mao, X.; Xu, J.; Wang, W.; Liang, C.; Hua, J.; Liu, J.; Zhang, B.; Meng, Q.; Yu, X.; Shi, S. Crosstalk between Cancer-Associated Fibroblasts and Immune Cells in the Tumor Microenvironment: New Findings and Future Perspectives. *Mol. Cancer* **2021**, *20*, 131. [CrossRef] [PubMed]
71. Ribeiro Franco, P.I.; Rodrigues, A.P.; Menezes, L.B.; Pacheco Miguel, M. Tumor Microenvironment Components: Allies of Cancer Progression. *Pathol. Res. Pract.* **2020**, *216*, 152729. [CrossRef]
72. Xiao, S.; Liu, L.; Lu, X.; Long, J.; Zhou, X.; Fang, M. The Prognostic Significance of Bromodomain PHD-Finger Transcription Factor in Colorectal Carcinoma and Association with Vimentin and E-Cadherin. *J. Cancer Res. Clin. Oncol.* **2015**, *141*, 1465–1474. [CrossRef] [PubMed]
73. Tang, N.; Marshall, W.F. Centrosome positioning in vertebrate development. *J. Cell Sci.* **2012**, *125*, 4951–4961. [CrossRef] [PubMed]
74. Das, L. Epigenetic Alterations Impede Epithelial-Mesenchymal Transition by Modulating Centrosome Amplification and Myc/RAS Axis in Triple Negative Breast Cancer Cells. *Sci. Rep.* **2023**, *13*, 2458. [CrossRef]
75. Olivares-Urbano, M.A.; Griñán-Lisón, C.; Marchal, J.A.; Núñez, M.I. CSC Radioresistance: A Therapeutic Challenge to Improve Radiotherapy Effectiveness in Cancer. *Cells* **2020**, *9*, 1651. [CrossRef]
76. Phi, L.T.H.; Sari, I.N.; Yang, Y.G.; Lee, S.H.; Jun, N.; Kim, K.S.; Lee, Y.K.; Kwon, H.Y. Cancer Stem Cells (CSCs) in Drug Resistance and Their Therapeutic Implications in Cancer Treatment. *Stem Cells Int.* **2018**, *2018*, 5416923. [CrossRef] [PubMed]
77. Gupta, G.; Merhej, G.; Saravanan, S.; Chen, H. Cancer Resistance to Immunotherapy: What Is the Role of Cancer Stem Cells? *Cancer Drug Resist.* **2022**, *5*, 981–994. [CrossRef] [PubMed]
78. Dai, J.; Su, Y.; Zhong, S.; Cong, L.; Liu, B.; Yang, J.; Tao, Y.; He, Z.; Chen, C.; Jiang, Y. Exosomes: Key Players in Cancer and Potential Therapeutic Strategy. *Signal Transduct. Target. Ther.* **2020**, *5*, 145. [CrossRef]
79. Fadavi, P.; Nafissi, N.; Mahdavi, S.R.; Jafarnejadi, B.; Javadinia, S.A. Outcome of Hypofractionated Breast Irradiation and Intraoperative Electron Boost in Early Breast Cancer: A Randomized Non-Inferiority Clinical Trial. *Cancer Rep.* **2021**, *4*, 1376. [CrossRef]
80. Ameri, A.; Norouzi, S.; Sourati, A.; Azghandi, S.; Novin, K.; Taghizadeh-Hesary, F. Randomized Trial on Acute Toxicities of Weekly vs Three-weekly CISPLATIN-BASED Chemoradiation in Head and Neck Cancer. *Cancer Rep.* **2022**, *5*, e1425. [CrossRef]
81. Borrego-Soto, G.; Ortiz-López, R.; Rojas-Martínez, A. Ionizing Radiation-Induced DNA Injury and Damage Detection in Patients with Breast Cancer. *Genet. Mol. Biol.* **2015**, *38*, 420–432. [CrossRef]
82. Syljuåsen, R.G. Cell Cycle Effects in Radiation Oncology. In *Radiation Oncology*; Wenz, F., Ed.; Springer International Publishing: Cham, Switzerland, 2019; pp. 1–8.

83. Suwa, T.; Kobayashi, M.; Nam, J.-M.; Harada, H. Tumor microenvironment and radioresistance. *Exp. Mol. Med.* **2021**, *53*, 1029–1035. [CrossRef] [PubMed]
84. Liu, C.; Lin, Q.; Yun, Z. Cellular and molecular mechanisms underlying oxygen-dependent radiosensitivity. *Radiat. Res.* **2015**, *183*, 487–496. [CrossRef]
85. You, L.; Wu, W.; Wang, X.; Fang, L.; Adam, V.; Nepovimova, E.; Wu, Q.; Kuca, K. The Role of Hypoxia-Inducible Factor 1 in Tumor Immune Evasion. *Med. Res. Rev.* **2021**, *41*, 1622–1643. [CrossRef] [PubMed]
86. Vaupel, P.; Multhoff, G. Hypoxia-/HIF-1α-Driven Factors of the Tumor Microenvironment Impeding Antitumor Immune Responses and Promoting Malignant Progression. *Adv. Exp. Med. Biol.* **2018**, *1072*, 171–175. [CrossRef]
87. Shurin, M.R.; Umansky, V. Cross-Talk between HIF and PD-1/PD-L1 Pathways in Carcinogenesis and Therapy. *J. Clin. Investig.* **2022**, *132*, e159473. [CrossRef] [PubMed]
88. Sethumadhavan, S.; Silva, M.; Philbrook, P.; Nguyen, T.; Hatfield, S.M.; Ohta, A.; Sitkovsky, M.V. Hypoxia and Hypoxia-Inducible Factor (HIF) Downregulate Antigen-Presenting MHC Class I Molecules Limiting Tumor Cell Recognition by T Cells. *PLoS ONE* **2017**, *12*, 0187314. [CrossRef]
89. Chiu, D.K.; Tse, A.P.; Xu, I.M.; Cui, J.; Lai, R.K.; Li, L.L.; Koh, H.Y.; Tsang, F.H.; Wei, L.L.; Wong, C.M. Hypoxia Inducible Factor HIF-1 Promotes Myeloid-Derived Suppressor Cells Accumulation through ENTPD2/CD39L1 in Hepatocellular Carcinoma. *Nat. Commun.* **2017**, *8*, 517. [CrossRef] [PubMed]
90. Groth, C.; Hu, X.; Weber, R.; Fleming, V.; Altevogt, P.; Utikal, J.; Umansky, V. Immunosuppression Mediated by Myeloid-Derived Suppressor Cells (MDSCs) during Tumour Progression. *Br. J. Cancer* **2019**, *120*, 16–25. [CrossRef] [PubMed]
91. Behnam, B.; Taghizadeh-Hesary, F. Mitochondrial Metabolism: A New Dimension of Personalized Oncology. *Cancers* **2023**, *15*, 4058. [CrossRef] [PubMed]

cancers

MDPI

Article

FBXO11 Mediates Ubiquitination of ZEB1 and Modulates Epithelial-to-Mesenchymal Transition in Lung Cancer Cells

Xinyue Zhao [1], Zhihui Han [1], Ruiying Liu [1], Zehao Li [1], Ling Mei [1] and Yue Jin [1,2,3,*]

1 Edmond H. Fischer Signal Transduction Laboratory, School of Life Sciences, Jilin University, Changchun 130012, China; zhaoxinyue20@mails.jlu.edu.cn (X.Z.); hanzhihui962024@163.com (Z.H.); liury22@mails.jlu.edu.cn (R.L.); zehao21@mails.jlu.edu.cn (Z.L.); meiling23@mails.jlu.edu.cn (L.M.)

2 National Engineering Laboratory of AIDS Vaccine, School of Life Sciences, Jilin University, Changchun 130012, China

3 Key Laboratory of Organ Regeneration & Transplantation of the Ministry of Education, Jilin University, Changchun 130061, China

* Correspondence: yuejin@jlu.edu.cn; Tel.: +86-18004458591

Simple Summary: Epithelial-to-mesenchymal transition (EMT) plays a critical role in cancer progression, contributing to the invasive and migratory abilities of tumor cells. In this study, we show that FBXO11 promotes the degradation of ZEB1, a key EMT regulator, via ubiquitination. Loss of FBXO11 increases ZEB1 levels, enhancing the invasiveness of lung cancer cells, while its overexpression reduces ZEB1 and suppresses invasion. Importantly, higher FBXO11 expression is associated with better prognosis in non-small cell lung cancer (NSCLC), highlighting its potential role as a therapeutic target for controlling EMT and cancer metastasis.

Abstract: Epithelial-to-mesenchymal transition (EMT) affects the invasion and migration of cancer cells. Here, we show that FBXO11 recognizes and promotes ubiquitin-mediated degradation of ZEB1. There is a strong association between FBXO11 and ZEB1 in non-small cell lung cancer (NSLC) in a clinical database. FBXO11 interacts with ZEB1, a core inducer of EMT. FBXO11 leads to increased ubiquitination and proteasomal degradation of ZEB1. Depletion of endogenous FBXO11 causes ZEB1 protein accumulation and EMT in A549 and H1299 cells, while overexpression of FBXO11 reduces ZEB1 protein abundance and cellular invasiveness. Importantly, the depletion of ZEB1 suppresses the increased migration and invasion of A549 and H1299 cells promoted by the depletion of FBXO11. The same results are shown in xenograft tumors. High FBXO11 expression is associated with a favorable prognosis in NSLC. Collectively, our study demonstrates that FBXO11 modulates EMT by mediating the stability of ZEB1 in lung cancer cells.

Keywords: ZEB1; FBXO11; lung cancer; EMT; invasion and metastasis

1. Introduction

According to GLOBOCAN's Global Cancer Statistics 2023, lung cancer is the number-one cause of death [1,2]. There are two main histological types of lung cancer, namely small cell lung cancer (SCLC) and non-small cell lung cancer (NSCLC). Of these, NSCLC accounts for about 85% [3–6]. Metastasis is the primary cause of death in patients with lung cancer [7]. Tumor cell invasion and metastasis are closely linked to epithelial-to-mesenchymal transition (EMT). EMT transforms polarized epithelial cells, typically anchored to the basement membrane, into mesenchymal cells with increased migration, invasiveness, and resistance to apoptosis [8]. ZEB1, also known as TCF8 or δEF1, is a key transcriptional regulator of EMT [9,10]. ZEB1 transcriptionally represses the expression of E-cadherin by binding to its promoter region [11]. Research shows that ZEB1 can induce EMT in breast cancer [10], osteosarcoma [12], lung cancer [13], melanoma [14], and other epithelioma, leading to tumor metastasis and promoting drug resistance. High expression of ZEB1 is related to the poor prognosis of cancer patients. Studies by Manshouri R et al. have shown that ZEB1 can become a therapeutic target for metastatic NSCLC. Various F-box proteins, including TrCP1/Fbxw1, Fbxw7, Ppa/Fbxl14, Fbxl5, Fbxo11, and Fbxo45, have been implicated in EMT by facilitating the degradation of EMT-related transcription factors (EMT-TFs) [15–20]. FBXO11, as a member of the SCF (Skp1-Cul1-F-box) ubiquitin ligase complex, exhibits E3 ubiquitin ligase activity and methyltransferase activity [21]. FBXO11 was first mentioned as a ubiquitin ligase in diffuse large B-cell lymphomas (DLBCLs), targeting BCL6 for its degradation and stabilization and, thus, inhibiting cell proliferation and inducing cell death [22]. FBXO11 can promote CDT2 polyubiquitylation, and degradation controls the timing of cell-cycle exit [23,24]. FBXO11 can regulate the invasive metastasis of tumor cells by associating with the EMT-related factor Snail, as well as the classical PI3k/Akt signaling pathway [25,26]. Recently, FBXO11 has been reported to be a major negative regulator of MHC class II through ubiquitin-dependent proteasomal degradation of CIITA in breast cancer [27]. In hepatocellular carcinoma, FBXO11 mediates heterogeneous ribonucleoprotein A2/B1 ubiquitination to regulate lipid metabolic reprogramming and promote tumorigenesis [28]. The modulation of cisplatin resistance by the miR-324-5pFBXO11 axis is observed in lung cancer [29]. In this study, we found that FBXO11 interacts with the ZEB1 CZF domain and promotes it ubiquitination and degradation. FBXO11 activity affects the stability and function of the ZEB1 protein, resulting in marked biological effects in vivo. This molecular interplay shapes the proliferative and metastatic competencies of lung cancer cells, thereby transforming their dissemination behavior within the biological milieu.

2. Materials and Methods

2.1. Cell Culture and Transfection

Human embryonic kidney cell HEK293T and human lung adenocarcinoma cell line A549 cells were cultured in DMEM (Gibco, Waltham, MA, USA, C11995500B) with 10% FBS (Kangyuan Biotechnology, Shanghai, China, KY-01000). Human lung adenocarcinoma cell line H1299 cells were grown in RPMI-1640 (Gibco) with 10% FBS. Transfections were performed using Lipofectamine 2000 (Invitrogen, Carlsbad, CA, USA) per the manufacturer's instructions.

2.2. Plasmids and Mutagenesis

ZEB1 and FBXO11 cDNA were synthesized and cloned into lentiviral vector pLVX-IRES-neo and pET28a by Miaoling Biology (Wuhan, China). FBXO11 was cloned into a pLVX-IRES-neo vector with Myc tagged. ZEB1 was cloned into a pLVX-IRES-neo vector with Flag tagged. FBXO11 and ZEB1 full-length and truncated mutants of ZEB1 were cloned into a pET28a vector.

2.3. Western Blotting and Immunoprecipitation

Western blotting was conducted as previously described [30]. Cell extracts were prepared with cold RIPA buffer (150 mM NaCl, 50 mM Tris pH 7.5, 1% NP-40, and 10% glycerol). Proteins were separated on 6–12% gels and transferred to PVDF membranes. Membranes (Millipore, Shanghai, China, IPVH00010) were probed with primary antibodies, including FBXO11 (Proteintech, Wuhan, China, #67365-1-Ig), ZEB1 (Santa Cruz Biotechnology, Oregon, USA, #515797), E-cadherin (Proteintech, #20874-1-AP), N-cadherin (BD Transduction Laboratories, #610920), GFP (Proteintech, #66002-1-Ig), and GAPDH (GAPDH; Bioss, Woburn, MA, USA, #0978M).

For immunoprecipitation assays, indicated plasmids were co-transfected with HEK293T cells in six-well plates. After 18 h, transfected cells were treated with 10 μM proteasome inhibitor MG132 (Sigma-Aldrich, St. Louis, MO, USA, C2211) for 6 h and lysed with cold RIPA lysate buffer to collect the supernatant and retain the input. The supernatants were combined with RIPA binding buffer (150 mM NaCl, 50 mM Tris PH

7.5, and 10% glycerol) and 20 μL protein A/G agarose beads (Invitrogen, 20421) and incubated at 4 °C for 6 h. The mixture was subjected to immunoprecipitation with either Flag (Proteintech, #66008-4-Ig) or Myc (Proteintech, #60003-2-Ig) antibodies and slowly overturned overnight at 4 °C. The agarose beads–antigen–antibody mixture was washed thoroughly with RIPA buffer, and Western blot analysis was conducted with indicated antibodies.

2.4. His-Tagged Protein Interaction Pull-Down Assay

FBXO11 wild-type or ZEB1 truncated cDNA was amplified by PCR and cloned into bacterial expression vector pET28a with an N-terminal His tag. The plasmids were transformed into BL21(DE3) (TransGen, Beijing, China, CD601-02). His-tagged FBXO11 protein or ZEB1 truncated proteins were induced by 0.4 mM IPTG at 21 °C for 16 h and purified by Ni-NTA beads (Qiagen, Hilden, Germany, 1018244). The His-tagged FBXO11 proteins or ZEB1-truncated proteins that bound to Ni-NTA beads were washed with PBS and collected at 4 °C. HEK293T cells transfected with ZEB1-Flag or FBXO11-Myc expression constructs were harvested in lysis buffer (50 mM $NaH_2PO_4.H_2O$ pH 8, 300 mM NaCl, 10 mM imidazole, protease inhibitor cocktail) and sonicated (power, 300 W; time, 10 s; interval, 50 s). The sonicated cell lysates were incubated with purified sFBXO11 or ZEB1-truncated proteins and Ni-NTA beads. Proteins that bound to Ni-NTA beads were eluted (50 mM $NaH_2PO_4.H_2O$, pH 8, 300 mM NaCl, 20 mM imidazole, protease inhibitor cocktail) and subjected to Western blot analysis with indicated antibodies.

2.5. Immunofluorescence

An immunofluorescence assay was performed as previously described [31]. Fluorescence photography was captured with a laser scanning confocal microscope (Leica Microsystems, Wetzlar, Germany, LSM 710). Primary antibodies included FBXO11 (Novussbio, Centennial, CO, USA 100-59826), ZEB1 (Santa Cruz, Dallas, TX, USA, #515797), and E-cadherin (Proteintech, 20874-1-AP). Secondary antibodies included CoraLite488-conjugated (Proteintech, SA00013-2), CoraLite594-conjugated (Proteintech, SA00013-3), and DAPI (Beyotime, Haimen, China) antibodies.

2.6. Lentiviral shRNA Depletion and qRT-PCR

To overexpress or deplete FBXO11 and ZEB1, cells were infected with lentiviral vectors or shRNAs. After puromycin selection, RNA was extracted using Trizol (Thermo Fisher Scientific, Waltham, MA, USA, 15596026). cDNA was synthesized using Takara RR047A-3. cDNA (Takara RR047A-3) was synthesized, and gene expression levels were measured by real-time PCR and normalized to β-actin. Primer and shRNA sequences are reported in Supplementary Tables S1 and S2.

2.7. In Vitro Ubiquitination Assay

Wild-type Ub, Ub-K48, and Ub-K63 plasmids with HA tags were co-transfected with the relevant plasmids. After 18 h, transfected cells were combined with 10 μM proteasome inhibitor MG132 for 6 h. The experimental manipulations were continued according to immunoprecipitation and Western blot analysis with indicated with Flag and HA (Proteintech, #81290-1-RR) antibodies.

2.8. Protein Degradation Assays

HEK293T cells were seeded in 6-well plates for 24 h and transfected with 0.5 μg of indicated expression plasmids. When indicated, 0.2 μg GFP was used as an internal transfection control. After 16 h, transfected cells were treated with 10 μM MG132 or 20 μM Chloroquine (MedChemExpress, Monmouth Junction, NJ, USA, HY-17589) for 8 h and collected with ice-cold whole-cell extraction buffer (25mM β-Glycerophospholipids PH 7.3, 2 mM EGTA, 10 mM EDTA, 10 mM β-mercaptoethanol, 0.1 M NaCl, 1% Triton X-100, protease inhibitor cocktail) for Western blot analysis.

2.9. ZEB1 Half-Life Assay

HEK293T cells were seeded in 6-well plates for 24 h and transfected with 0.5 μg of indicated expression plasmids and 0.2 μg GFP as an internal transfection control. After 24 h, transfected cells were treated with 50 μg/mL cycloheximide (Aladdin, Shanghai, China, C112766) for 0, 2, 4, 6, 8, or 10 h and collected with ice-cold whole-cell extraction buffer for Western blot analysis.

2.10. Transwell Invasion Assay

Transwell chamber membranes (24-well; 8 μM, Costar, Costar, Corning, NY, USA, #3422) were coated with fibronectin (Sigma-Aldrich). Cells (5×10^4) were plated in the top chamber and incubated at 37 °C overnight. Invaded cells were fixed and stained with 1% crystal violet (Solarbio, Beijing, China). Cell numbers were then counted under a microscope.

2.11. Wound-Healing Migration Assay

Cells were seeded in 6-well plates and grown to confluence. A wound was created by scraping with a 200 μL tip. Cells were washed with PBS and cultured in serum-free medium. Wound areas were photographed at 0 and 48 h, and the healing sizes were measured.

2.12. Mouse Subcutaneous Tumor Formation Assay

This study followed the Guide for the Care and Use of Laboratory Animals (Eighth Edition) and was approved by the Animal Ethics Committee of Changchun Wish Technology Company. After stable expression and screening of the corresponding plasmid cDNA in A549 cells infected with lentivirus, the cells were diluted to 5×10^6. Cells were injected subcutaneously into the hindlimbs of 4–6-week-old nude mice (Balb/c-nu/nu). Tumor growth was observed and recorded. After growth to three weeks, the mice were sacrificed and dissected. Tumors were fixed with 4% PFA and paraffin-embedded, and 5 μM thick tumor sections were made. H&E staining was performed, and the sections were observed and photographed under a microscope.

2.13. Molecular Docking

We uploaded the FBXO11 protein sequence and the ZEB1 CZF domain (725-1125 amino acids) truncated protein sequence to LiHDOCK SERVER software (http://huanglab.phys.hust.edu.cn/, accessed on 13 October 2023) for protein–protein docking. Then, the docking combination with the best docking effect (model No. 1) was selected based on the docking score and confidence score and analyzed using PyMOL.

2.14. Kaplan–Meier Plot

We selected patient cohorts from the KM-PLOTTER website (http://kmplot.com/). We selected histological data from 1161 patients with lung adenocarcinoma.

2.15. Statistical Analysis

Patient sample data were downloaded from the TCGA database and grouped according to high and low FBXO11 expression. To investigate FBXO11 expression in the LUAD, we applied independent-sample *t*-tests to unpaired samples and paired *t*-tests to paired samples. Categorical data, including sex and tumor differentiation, were analyzed using chi-square tests. All analyses were two-sided and conducted with R 3.2.0 and SPSS 16.0.2. Statistical significance was set at $p < 0.05$.

3. Results

3.1. ZEB1 as the Major Transcription Factor Induces EMT in Lung Cancer Cells

To clarify the major types of transcription factors that induce EMT in lung cancer cells, we used nickel chloride ($NiCl_2$) to induce EMT in human non-small cell lung cancer cell lines A549 and H1299. $NiCl_2$ induces epithelial–mesenchymal transition (EMT) and enhances cellular invasiveness by generating reactive oxygen species (ROS) and altering DNA methylation, inhibiting the expression of E-cadherin and promoting the upregulation of N-cadherin and vimentin [32]. This mechanism has been validated in a variety of cell models, including lung cancer and renal tubular epithelial cells [33,34]. The expression of E-cadherin gradually decreased with increasing $NiCl_2$ concentration, while ZEB1 expression was upregulated. It is noteworthy that the protein expression of Snail and Slug did not change (Figure 1A). The cell morphology shifted towards mesenchymal cell morphology with 2 mM $NiCl_2$, while the knockdown of ZEB1 was able to resist the induction effect of $NiCl_2$ (Figure 1B). Our previous study demonstrated that FBXO11 ubiquitinates and degrades Snail through the proteasome, thereby blocking Snail-induced EMT and inhibiting tumor metastasis in breast cancer [25,26]. Here, we speculated that FBXO11 might have a similar regulatory effect on ZEB1 in lung cancer cells. Using GEPIA, we performed gene correlation analysis using Transcripts Per Million (TPM) for lung cancer data in the TCGA/GTEx database and found that the ZEB1 gene was positively correlated with the FBXO11 gene (Figure 1C). To explore the role of FBXO11 in the development, progression, and prognosis of cancer, we investigated its expression in normal tissues and tumors in the TCGA and GETx databases (http://gepia.cancer-pku.cn) and found that FBXO11 was highly expressed in a variety of normal tissue compared to tumors, including BRCA, COAD, LUAD, and READ (Figure 1D).

We screened lung adenocarcinoma samples against FBXO11 expression in the TCGA clinical database. FBXO11 expression was not significantly correlated with the ages of the patients, sex, radiation therapy, or smoking history. It is noteworthy that the impact of FBXO11 expression on tumor grade is more likely to be seen in the early stages (I–II) of the tumor, when cancer cells are about to develop and have spread to a small extent (Table 1).

Table 1. Correlation of FBXO11 expression with clinicopathological covariates in lung cancer patients.

Characteristic	FBXO11—High (n = 286)	FBXO11—Low (n = 286)	p Value *
Age (years), mean (SD)	66 (60.73)	66 (57.72)	0.407 †
Sex (n(%))			0.737
Women	153 (53.5)	157 (54.9)	
Men	133 (46.5)	129 (45.1)	
Radiation therapy			0.385
No	224 (78.4)	195 (68.2)	
YES	33 (11.5)	36 (12.6)	
Missing	29 (10.1)	55 (19.2)	
Smoking history			0.718
≤2	100 (35)	108 (37.8)	
>2	177 (61.9)	167 (58.4)	
Missing	9 (3.1)	11 (3.8)	
Tumor stage			0.146 ‡
i	169 (59.1)	141 (49.4)	
ii	56 (19.6)	77 (26.9)	
iii	47 (16.4)	46 (16.1)	
iv	12 (4.2)	15 (5.2)	
Missing	2 (0.7)	7 (2.4)	

* $\chi 2$ test or Fisher's exact test. † Student's t test. ‡ Mann–Whitney U test (non-parametric). Missing values are excluded for all statistical tests.

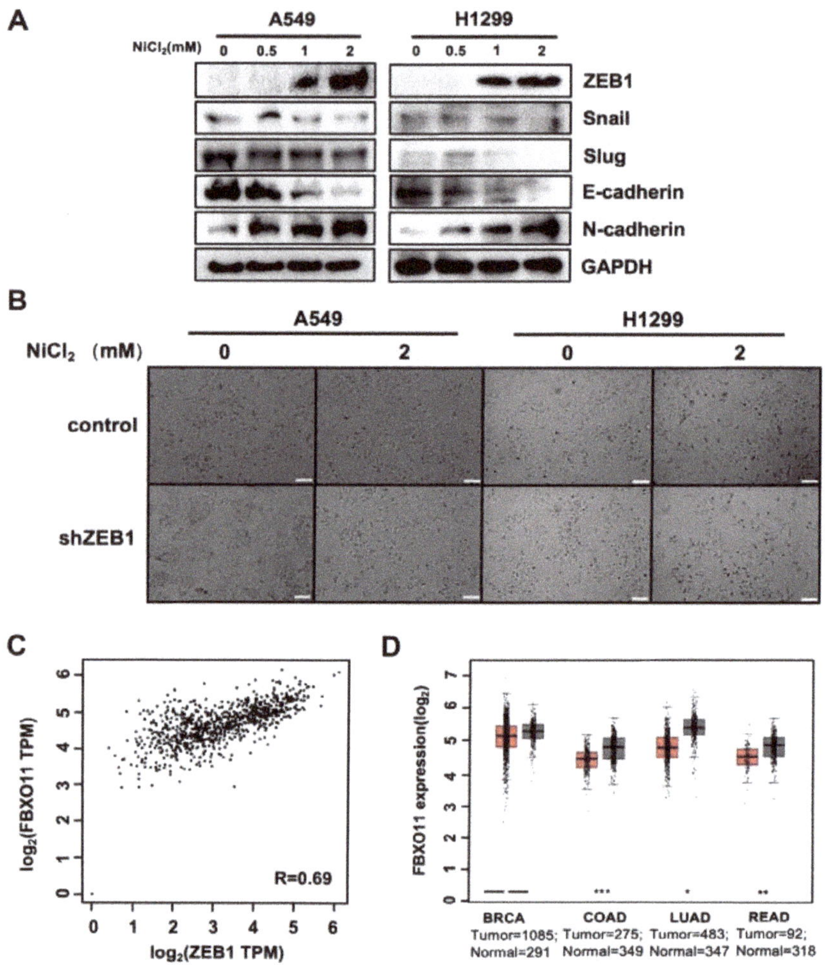

Figure 1. EMT is mainly induced in lung cancer cells by transcription factor ZEB1. (**A**) ZEB1 expression is increased in $NiCl_2$-induced EMT. A549 and H1299 cells were treated with different concentrations of $NiCl_2$ for 2 d. Immunoblotting assays were performed to detect the expression of each EMT marker. (**B**) Knockdown of ZEB1 in cells resistant to the induction effect of $NiCl_2$. A549 and H1299 shNC or shZEB1 cells were treated with 2 mM $NiCl_2$, and changes in cell morphology were observed under a microscope. Scale bar: 100 μm. (**C**) FBXO11 is positively correlated with ZEB1. Correlation analysis of LUAD data for TCGA and GTEx on the GEPIA website showed a correlation coefficient of R = 0.69 ($p = 0$; $p < 0.01$). TPM: transcripts per million. (**D**) FBXO11 exhibited higher expression in the normal tissue samples. The GEPIA database was used for FBXO11 expression in normal tissues and tumors (* $p < 0.05$, ** $p < 0.01$, *** $p < 0.001$).

3.2. FBXO11 Associates with ZEB1

To investigate the interaction between FBXO11 and ZEB1, we first transfected the Myc-tagged FBXO11 and Flag-tagged ZEB1 expression constructs into HEK293T cells. Immunoprecipitation of Myc-tagged FBXO11 pulled down Flag-tagged ZEB1. In a reciprocal assay, immunoprecipitation of Flag-tagged ZEB1 also pulled down Myc-tagged FBXO11 (Figure 2A). An immunofluorescence assay further confirmed colocalization of FBXO11 and ZEB1 in HEK293T cells within the nucleus (Figure 2B). Next, we validated the protein interactions and mapped interacting domains by in vitro His pulldown assay.

FBXO11 protein consists of an N-terminal F-box motif, a C-terminal zinc-finger-like UBR domain, and several repetitive cysteine-enriched domains located in the center [15]. The His pulldown assays showed the interaction between FBXO11 and ZEB1 (Figure 2C). ZEB1 consists of a central homeodomain (HD) and zinc-finger clusters at the N terminal (NZF) and C terminal (CZF) [10]. ZEB1 and its truncated mutants were expressed as His fusion protein. The ZEB1 C zinc-finger domain is indispensable for binding to FBXO11 by in vitro his pulldown assay (Figure 2D). After determining the binding domain, we simulated the interaction pattern of FBXO11 and ZEB1 C zinc finger using Z dock. The docking results show that the Z-dock docking fraction of FBXO11 and ZEB1 is 1742.563. Residues near the protein–protein interaction interface can form hydrogen bonds, which help stabilize the complex. PyMOL 2.3.0 was used to analyze the docking interaction patterns (Figure 2E). Overall, these results indicate that FBXO11 interacts with ZEB1 in vitro.

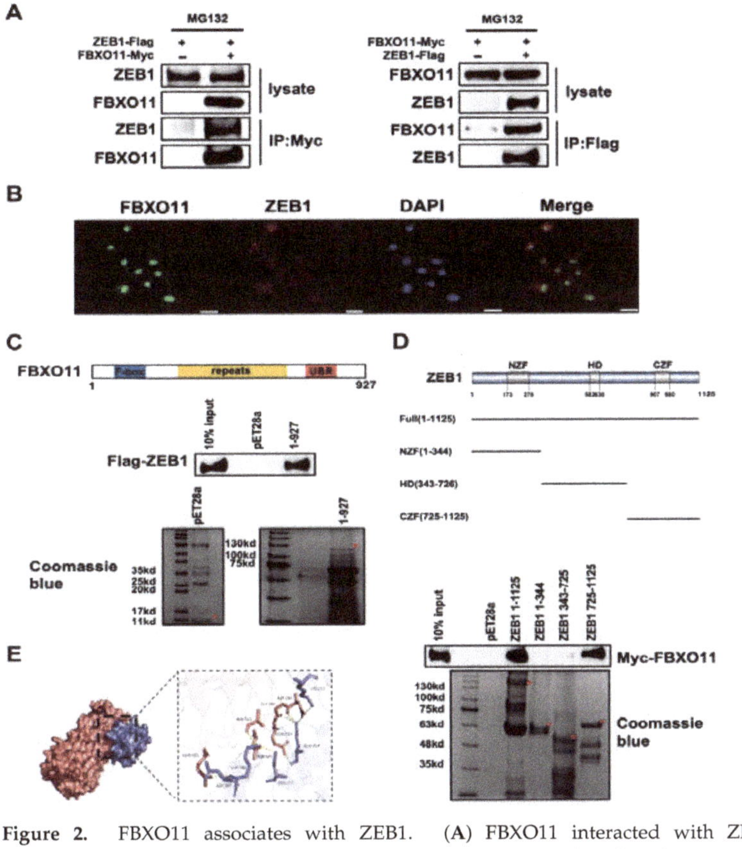

Figure 2. FBXO11 associates with ZEB1. (**A**) FBXO11 interacted with ZEB1 in a co-immunoprecipitation (co-IP) assay. HEK293T cells were transfected with Myc-tagged FBXO11 and Flag-tagged ZEB1 as indicated. Cell lysates were immunoprecipitated with either anti-Myc or anti-Flag antibodies and immunoblotted with anti-ZEB1 and anti-FBXO11 antibodies. (**B**) FBXO11 and ZEB1 co-location in the nucleus. Immunofluorescence assay probe colocalization of FBXO11 (green) and ZEB1 (red). Scale bar: 50 μm. (**C**) His pulldown assays showing FBXO11 bound to ZEB1 protein. A Coomassie blue staining image of PAGE gel is shown below to confirm the expression of pET28a and FBXO11. (**D**) His pulldown assays showing the CZF domain of ZEB1 bound to FBXO11 protein. A Coomassie blue staining image of PAGE gel is shown below to confirm the expression of various forms of ZEB1 proteins. (**E**) Molecular docking of FBXO11 protein and the ZEB1 CZF domain (725-1125 amino acids) truncated protein.

3.3. FBXO11 Stabilizes ZEB1 though Ubiquitination Acticity

Given that FBXO11 is an E3 ubiquitin ligase, we investigated its role in ZEB1 ubiquitination and degradation. Co-transfection with FBXO11 led to ZEB1 degradation, which was blocked by proteasome inhibitor MG132 but not by chloroquine, demonstrating that overexpression of FBXO11 caused the proteasomal degradation of ZEB1 (Figure 3A). FBXO11 overexpression significantly increased ZEB1 ubiquitination, particularly K48-linked ubiquitination (Figure 3B). As FBXO11 levels increased, ZEB1 protein levels decreased, indicating concentration-dependent degradation by FBXO11 (Figure 3C). FBXO11 also accelerated ZEB1 turnover but not that of GFP (Figure 3D). Taken together, the results suggest that FBXO11 promotes polyubiquitination and the degradation of ZEB1 protein.

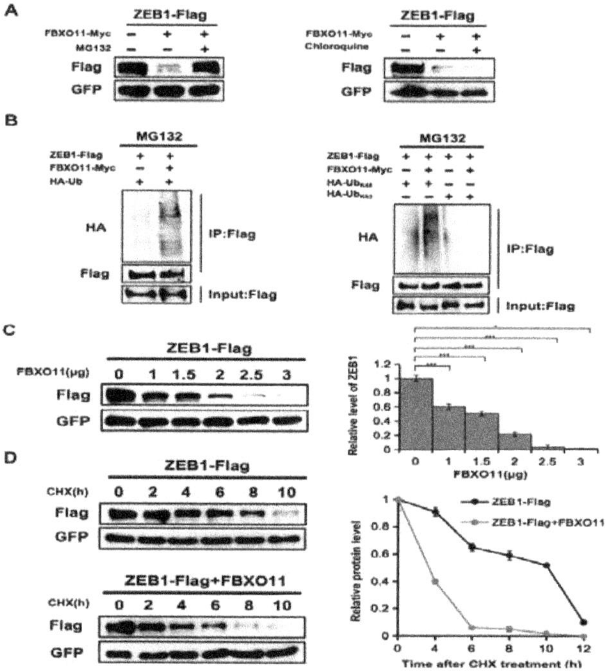

Figure 3. FBXO11 stabilizes ZEB1 though ubiquitination activity. (**A**) FBXO11 promotes ZEB1 proteasomal degradation. ZEB1-Flag was cotransfected with FBXO11-Myc in HEK293T cells, together with GFP, as well as an empty vector, and treated with DMSO, MG132, or chloroquine as indicated. The expressions of ZEB1 and GFP were assessed by Western blot. (**B**) FBXO11 promotes K48 polyubiquitinated chain generation of ZEB1 protein. In cellular ubiquitination assays, FBXO11-Myc was co-transfected with ZEB1-Flag plasmids or with HA-Ub-K63 and HA-Ub-K48 plasmids. Western blot was performed, and cell lysates were immunoprecipitated with anti-Flag antibody, then detected by anti-Ub antibody to poly-Ub levels. (**C**) FBXO11 degrades ZEB1 protein in a concentration-dependent manner. HEK293T cells were transfected with ZEB1-Flag, GFP, or FBXO11-Myc for 48 h, and cell lysates were immunoblotted and screened with anti-Flag antibody. Quantities were calculated as fold changes of 0 μg FBXO11. Error bars represent the SD from 3 independent experiments. * indicates $p < 0.05$, and *** indicates $p < 0.001$. Each experiment was repeated at least three times. (**D**) Exogenous FBXO11 accelerates ZEB1 protein turnover. HEK293T cells were transfected with ZEB1-Flag and GFP, in combination with FBXO11-Myc, and treated with cycloheximide (CHX) as indicated. Cell lysates were subjected to western blot analysis with anti-ZEB1 and anti-GFP antibodies. The graph shows the quantification of ZEB1 protein levels (based on the band intensity from the gels) normalized to GFP over the time course. The ZEB1 protein level at the 0 h time point of CHX treatment was set as 100%. The experiment was repeated three times, and a representative experiment is presented.

3.4. FBXO11 Regulates the Expression of EMT-Related Factors

To explore the role of FBXO11 in the modulation of the migration and invasion of lung cancer cells, we examined the expression of genes that regulate EMT in A549 and H1299 cells. The depletion of FBXO11 caused marked upregulation of ZEB1 protein levels and downregulation of epithelial marker E-cadherin. Mesenchymal marker N-cadherin was increased (Figure 4A). The conversion of E-cadherin and N-cadherin expression is one of the hallmarks of EMT. Similar results were observed in H1299 lung cancer cells (Figure 4A). Immunostaining also showed that FBXO11 depletion increased the expression of ZEB1 and decreased E-cadherin (Figure 4B). When A549 and H1299 cells were transduced with lentivirus-expressing FBXO11, endogenous ZEB1 and N-cadherin protein and mRNA levels decreased, while E-cadherin protein and mRNA expression increased (Figure 4D). The upregulation of E-cadherin is indicative of EMT and is expected to decrease cell motility and invasion. As expected, the overexpression of FBXO11 in A549 and H1299 cells strengthened intercellular adhesion, resulting in a more clustered morphology (Figure 4C). Concomitantly, in a wound-healing migration assay (Figure 4E) and transwell invasion assay (Figure 4F), the migratory and invasive abilities of the cells were weakened. These results suggest that FBXO11 can regulate the expression of EMT-related factors and affect the migration and invasion in lung cancer cells.

3.5. The FBXO11–ZEB1 Axis Regulates the EMT Pathway in LUAD

Since ZEB1 is a core regulator of EMT, we investigated whether FBXO11 depletion enhanced EMT through the increased expression of ZEB1. Compared with FBXO11 knockdown alone, FBXO11 and ZEB1 double knockdown increased E-cadherin and decreased N-cadherin protein expression, which is similar to the result of ZEB1 knockdown alone (Figure 5A). The depletion of FBXO11 in A549 and H1299 cells gave rise to a predominance of dispersed and elongated single cells (Figure 5B). Similar to ZEB1 knockdown alone, FBXO11 and ZEB1 double knockdown cells showed tight connections and regular arrangement (Figure 5B). Consistent with protein expression and cell morphology, FBXO11 and ZEB1 double knockdown caused decreased cell mobility (Figure 5C) and aggressiveness (Figure 5D). Moreover, ZEB1 knockdown alone showed the lowest cellular mobility and aggressiveness. Together, these results demonstrate that FBXO11 regulated the migration and invasion of A549 cells through ZEB1.

3.6. FBXO11 Inhibits LUAD Progression by Stabilizing ZEB1

To explore the physiological role of FBXO11 and ZEB1 in vivo, we transplanted A549 cells into nude mice (Balb/c-nu/nu) with an immunodeficient system suitable for tumor transplantation and immunology research [35] (Figure 6A). Tumors with FBXO11 deletion showed significant local invasion, while the results for tumors with FBXO11 and ZEB1 double knockdown were similar to those of tumors with ZEB1 single knockdown, with clear tumor boundaries (Figure 6B). Xenograft tumors derived from control A549 cells and FBXO11 overexpression cells both formed well-defined envelope tumors (Figure 6C). These results show that FBXO11 can affect the invasion and metastasis of lung cancer cells by regulating ZEB1 expression both in vivo and in vitro. In the lung cancer database organized by Kaplan–Meier plot, analysis demonstrated that ZEB1 expression levels were negatively correlated with patient survival, and patients with high levels of FBXO11 expression in their tumors had higher overall survival rates and survived longer than patients with low levels of FBXO11 expression (Figure 6D). This is in line with the trend of our findings. Taken together, we confirm that EMT transcription factor ZEB1 plays a major role in the invasion and metastasis of lung cancer and that FBXO11 can ubiquitinate ZEB1 and be recognized by the proteasome for degradation, thereby inhibiting the invasion and metastasis of lung cancer cells (Figure 7).

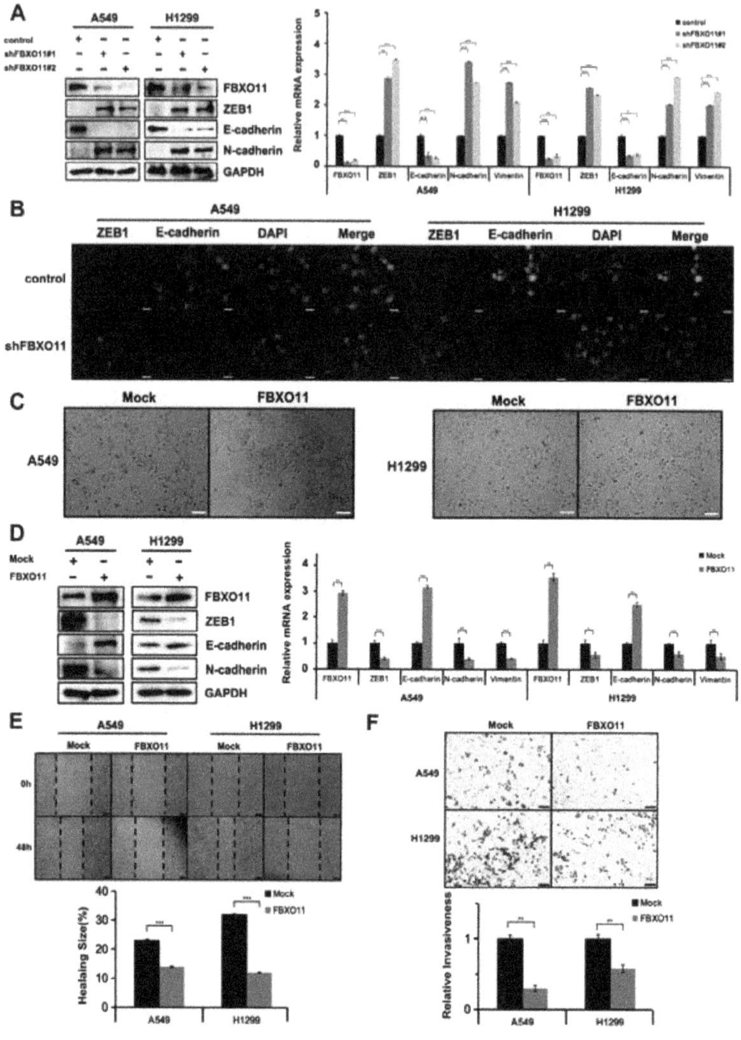

Figure 4. FBXO11 affects the expression of EMT-related factors. (**A**) Endogenous FBXO11 knockdown changes the expression of ZEB1 and EMT marker genes in lung adenocarcinoma cells. Immunoblotting analysis and quantitative RT-PCR analysis of ZEB1 and EMT markers in A549 and H1299 cells was transduced with lentiviral shRNAs targeting control or FBXO11. Error bars represent the SD from 3 independent experiments. * indicates $p < 0.05$, ** indicates $p < 0.01$, and *** indicates $p < 0.001$. (**B**) Immunofluorescence analysis of ZEB1 and E-cadherin protein expression in control and shFBXO11 of A549 and H1299 cells (ZEB1, red; E-cadherin, green; DAPI, blue). Scale bar: 50 µm. (**C**) FBXO11 overexpression affects changes cells into the mesenchymal phenotype. Cell morphology in A549 and H1299 cells with overexpressed Mock or FBXO11. Scale bars: 100 µm. (**D**) The overexpression of FBXO11 alters the expression of ZEB1 and EMT marker genes in lung adenocarcinoma cells. Immunoblotting and quantitative RT-PCR analysis of ZEB1 and EMT markers in A549 and H1299 cells transduced with Mock or FBXO11. Error bars represent the SD from 3 independent experiments. ** indicates $p < 0.01$, and *** indicates $p < 0.001$. (**E**) Wound-healing assay showing the migration of A549 and H1299 cells after transfection with Mock or FBXO11-Myc. Representative images are shown 0 and 48 h post scratch ($n = 3$). Scale bars: 100 µm. (**F**) Transwell assay showing the invasion of A549 and H1299 cells transfected with Mock or FBXO11-Myc ($n = 3$). Scale bars: 100 µm.

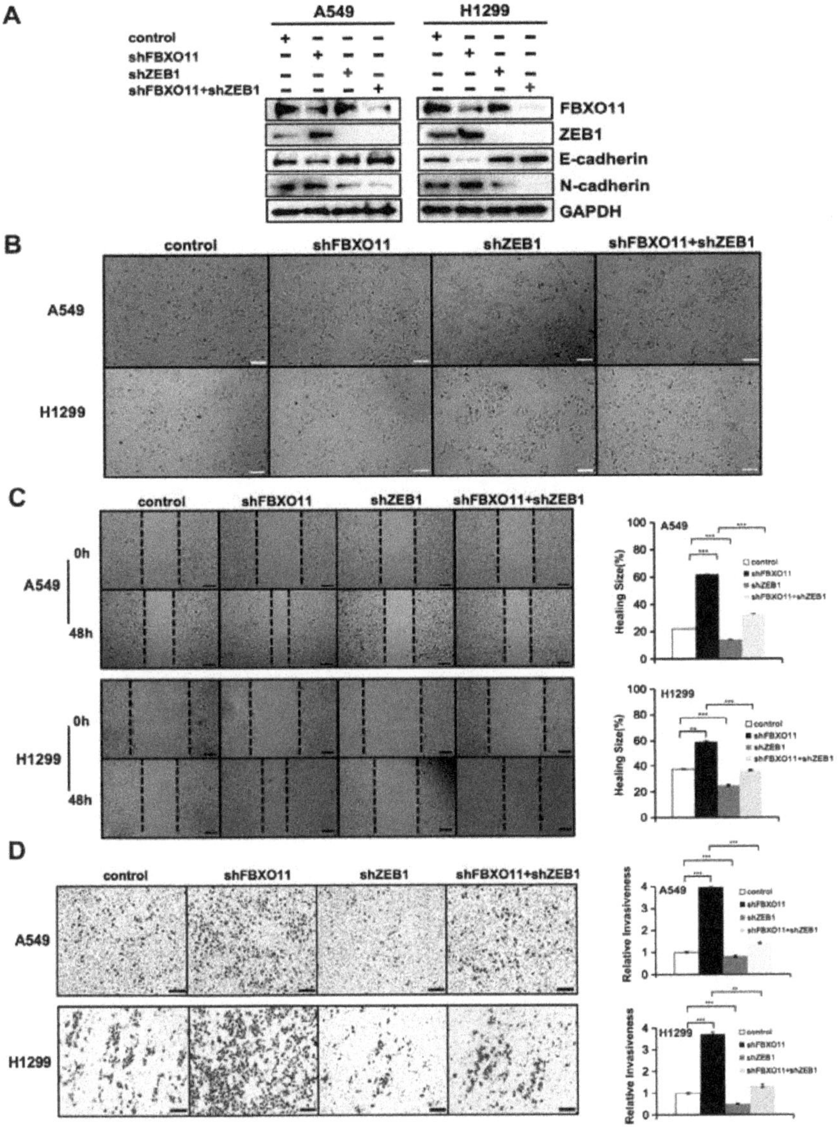

Figure 5. The FBXO11–ZEB1 axis regulates the EMT pathway in LUAD. (**A**) Detection of FBXO11 or ZEB1 in A549 and H1299 cells expressing the indicated shRNAs. (**B**) The FBXO11–ZEB1 axis regulates epithelial–mesenchymal cell morphology in lung cancer cells. Morphological changes in A549 and H1299 control, shFBXO11, and shZEB1 and co-knockdown of ZEB1 and FBXO11 cells. Scale bars: 100 μm. (**C**) FBXO11 affects the migratory ability of A549 and H1299 cells via ZEB1. Scratching experiments were performed to analyze changes in the migratory capacity of A549 and H1299 control, shFBXO11, and shZEB1 and co-knockdown of ZEB1 and FBXO11 cells. Scale bar: 100 μm. The area of cell invasion was counted (n = 6). ** indicates $p < 0.01$, *** indicates $p < 0.001$. (**D**) Cell invasiveness is moderated in vitro via the FBXO11–ZEB1 axis. A transwell assay to was conducted to analyze changes in the invasive capacity of A549 and H1299 control, shFBXO11, and shZEB1 and co-knockdown of ZEB1 and FBXO11 cells. Scale bar: 100 μm. The count of cells crossing the basement membrane was n = 6. ** indicates $p < 0.01$, and *** indicates $p < 0.001$.

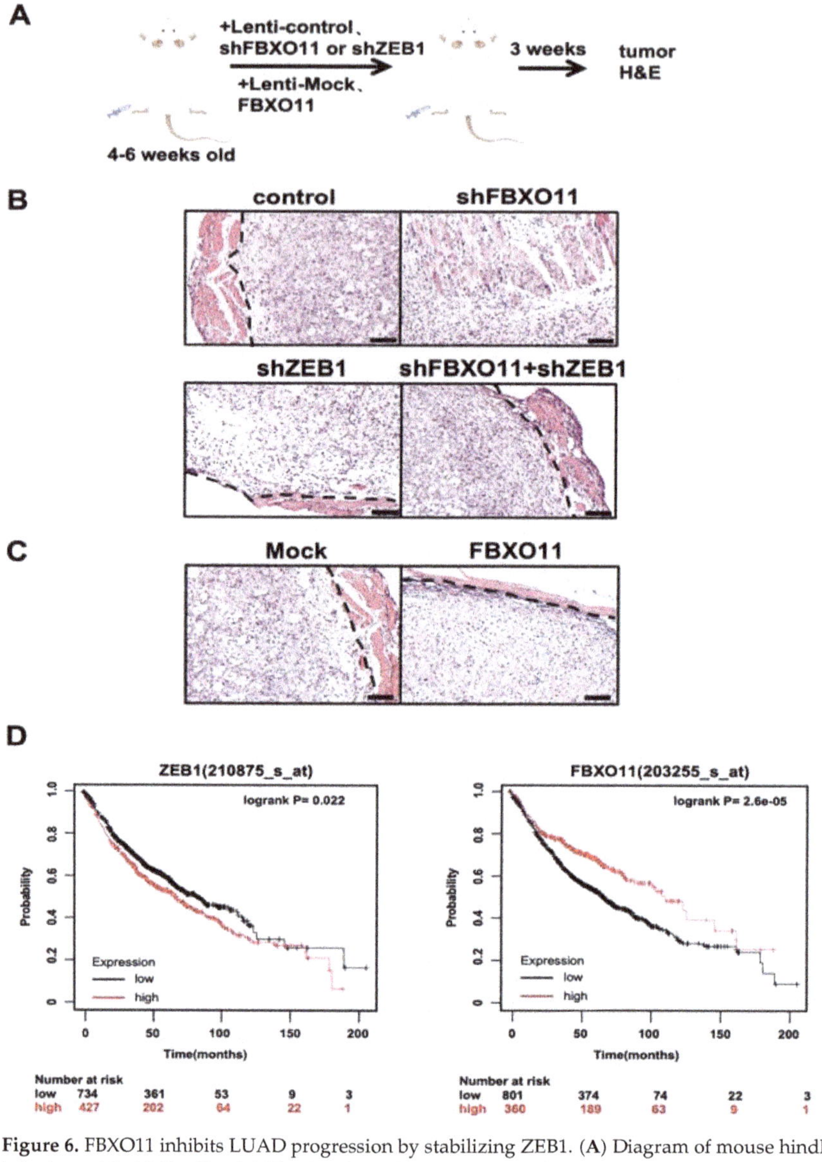

Figure 6. FBXO11 inhibits LUAD progression by stabilizing ZEB1. (**A**) Diagram of mouse hindlimb injection with A549 cells. (**B**) The FBXO11–ZEB1 axis controls tumor cell infiltration. Xenograft experiments were performed to analyze changes in the invasive ability of control, shFBXO11, shZEB1, co-knockdown, ZEB1, and FBXO11 cells in mice. Scale bars: 50 μm. (**C**) Xenograft experiments were performed to analyze changes in the invasive ability of A549, Mock, and FBXO11 cells in mice. Scale bars: 50 μm. (**D**) High ZEB1 and low FBXO11 expression predict poor prognosis. Kaplan–Meier plot of overall survival of lung cancer patients stratified by the expression of ZEB1 and FBXO11 genes. Data were obtained from KMplot.com.

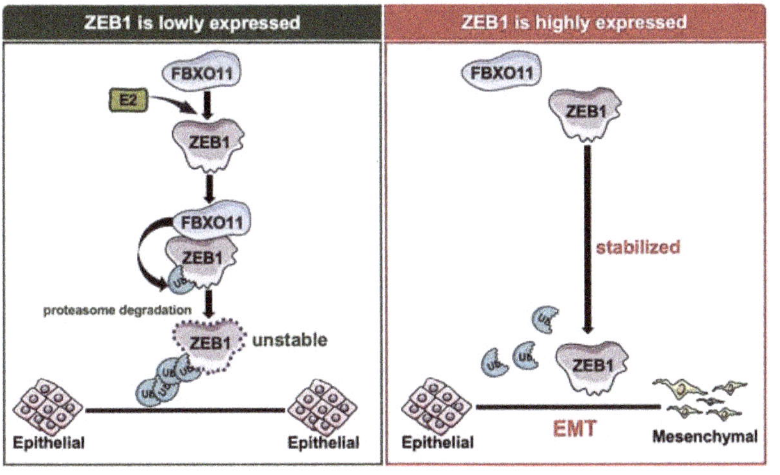

Figure 7. FBXO11 regulates EMT and metastasis through ZEB1 in LUAD. FBXO11-dependent ubiquitination and ZEB1 protein stability regulate the invasion and metastasis of lung cancer cells.

4. Discussion

Lung cancer is the leading cause of cancer death and has one of the lowest 5-year survival rates among all cancer types [36,37]. It is often accompanied by metastases in late stages [38,39]. During tumorigenesis, progression, and metastasis, cancer cells invade the stroma, enter the bloodstream, and colonize distant organs. It is well-established that epithelial–mesenchymal transition (EMT) and mesenchymal–epithelial transition (MET) are critical for cancer cell dissemination [40]. Classical EMT transcription factors like Snail, Slug, Twist, and ZEB1/2 are upregulated in cancer cells, which enhances their invasive and metastatic potential [41]. ZEB1, in particular, is highly expressed in various cancers and plays a crucial role in tumor progression and metastasis. ZEB1 interacts with the tumor microenvironment, including immune cells, and exhibits varying degrees of radioresistance and drug resistance [42,43]. The SCF (Skp1-Cul1-F-box) ubiquitin ligase complex is the largest family of E3 ubiquitin ligases that mediate the ubiquitination of post-translationally modified target proteins or substrates and their degradation by the proteasome [23,44,45]. F-box proteins are a large family of proteins, of which FBXO11 is a member, and act as an E3 ligase that can play a role in carcinogenesis by regulating oncogenes or tumor suppressors [46–48].

This study systematically investigated the molecular mechanisms through which E3 ubiquitin ligase FBXO11 mediates the ubiquitination and degradation of ZEB1, thereby regulating the invasive metastasis of lung cancer cells. We demonstrated that the expression of the FBXO11–ZEB1 axis is strongly associated with the incidence and prognosis of non-small cell lung cancer (NSCLC), the most common type of lung cancer, and has potential as a therapeutic target. Specifically, EMT was induced in human NSCLC A549 cells by $NiCl_2$, leading to the upregulation of ZEB1, while other EMT drivers remained unchanged, underscoring ZEB1's primary role in the EMT pathway in lung adenocarcinoma cells. Our findings reveal that ZEB1 interacts with FBXO11, which enhances the polyubiquitination and proteasomal degradation of ZEB1 and reduces the intracellular half-life of ZEB1. Low FBXO11 expression in lung adenocarcinoma cells increases ZEB1 protein levels, decreases epithelial marker E-cadherin, and increases mesenchymal markers N-cadherin and vimentin, promoting EMT and enhancing cell migration and invasion. The overexpression of FBXO11 reduces ZEBs1 protein levels and maintains epithelial cell status, suggesting a better clinical prognosis. FBXO11 was also found to inhibit lung cancer cell migration and invasion in vivo in xenograft mice.

5. Conclusions

Our findings elucidate the roles played by FBXO11 and ZEB1 in lung adenocarcinoma from a fundamental molecular perspective. Future research should focus on elucidating the role of the FBXO11–ZEB1 axis in vivo, particularly in human lung cancer. This includes the use of animal models to study resistance and evaluate the therapeutic potential and off-target effects. Such a comprehensive could further validate our findings and explore the clinical implications of the interaction between FBXO11 and ZEB1. In addition, identifying small-molecule inhibitors of FBXO11 and studying their efficacy in preclinical models could pave the way for novel targeted therapies. Our ultimate goal is to develop effective therapies that improve survival and reduce mortality in lung cancer patients.

Supplementary Materials: The following supporting information can be downloaded at https://www.mdpi.com/article/10.3390/cancers16193269/s1: Figure S1: Morphological changes in A549 cells were induced by $NiCl_2$; Figure S2: FBXO11 and UBR5 are not colocalized in the nucleus. Immunofluorescence assay probing the colocalization of FBXO11 (Green) and UBR5 (Red). Scale bar: 50 μm. Table S1: q-PCR primers; Table S2: Lentiviral packaging sequence; Original blot data file.

Author Contributions: X.Z.: investigation, methodology, data curation, and writing—original draft; Z.H.: formal analysis and software; R.L.: data curation and visualization; Z.L.: resources; L.M.: validation; Y.J.: conceptualization, project administration, funding acquisition, management, and writing—review and editing. All authors have read and agreed to the published version of the manuscript.

Funding: This work was supported by grants from the Jilin Provincial Scientific and Technological Development Program (20220101276JC and 20210402018GH).

Institutional Review Board Statement: The animal study protocol was approved by the Ethics Committee of the Changchun Wish Technology Company (approval number: 20231129-01; approved on 24 November 2023).

Informed Consent Statement: Not applicable.

Data Availability Statement: The datasets used and analyzed in this paper are available from the corresponding author upon reasonable request.

Conflicts of Interest: The authors declare no conflicts of interest.

References

1. Wang, C.D.; Wu, Y.X.; Shao, J.; Zhang, L.; Li, H.; Wang, Q.; Liu, X.; Huang, Z.; Zhou, Y.; Xie, Y.; et al. Clinicopathological variables influencing overall survival, recurrence and post-recurrence survival in resected stage I non-small-cell lung cancer. *BMC Cancer* **2020**, *20*, 150. [CrossRef] [PubMed]
2. Siegel, R.L.; Miller, K.D.; Wagle, N.S.; Jemal, A.; Stiehler, M.F.; Ou, Y.; Ma, J.; Islami, F.; Xu, J.; Cheng, C.; et al. Cancer statistics, 2023. *CA Cancer J. Clin.* **2023**, *73*, 17–48. [CrossRef] [PubMed]
3. Ruiz-Cordero, R.; Devine, W.P. Targeted Therapy and Checkpoint Immunotherapy in Lung Cancer. *Surg. Pathol. Clin.* **2020**, *13*, 17–33. [CrossRef]
4. Rami-Porta, R.; Bolejack, V.; Giroux, D.J.; Chansky, K.; Crowley, J.; Asamura, H.; Detterbeck, F.C.; Rusch, V.W.; Tsuboi, M.; Goldstraw, P.; et al. The IASLC lung cancer staging project: The new database to inform the eighth edition of the TNM classification of lung cancer. *J. Thorac. Oncol.* **2014**, *9*, 1618–1624. [CrossRef] [PubMed]
5. Chansky, K.; Detterbeck, F.C.; Nicholson, A.G.; Rusch, V.W.; Vallieres, E.; Groome, P.; Kennedy, C.; Krasnik, M.; Peake, M.; Shemanski, L.; et al. The IASLC Lung Cancer Staging Project: External Validation of the Revision of the TNM Stage Groupings in the Eighth Edition of the TNM Classification of Lung Cancer. *J. Thorac. Oncol.* **2017**, *12*, 1109–1121. [CrossRef] [PubMed]
6. Wahbah, M.; Boroumand, N.; Castro, C.; El-Zeky, F.; Dean, S.; de la Cruz, M.; Goel, A.; Chung, J.; Pusztai, L.; Albo, D.; et al. Changing trends in the distribution of the histologic types of lung cancer: A review of 4439 cases. *Ann. Diagn. Pathol.* **2007**, *11*, 89–96. [CrossRef]
7. Van't Veer, L.J. Road map to metastasis. *Nat. Med.* **2003**, *9*, 999–1000. [CrossRef]
8. Kalluri, R.; Neilson, E.G.; Zeisberg, M.; Kanagawa, M.; von Levetzow, C.; Li, Y.; Campbell, H.; Xiao, Y.; Schnaper, H.W.; Ash, S.R.; et al. Epithelial-mesenchymal transition and its implications for fibrosis. *J. Clin. Investig.* **2003**, *112*, 1776–1784. [CrossRef]
9. Liskova, P.; Tuft, S.J.; Gwilliam, R.; Ebenezer, N.D.; Jirsova, K.; Prescott, Q.; Bhattacharya, S.S.; Delp, M.; Stankovska, M.; Hegedus, L.; et al. Novel Mutations in the ZEB1 Gene Identified in Czech and British Patients With Posterior Polymorphous Corneal Dystrophy. *Hum. Mutat.* **2007**, *28*, 638. [CrossRef]

10. Wu, H.T.; Zhong, H.T.; Li, G.W.; Zhang, Y.; Liu, M.; Zhu, L.; Li, Z.; Wang, X.; Chen, P.; Zhuang, L.; et al. Oncogenic functions of the EMT-related transcription factor ZEB1 in breast cancer. *J. Transl. Med.* **2020**, *18*, 51. [CrossRef]
11. Eger, A.; Aigner, K.; Sonderegger, S.; Dampier, B.; Oehler, S.; Schreiber, M.; Mikula, M.; Schwarz, H.; Grillari, J.; Nagy, Z.; et al. DeltaEF1 is a transcriptional repressor of E-cadherin and regulates epithelial plasticity in breast cancer cells. *Oncogene* **2005**, *24*, 2375–2385. [CrossRef] [PubMed]
12. Jiang, M.; Jike, Y.; Liu, K.; Gan, F.; Zhang, K.; Xie, M.; Zhang, J.; Chen, C.; Zou, X.; Jiang, X.; et al. Exosome-mediated miR-144-3p promotes ferroptosis to inhibit osteosarcoma proliferation, migration, and invasion through regulating ZEB1. *Mol. Cancer* **2023**, *22*, 113. [CrossRef] [PubMed]
13. Banerjee, P.; Xiao, G.Y.; Tan, X.; Zheng, V.J.; Shi, L.; Rabassedas, M.N.B.; Guo, H.F.; Liu, X.; Yu, J.; Diao, L.; et al. The EMT activator ZEB1 accelerates endosomal trafficking to establish a polarity axis in lung adenocarcinoma cells. *Nat. Commun.* **2021**, *12*, 6354. [CrossRef] [PubMed]
14. Chen, Y.; Lu, X.; Montoya-Durango, D.E.; Liu, Y.H.; Dean, K.C.; Darling, D.S.; Kaplan, H.J.; Dean, D.C.; Gao, L.; Liu, Y. ZEB1 Regulates Multiple Oncogenic Components Involved in Uveal Melanoma Progression. *Sci. Rep.* **2017**, *7*, 45. [CrossRef]
15. Díaz, V.M.; García de Herreros, A. F-box proteins: Keeping the epithelial-to-mesenchymal transition (EMT) in check. *Semin. Cancer Biol.* **2016**, *36*, 71–79. [CrossRef]
16. Mao, J.H.; Kim, I.J.; Wu, D.; Climent, J.; Kang, H.C.; DelRosario, R.; Balmain, A.; Reilly, R.; Lu, Y.; Qiu, Z.; et al. FBXW7 targets mTOR for degradation and cooperates with PTEN in tumor suppression. *Science* **2008**, *321*, 1499–1502. [CrossRef]
17. Lander, R.; Nordin, K.; LaBonne, C. The F-box protein Ppa is a common regulator of core EMT factors Twist, Snail, Slug, and Sip1. *J. Cell Biol.* **2011**, *194*, 17–25. [CrossRef]
18. Vernon, A.E.; LaBonne, C. Slug stability is dynamically regulated during neural crest development by the F-box protein Ppa. *Development* **2006**, *133*, 3359–3370. [CrossRef]
19. Vinas-Castells, R.; Frias, A.; Robles-Lanuza, E.; Zhang, K.; Longmore, G.D.; García de Herreros, A.; Batlle, E.; Postigo, A.; Nieto, M.A.; Cano, A.; et al. Nuclear ubiquitination by FBXL5 modulates Snail1 DNA binding and stability. *Nucleic Acids Res.* **2014**, *42*, 1079–1094. [CrossRef]
20. Xu, M.; Zhu, C.; Zhao, X.; Chen, C.; Zhang, H.; Yuan, H.; Qian, Y.; Liu, J.; Wang, Z.; Li, M.; et al. Atypical ubiquitin E3 ligase complex Skp1-Pam-Fbxo45 controls the core epithelial-to-mesenchymal transition-inducing transcription factors. *Oncotarget* **2015**, *6*, 979–994. [CrossRef]
21. Chandra, D.S.; Nathubhai, K.N.; Kumar, A. Molecular dynamics simulations elucidate the mode of protein recognition by Skp1 and the F-box domain in the SCF complex. *Proteins* **2016**, *84*, 159–171. [CrossRef] [PubMed]
22. Duan, S.S.; Cermak, L.; Vangala, D.; Rahman, S.; Nawaz, Z.; Jung, K.H.; Qiu, X.; Lou, H.J.; Cheng, M.; Perna, F.; et al. FBXO11 targets BCL6 for degradation and is inactivated in diffuse large B-cell lymphomas. *Nature* **2012**, *481*, 90–94. [CrossRef] [PubMed]
23. Rossi, M.; Duan, S.; Jeong, Y.T.; Dai, Q.; Zhang, J.; Chung, S.Y.; Xing, L.; Lu, M.; Feng, Z.; Hu, H.; et al. Regulation of the CRL4Cdt2 ubiquitin ligase and cell cycle exit by the SCF Fbxo11 ubiquitin ligase. *Mol. Cell* **2013**, *49*, 1159–1166. [CrossRef] [PubMed]
24. Abbas, T.; Keaton, M.; Dutta, A.; Liu, S.; Wang, Z.; Wang, H.; Ranjan, A.; Wei, G.; Zhang, S.; Lin, C.; et al. Regulation of TGF-β signaling, exit from the cell cycle, and cellular migration through cullin cross-regulation SCF-FBXO11 turns off CRL4-Cdt2. *Cell Cycle* **2013**, *12*, 2175–2182. [CrossRef]
25. Jin, Y.; Anitha, K.; Zhang, Q.; Swensen, J.; Long, D.T.; Reese, J.C.; Wang, L.; Hu, W.; Zhou, J.; Zhang, Y.; et al. FBXO11 promotes ubiquitination of the Snail family of transcription factors in cancer progression and epidermal development. *Cancer Lett.* **2015**, *362*, 70–82. [CrossRef]
26. Zheng, H.Q.; Shen, M.H.; Zhao, C.; Li, S.; Cheng, J.Q.; Du, Y.; Li, X.; Wang, Y.; Wei, Y.; Li, H.; et al. PKD1 phosphorylation-dependent degradation of SNAIL by SCF FBXO11 regulates epithelial-mesenchymal transition and metastasis. *Cancer Cell* **2014**, *26*, 358–373. [CrossRef]
27. Kasuga, Y.; Ouda, R.; Watanabe, M.; Goto, A.; Kikuchi, R.; Takahashi, N.; Ito, S.; Yamada, T.; Sugiyama, Y.; Saito, K.; et al. FBXO11 constitutes a major negative regulator of MHC class II through ubiquitin-dependent proteasomal degradation of CIITA. *Proc. Natl. Acad. Sci. USA* **2023**, *120*, e2218955120. [CrossRef]
28. Zhang, H.; Xia, P.; Yang, Z.S.; Liu, X.; Wang, W.; Chen, L.; Zhang, M.; Sun, Y.; Zhou, H.; Guo, X.; et al. Cullin-associated and neddylation-dissociated 1 regulate reprogramming of lipid metabolism through SKP1-Cullin-1-F-boxFBXO1-mediated heterogeneous nuclear ribonucleoprotein A2/B1 ubiquitination and promote hepatocellular carcinoma. *Clin. Transl. Med.* **2023**, *13*, e1443. [CrossRef]
29. Ba, Z.C.; Zhou, Y.F.; Sun, M.; Zhang, S.; Zhao, Y.; Li, W.; Zhang, Q.; Zhou, X.; Liu, Y.; Wang, H.; et al. miR-324-5p upregulation potentiates resistance to cisplatin by targeting FBXO11 signalling in non-small cell lung cancer cells. *J. Biochem.* **2019**, *166*, 517–527. [CrossRef]
30. Sonego, M.; Pellarin, I.; Costa, A.; Vinciguerra, G.L.R.; Coan, M.; Kraut, A.; D'Andrea, S.; Dall'Acqua, A.; Castillo-Tong, D.C.; Califano, D.; et al. USP1 links platinum resistance to cancer cell dissemination by regulating Snail stability. *Sci. Adv.* **2019**, *5*, eaav3235. [CrossRef]
31. Xu, X.; Zhuang, X.; Yu, H.; Li, P.; Li, X.; Lin, H.; Teoh, J.P.; Chen, Y.; Yang, Y.; Chen, W.; et al. FSH induces EMT in ovarian cancer via ALKBH5-regulated Snail m6A demethylation. *Theranostics* **2023**, *13*, 1092–1104. [CrossRef] [PubMed]
32. Wu, C.H.; Tang, S.C.; Wang, P.H.; Lee, H.; Ko, J.L. Nickel-induced epithelial-mesenchymal transition by reactive oxygen species generation and E-cadherin promoter hypermethylation. *J. Biol. Chem.* **2012**, *287*, 25292–25302. [CrossRef]

33. Chen, H.; Ke, Q.; Costa, M.; Zhang, P.; Yan, Y.; Bai, W.; Zhang, Z.; Liu, L.; Huang, C.; Shi, X.; et al. Nickel ions increase the invasiveness of human lung cancer cells via TGF-β signaling pathway. *Toxicol. Appl. Pharmacol.* **2006**, *210*, 148–155.
34. Zhang, Q.; Bhattacharya, S.; Andersen, M.E.; Conolly, R.B.; Clewell, H.J.; John, B.T.; Zang, Y.; Teng, C.; Wang, L.; Chen, H.; et al. Nickel-induced epithelial-mesenchymal transition via downregulation of E-cadherin in human bronchial epithelial cells. *Toxicology* **2013**, *305*, 146–153.
35. Ito, M.; Hiramatsu, H.; Kobayashi, K.; Suzue, K.; Kawahata, M.; Hioki, K.; Ueyama, Y.; Sugawara, M.; Nakamura, T.; Tsujimura, K.; et al. NOD/SCID/γc null mouse: An excellent recipient mouse model for engraftment of human cells. *Blood* **2002**, *100*, 3175–3182. [CrossRef]
36. Maiuthed, A.; Chantarawong, W.; Promjantuek, W.; Rattanaburi, P.; Wattanapan, P.; Worawichawong, S.; Suwankulanan, S.; Jitariu, A.A.; Berindan-Neagoe, I.; Crea, F.; et al. Lung Cancer Stem Cells and Cancer Stem Cell-targeting Natural Compounds. *Anticancer Res.* **2018**, *38*, 3797–3809. [CrossRef]
37. Sung, H.; Ferlay, J.; Siegel, R.L.; Laversanne, M.; Soerjomataram, I.; Jemal, A.; Bray, F.; Giovannucci, E.; Misghinna, M.; Lortet-Tieulent, J.; et al. Global Cancer Statistics 2020: GLOBOCAN estimates of incidence and mortality worldwide for 36 cancers in 185 countries. *CA Cancer J. Clin.* **2021**, *71*, 209–249. [CrossRef]
38. Thai, A.A.; Solomon, B.J.; Sequist, L.V.; Gainor, J.F.; Heist, R.S.; Shaw, A.T.; Lim, J.K.; McGranahan, N.; Swanton, C.; Landau, D.A.; et al. Lung cancer. *Lancet* **2021**, *398*, 535–554. [CrossRef]
39. Toyokawa, G.; Yamada, Y.; Tagawa, T.; Takahashi, F.; Yamasaki, Y.; Hirai, F.; Taira, T.; Miyazawa, Y.; Kodama, Y.; Shoji, F.; et al. Significance of spread through air spaces in early-stage lung adenocarcinomas undergoing limited resection. *Thorac. Cancer* **2018**, *9*, 1255–1261. [CrossRef]
40. Bakir, B.; Chiarella, A.M.; Pitarresi, J.R.; Rustgi, A.K. EMT, MET, Plasticity, and Tumor Metastasis. *Trends Cell Biol.* **2020**, *30*, 764–776. [CrossRef]
41. Beerling, E.; Seinstra, D.; de Wit, E.; Kester, L.; van der Velden, D.; Maynard, C.; Schafer, R.; van Diest, P.; Voest, E.; van Oudenaarden, A.; et al. Plasticity between epithelial and mesenchymal states unlinks EMT from metastasis-enhancing stem cell capacity. *Cell Rep.* **2016**, *14*, 2281–2288. [CrossRef] [PubMed]
42. Chen, L.O.; Gibbons, D.L.; Goswami, S.; Cortez, M.A.; Ahn, Y.H.; Byers, L.A.; Zhang, X.; Yi, X.; Dwyer, D.; Lin, W.; et al. Metastasis is regulated via microRNA-200/Z EB1 axis control of tumour cell PD-L1 expression and intratumoral immunosuppression. *Nat. Commun.* **2014**, *5*, 5241. [CrossRef] [PubMed]
43. Zhang, P.J.; Wei, Y.K.; Wang, L.; Debeb, B.G.; Yuan, Y.; Zhang, J.; Zhou, X.; Li, H.; Xu, W.; Liu, Y.; et al. ATM-mediated stabilization of ZEB1 promotes DNA damage response and radioresistance through CHK. *Nat. Cell Biol.* **2014**, *16*, 864–875. [CrossRef] [PubMed]
44. Peng, D.H.; Kundu, S.T.; Fradette, J.J.; Diao, L.; Tong, P.; Byers, L.A.; Wang, J.; Zhang, H.; Liu, H.; Wu, Y.; et al. ZEB1 suppression sensitizes KRAS mutant cancers to MEK inhibition by an IL17RD-dependent mechanism. *Sci. Transl. Med.* **2019**, *11*, eaaq1238. [CrossRef]
45. Zheng, N.; Schulman, B.A.; Song, L.; Miller, J.J.; Jeffrey, P.D.; Wang, P.; Chu, C.; Koepp, D.M.; Elledge, S.J.; Pagano, M.; et al. Structure of the Cul1-Rbx1-Skp1-F boxSkp2 SCF ubiquitin ligase complex. *Nature* **2002**, *416*, 703–709. [CrossRef]
46. Cardozo, T.; Pagano, M. The SCF ubiquitin ligase: Insights into a molecular machine. *Nat. Rev. Mol. Cell Biol.* **2004**, *5*, 739–751. [CrossRef]
47. Skaar, J.R.; Pagan, J.K.; Pagano, M. SnapShot: F box proteins I. *Cell* **2009**, *137*, 1160–1160.e1. [CrossRef]
48. Nelson, D.E.; Randle, S.J.; Laman, H. Beyond ubiquitination: The atypical functions of Fbxo7 and other F-box proteins. *Open Biol.* **2013**, *3*, 130131. [CrossRef]

cancers

MDPI

Review

Emerging Paradigms in Cancer Metastasis: Ghost Mitochondria, Vasculogenic Mimicry, and Polyploid Giant Cancer Cells

Mateusz Krotofil [1], Maciej Tota [1], Jakub Siednienko [2] and Piotr Donizy [1,3,*]

[1] Department of Clinical and Experimental Pathology, Wroclaw Medical University, Borowska 213, 50-556 Wroclaw, Poland
[2] Department of Experimental Oncology, Ludwik Hirszfeld Institute of Immunology and Experimental Therapy, Polish Academy of Sciences, 53-114 Wroclaw, Poland
[3] Department of Pathology and Clinical Cytology, Jan Mikulicz-Radecki University Hospital, 50-556 Wroclaw, Poland
* Correspondence: piotr.donizy@umw.edu.pl; Tel.: +48-717-343-955

Simple Summary: Metastatic disease is the leading cause of cancer-related morbidity and mortality, accounting for over 80% of cancer deaths. Recent studies have introduced and refined several theories on cancer metastasis, including ghost mitochondria (GM), vasculogenic mimicry (VM), and the formation of polyploid giant cancer cells (PGCCs). Understanding the complex processes of cancer metastasis is crucial for developing effective treatment options.

Abstract: The capacity of cancer cells to migrate from a primary tumor, disseminate throughout the body, and eventually establish secondary tumors is a fundamental aspect of metastasis. A detailed understanding of the cellular and molecular mechanisms underpinning this multifaceted process would facilitate the rational development of therapies aimed at treating metastatic disease. Although various hypotheses and models have been proposed, no single concept fully explains the mechanism of metastasis or integrates all observations and experimental findings. Recent advancements in metastasis research have refined existing theories and introduced new ones. This review evaluates several novel/emerging theories, focusing on ghost mitochondria (GM), vasculogenic mimicry (VM), and polyploid giant cancer cells (PGCCs).

Keywords: metastasis; cancer; ghost mitochondria; vasculogenic mimicry; polyploid giant cancer cells; GM; VM; PGCCs

1. Introduction

Metastasis is a complex process in cancer biology that is not fully understood. Disseminated cancer is often a fatal disease and represents a significant global health challenge [1–3]. Among a cohort of 1,030,937 metastatic cancer survivors in the United States from the period 1992–2019, 82.6% of the patients (n = 688,529) died from the cancer for which they were diagnosed. The median survival duration for patients living with metastases is 10 months [4]. The primary hypothesis explaining the formation of nodal and distant organ metastases is based on direct invasion by tumor cells into lymphatic and/or blood vessels and the generation of circulating tumor cells (CTCs) within the lymphatic and blood vessel systems [5,6]. However, the process of CTC colonization of regional lymphatic nodes and distant organs seems to be disproportionately insufficient [7–9]. Studies on metastasis in patients with melanoma have shown that only 2.5% of circulating cancer cells are able to survive and form micrometastases [9]. Moreover, the formation of micrometastases does not always result in the development of metastatic tumors, as only 1% of micrometastases progress to disseminated disease [9]. These discoveries are inconsistent with the clinically observed high efficiency of the metastatic process, which occurs even in patients who undergo aggressive treatment [10,11]. In our work, we review the literature to reveal potential novel concepts related to cancer cell metastasis.

2. "Ghost" Mitochondria (GM)—Cancer Cell Communication or Independent Metastatic Route?

The role of mitochondria in cancer was first identified by Otto Warburg, who hypothesized that there is a mitochondrial defect in cancer cells [12]. Research has shown that this defect should be considered not only an important factor of cancer cell metabolism but also an essential part of carcinogenesis and metastasis. Thomas Seyfried proposed that initial changes in cancer cells are not associated with mutations in genes but with defective mitochondrial metabolism [13]. His hypothesis was based on numerous in vitro studies that described the suppression of carcinogenesis in hybrid tumor cells after the transplantation of normal cell cytoplasm or mitochondria [14–17]. Other studies revealed that the transfer of mtDNA from cells with high metastatic potential to cells with low metastatic potential can generate high metastatic potential in recipient cells [18].

Mitochondria in cancer cells exhibit remarkable adaptability to endure adverse conditions, including hypoxia, nutrient deficiency, and various therapeutics. Consequently, they are integral to tumor development, requiring a capacity for adjustment in response to both cellular and environmental fluctuations. In addition to their role in bioenergetics, numerous other facets of mitochondrial biology are associated with cellular transformation. These processes encompass mitochondrial biogenesis and turnover, metabolic pathways, dynamics of fission and fusion, regulation of oxidative stress, susceptibility to cell death, and signaling mechanisms [19].

Cancer cells are able to produce various types of microparticles containing extracellular vesicles (EVs) surrounded by a lipid bilayer and nonmembranous particles [20–22]. EVs can be divided into large exopheres, which can contain cellular organelles, and smaller microvesicles (>150 nm) and exosomes (30–150 nm) [22,23]. EVs containing mitochondria can not only appear locally but also be released into the bloodstream by both healthy and injured cells [24–26]. The number of excreted mitochondria is elevated in injured cells, including cancer cells [24,26]. Additionally, cancer cells release small exosomes that contain mitochondrial proteins or mtDNA. These exosomes are present in blood and may serve as potential new cancer markers detectable in blood samples [27,28]. Microparticles that carry mitochondria often present surface markers of platelets or immune cells to promote inflammation [26,29]. Circulating mitochondria are involved in immune response modulation and are important in intracellular signaling pathways [30]. Cancer cells can receive healthy mitochondria from tumor microenvironment (TME) cells, which may have a therapeutic effect and suppress tumor growth [14,31,32]. However, mitochondrial transfer can also enhance cancer cell metabolism and is associated with progression and metastasis [33,34]. Studies of mitochondrial transfer in the TME revealed that prostate cancer cells can recruit mitochondria from stromal fibroblasts [21]. Cancer cells can receive mitochondria from tumor-associated stromal cells even if functional mitochondria are already present in their cytoplasm. The process of mitochondrial transplantation between cells is locally regulated. Only fibroblasts reprogrammed by cancer cells to be highly glycolytic can transfer their mitochondria, and only selected cancer cells can act as recipients [21]. The authors suggested that the receiving mechanisms may depend on lactate excreted by cancer-associated fibroblasts, which triggers changes in cancer cell metabolism and activates signaling pathways associated with mitochondrial uptake [21]. Mitochondrial transplantation is vital for cancer cells to produce sufficient energy for active proliferation [21,32,35]. Studies of cancer mitochondrial transfer are consistent with discoveries claiming that functional mitochondria and oxidative energy production are essential for tumor cell proliferation [24,26,36,37]. Alterations in energy metabolism or the function of mitochondria may inhibit the anticancer capabilities of immune cells and promote the immune escape of cancer cells within the TME [38,39]. Recent research suggests that cancer cells may have the ability to appropriate mitochondria from non-tumor cells within the TME [40]. Mitochondrial transfer is considered a form of intercellular communication. It has been posited that cancer cells could acquire mitochondria from T cells through nanotubes, thereby enhancing their own cellular strength and facilitating immune escape.

Interestingly, mitochondrial transfer by platelets is performed in different physiological and pathological states, e.g., wound healing. Activated platelets release mitochondria to fibroblasts in wounds to enhance metabolic processes and promote angiogenesis [41]. Studies have revealed that mitochondrial transplantation between platelets occurs in cancer and can potentially be associated with metastasis and proliferation. Zhang et al. [42] reported that osteosarcoma cells are reprogrammed into a metastatic state through the acquisition of platelet mitochondria via the PINK1/Parkin-Mfn2 pathway. Platelet mitochondria promote lung metastasis by regulating the GSH/GSSG ratio and reactive oxygen species (ROS) in cancer cells. Mitochondrial transfer between cells possibly plays an important role in local intracellular interactions and is potentially a novel target in cancer therapy.

However, cancer cells contain not only functional but also damaged and nonfunctional cancer mitochondria called "ghost" mitochondria (GM) [43]. The term GM refers to spherical mitochondria that arise from the downregulation of Mic60 in cancer cells [44]. Mic60 is an essential component of the mitochondrial contact site and cristae organizing system (MICOS), which is pivotal for the development of the inner mitochondrial membrane and the formation of functional cristae [45]. Impaired MICOS function results in the formation of mitochondria, which are unable to fulfil their major functions: effective ATP production and the regulation of cell death [43,45]. The presence of the GM may seem contradictory to the energy demands of cancer cells, but it is linked to several proinflammatory signaling pathways, the regulation of nuclear gene expression, and the evasion of programmed cell death mechanisms [43]. Moreover, the presence of the GM can possibly influence cancer metastasis because there seems to be a connection between the downregulation of Mic60 and the ability of cells to produce metastases [43].

Oxidative phosphorylation (OXPHOS) significantly contributes to the advancement of various cancer cells. It not only supplies adequate energy essential for the survival of tumor tissue but also modulates the conditions necessary for tumor proliferation, invasion, and metastasis. Furthermore, modifications in OXPHOS can adversely affect the immune functionality of immune cells within the TME, resulting in immune escape. Mic60-low tumors demonstrate a significant decline in mitochondrial functionality, which is paradoxically correlated with an enhanced tendency for metastasis. An elevated expression of the Mic60-low gene signature has been linked to reduced patient survival in various cancer types, including kidney cancer, uveal melanoma, testicular germ cell tumors, and thymomas [46].

The role of the GM in cancer cells remains unclear and should be investigated in future studies. We hypothesize that the mitochondrial transfer processes mentioned above could also involve the pathological GM. The GM can be transported in EVs to distant tissues, where it can serve as a signal to activate and proliferate dormant cells. The second possible hypothesis involves the interaction of the GM with immune cells and metastatic niche cells, which could be essential for preparing the TME for migrating cancer cells. Studies have not yet clarified the quantity and proportions of the GM compared with those of functional mitochondria in primary and metastatic cancer cells. Identifying markers for cell-free GM could help determine the number of circulating mitochondria and their potential role in assessing the risk of metastasis in patients with various types of cancer. Given the evidence suggesting the inefficiency of metastasis formation solely through circulating tumor cells [7,9], it is also important to explore the possibility that metastasis could spread via smaller entities, such as mitochondria. Perhaps cytokinesis in highly proliferative cancer cells (especially in tumors with an increased number of pathological mitoses) could also be a source of mitochondria released from dividing cells. The GM can enter previously healthy cells of distant organs and alter their metabolism and gene expression to produce a cancer phenotype (Figures 1 and 2). Further studies are needed to investigate this hypothesis and identify the potential signaling pathways involved.

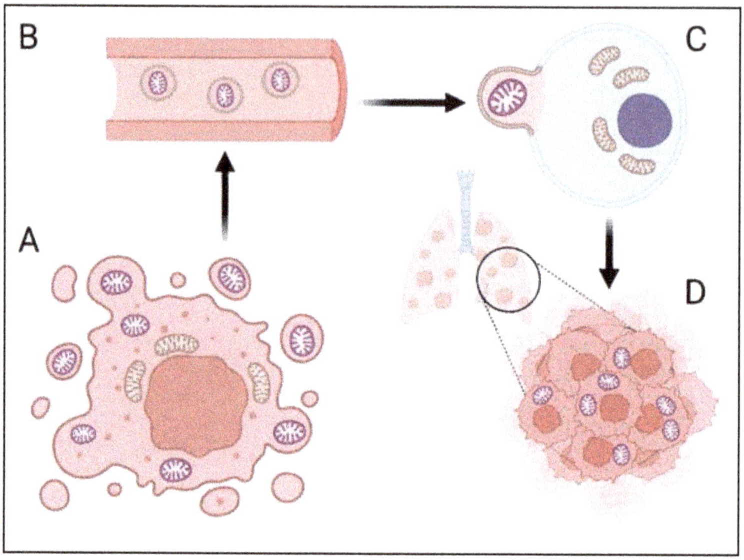

Figure 1. The possible role of circulating "ghost" mitochondria (GM) in metastasis. (**A**) The GM is released by cancer cells into the tumor microenvironment (TME). Vesicles with mitochondria act as local signaling factors but can also enter the circulation. (**B**) Circulating GM enters sites of cancer metastasis through the bloodstream. (**C**) Vesicles with a GM fuse with the membrane of niche cells and modify their metabolism and gene expression to create a TME that is optimal for metastasis. (**D**) Previously healthy cells with internalized GM undergo dysregulation of cell death control and can present a cancer phenotype.

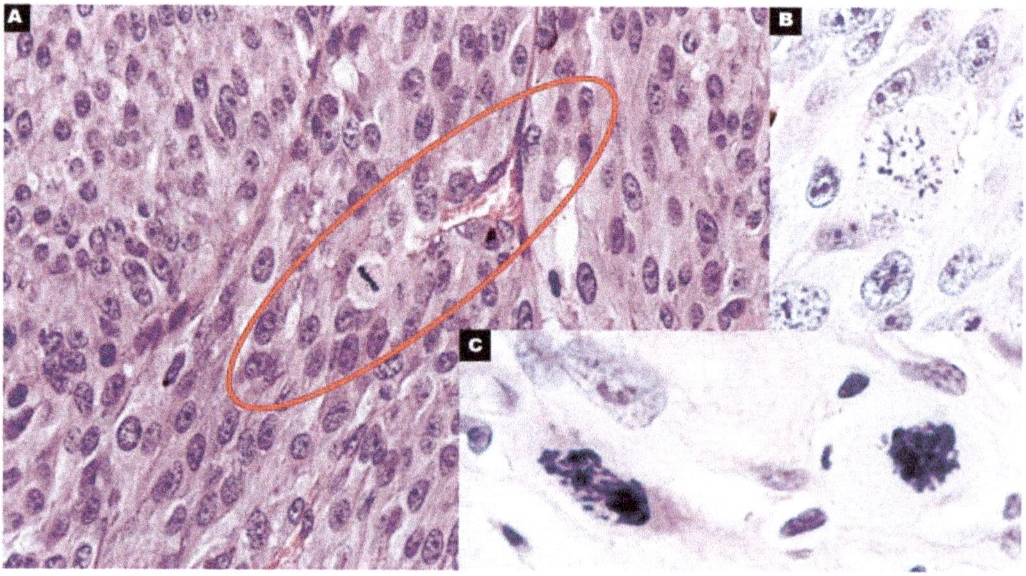

Figure 2. Melanoma with a single normal mitotic figure near small blood vessels with erythrocytes ((**A**), 400×) was observed. Pathologic mitoses with severe abnormal/asymmetrical morphology of chromatin in uterine leiomyosarcoma as a possible source of mitochondria during cytokinesis ((**B**,**C**); 600×).

3. Vasculogenic Mimicry (VM)—Need for Nutrients or Method of Metastasis on the Basis of Interactions with Blood Cells?

Blood flow in cancer is crucial for supplying proliferating cancer cells with the necessary nutrients and oxygen. Angiogenesis is one of the best-examined processes for providing the blood supply to the TME. However, other mechanisms also exist, including vasculogenic mimicry (VM). VM was first described by Maniotis et al. and refers to a process by which cancer cells imitate endothelial cells to form the vasculature [47]. In VM, cancer cells can create vessel-like structures without the presence of endothelial cells [48]. Mechanisms underlying the VM formation are complex and associated with hypoxia. The decreased oxygen supply to the tumor leads to the activation of the hypoxia-inducible factor (HIF) pathway in cancer stem cells (CSCs) [49]. CSCs are a population of cancer cells that possess two features typical to stem cells: self-renewal and differentiation potential [50]. The activation of HIF-1 cascade in CSCs is responsible for the activation of pathways leading to epithelial-to-endothelial transition (EET) [49,51]. EET is a process of differentiation of CSCs into cells with endothelial characteristics [51]. The extracellular matrix remodeling and the EET are major processes leading to the formation of vessel-like structures in VM [49]. Two types of VM have been described: the tubular type and the patterned matrix type [52]. In the tubular type, cancer cells form vessel-like structures very similar to those of normal microvasculature [52,53]. The patterned matrix type is associated with the formation of morphologically dissimilar vessel-like structures surrounded by a matrix and clusters of cancer cells [52,53]. Regardless of the type, vessel-like canals originate from blood vessels and provide slow blood flow into cancer tissue, which enables cancer cells to interact with blood components [52]. The VM process is associated with the presence of PGCCs, which can create tubular vascular-like channels [54]. VM and the formation of PGCCs require similar hypoxia-induced pathways, and the stemness phenotype of PGCCs enables them to perform EET [54,55]. Studies of PGCCs have revealed that their stem-like features permit the cells to differentiate into erythrocyte-like cells which can enter the bloodstream [54]. VM has been observed in various cancer types, including melanoma, breast cancer, gastrointestinal cancers, and glioblastoma [47,53,56–58]. The presence of VM is associated with poor clinical outcomes and is characterized by increased invasion, metastasis, and resistance to antiangiogenic therapies [48,53]. In VM, cancer cells express VE-cadherin, which participates in the formation of vessel-like structures and activates pathways that promote survival and proliferation [53,59]. The structures formed during the VM process allow cancer cells to enter the bloodstream and interact with blood cells.

Interactions between different blood cells and cancer cells have important influences on CTC survival and the metastatic potential of tumors [60]. Studies have shown that cancer cells can interact with platelets to avoid the immune system and facilitate metastasis formation [61,62]. Once cancer cells enter the bloodstream, they can activate platelets by secreting different factors, including tissue factor and thrombin [63,64]. Activated platelets attach to CTCs and cover their surface, creating a protective layer and enabling cell interactions [60,63,65]. Platelets are able to both promote and inhibit cancer metastasis [63]. Prometastatic platelet function is associated with facilitating cancer cell immune escape, increasing adhesion ability and inhibiting cancer cell anoikis (a form of programmed cell death that occurs when cells detach from the surrounding extracellular matrix) [63]. Immune escape is connected mainly with the modification of antigens presented by cancer cells, which protects them from elimination by NK cells [63,66]. Platelets can also impair tumor progression and metastasis via modification of the cell cycle and apoptosis [63].

Cancer cells can also influence erythrocytes by altering their immunological function [67]. DNA fragments, likely originating from cancer cells, have been detected in mature red blood cells [68]. Tumor cells are able to produce vesicles containing RNA or DNA (both nuclear and mitochondrial) that can be transferred to other cells and translated into proteins, altering their function [69–72]. We postulate that VM and direct contact between tumor cells and the blood are involved in the metastasis process. Cancer cells possibly recruit platelets and erythrocytes to protect migrating cells from the immune system. Another potential mechanism involves these blood cells acting as vectors for cancer genetic material, transporting it to distant sites, such as bone marrow. This material could help prepare niches for future colonization by metastatic cancer cells, or it could be directly transferred to stem cells, which may then be altered to express cancer phenotypes. Studies on mouse models have not demonstrated the degree of deformability in cancer cells that is essential for their entry into the circulation. However, observations of human prostate cells suggest that circulating tumor cells (CTCs) exhibit a size and deformability similar to those of blood cells [73]. This finding raises the possibility that CTCs, which are responsible for metastasis, could actually be blood cells with cancer-like features. Moreover, studies that are used to diagnose the presence of CTCs in the blood often rely on the detection of surface antigens present in cancer cells and absent in healthy blood cells [73]. More studies, including studies on the greater quantity of antigens associated with the examination of cellular morphology, should be performed to determine the exact origin of CTCs. The potential role of stem cells in carcinogenesis has been previously proposed, with the idea that stem cells could differentiate into cancer cells [74]. Changes in the bone microenvironment in premetastatic disease, including changes in stem cell proportions, have been observed in previous studies, which suggests that cancer can modify the microenvironment before the development of metastasis [75] (Figures 3 and 4). In our opinion, further research is needed to investigate whether the indirect evidence of the use of blood cells to disseminate and communicate with distant organs in cancer supports our hypothesis [76,77].

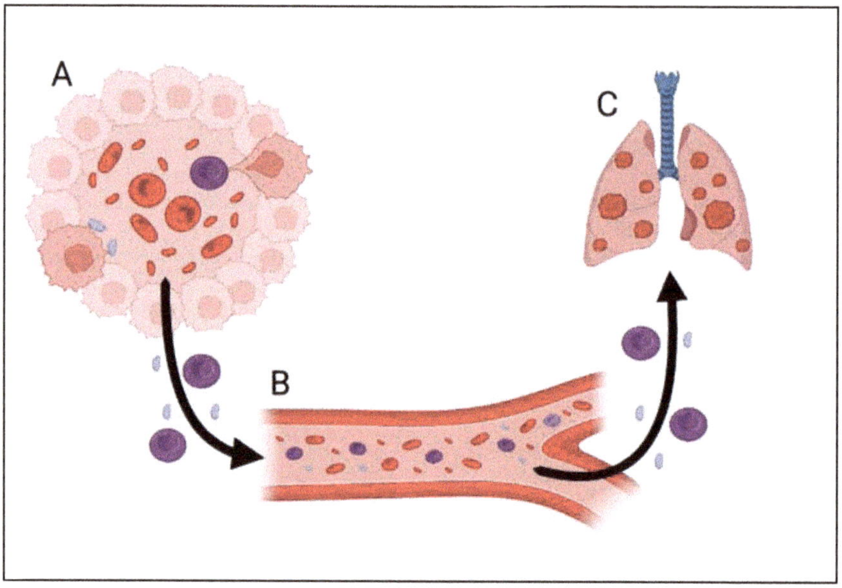

Figure 3. The proposed mechanism of cancer metastasis through vasculogenic mimicry. (**A**) "Education" of blood cells in vessel-like structures in cancer tissue. Cancer cells modify the function of blood cells and implant DNA into erythrocytes and platelets. (**B**) "Modified" platelets and erythrocytes are released into the circulation and migrate into the site of metastasis. (**C**) Metastasis formation occurs via the migration of "modified" cells as a result of interactions with niche cells, the activation of dormant cancer cells or the internalization of genetic material from cancer cells into healthy cells (acting as a "vector").

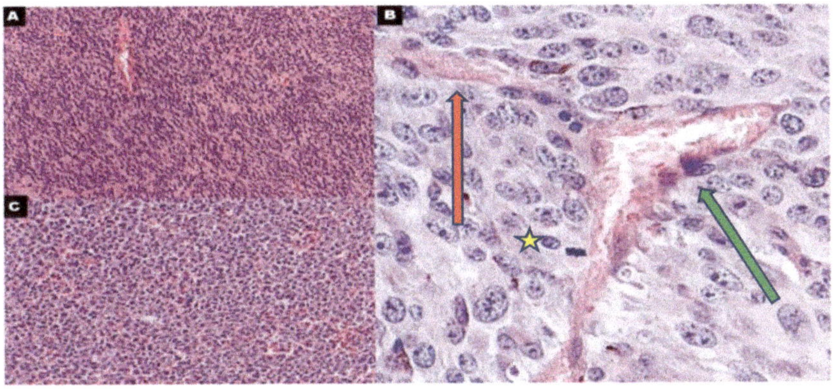

Figure 4. A regular blood vessel lined by endothelial cells in uveal melanoma ((**A**); 400×). Connection between the regular blood vessel lined by endothelial cells (green arrow) and the vascular channel partially lined by tumoral cells (red arrow) in uveal melanoma; asterisk: tumoral cell during mitosis in direct connection with the lumen of the vessel ((**B**); 600×). Dispersed vasculogenic mimicry (vascular channels lined only by tumor cells) in metastatic cutaneous melanoma ((**C**); 400×).

4. Polyploid Giant Cancer Cells—Dormant Locators or Migrating Pioneers

Polyploid giant cells (PGCs) are formed through the fusion of multiple mononuclear cells. The formation of PGCs is a common feature of the monocyte lineage and can occur in both physiological conditions (e.g., osteoclasts) and pathological conditions, such as Langhans giant cells in tuberculosis [78]. Polyploid giant cancer cells (PGCCs) are a subtype of PGC composed of fused cancer cells and are commonly observed in various cancer types, including melanoma, urothelial carcinoma, renal cell carcinoma, breast carcinoma, ovarian carcinoma, pancreatic adenocarcinoma, and prostate carcinoma [79]. PGCCs are a heterogenous group of cancer cells with diversified roles in cancer dormancy and metastasis, and their numbers are correlated with disease progression [80,81].

Different functions of PGCCs are associated with stemness features. Studies revealed that progeny cells of PGCCs possess stem cell markers including Oct-4, Sox-2, NANOG, CD44, and CD133 [54,82]. PGCCs exhibit characteristics similar to CSCs and have the ability to express CSC-related markers such as CD44 and CD133. Tumors derived from PGCCs express proteins associated with EMT, such as N-cadherin, Snail/Slug, and Twist. These proteins contribute to the increased metastatic and invasive capabilities of cancer cells [83].

The process of PGCC formation is initiated by cellular stress factors such as hypoxia, chemotherapy, radiotherapy, ROS production, viral infections, and other factors [79,84]. Hypoxia is the best known inductor and the hypoxia mimicking $CoCl_2$ is commonly used for in vitro PGCC formation [85]. The exact mechanisms of PGCC induction are not fully understood; however, studies suggest similar pathways that are associated with VM [54,55]. Studies have revealed that important events leading to PGCC creation are connected with HIF-1 pathway activation and DNA damage [54,55,84].

PGCCs remain arrested in the G2/M phase of the cell cycle [81,86]. Cell cycle arrest is caused by impaired function of the cell cycle regulatory proteins CDC25B and CDC25C. During mitosis initiation, these proteins are normally translocated from the cytoplasm into the nucleus by cyclin B1. In dormant PGCCs, cyclin B1 is downregulated; however, it can also be highly expressed in PGCCs, which is associated with reactivation and production of daughter cells [81,85–87]. Studies of high-grade serous ovarian carcinoma revealed that PGCCs are able to undergo microevolution, which is responsible for resistance to chemotherapy and adaptation to metastasis [88,89]. During their dormant phase, PGCCs undergo endomitotic nuclear divisions without cytokinesis, increasing the quantity of genetic material [88]. Some factors, e.g., the cellular stress caused by radiotherapy or chemotherapy, can activate dormant PGCCs [90–93]. The activated PGCCs divide by neosis, which consists of amitotic nuclear fragmentation and cytokinesis, to produce blastomere-like organoids filled with daughter cells [90,92]. Moreover the stemness of PGCCs enables them to differentiate into different cellular lines, including cells of each germ layer, as a response to signalizing factors in the TME [54,55,82]. Fecundity daughter cells can be released into the TME, where they undergo mitotic divisions, resulting in tumor formation [88,92,94]. While PGCCs are widely described as dormant cells formed by fused cancer cells [64,76], there is also a population of PGCCs that express surface proteins typical of immune cells (e.g., macrophages), allowing them to migrate through blood vessels and contribute to metastasis [95].

Macrophages constitute another significant cell population associated with cancer metastasis and the possible formation of PGCCs [96]. As innate immune cells with phagocytic capabilities, macrophages play dual roles in cancer, contributing to both antitumor immunity and disease progression. The role of macrophages in cancer is diverse; for that reason, three novel types of macrophages have been identified [97]. The first type is regular tumor-associated macrophages (TAMs), which are important elements of the TME. The second type is the population of giant cells that contain cancer proteins and DNA fragments but also express the macrophage-specific protein CD14; these cells are referred to as cancer-associated macrophage-like cells (CAMLs) [97]. CAMLs carry the DNA and proteome from phagocytosed cancer cells and can migrate into blood vessels to perform immune functions. The third type of cancer-associated macrophage is giant hybrid cells (GHCs), which are formed through the fusion of macrophages and cancer cells and contain functional cancer cell nuclei. GHCs leave the TME in large numbers and enter the circulation as circulating hybrid cells (CHCs) [95] (Figures 5–7). The immunotolerance provided by macrophage surface markers and the ability to promote neovascularization through the expression of TIE-2/CD146 support the hypothesis that PGCCs with a macrophage phenotype could serve as an effective mechanism for cancer dissemination [95].

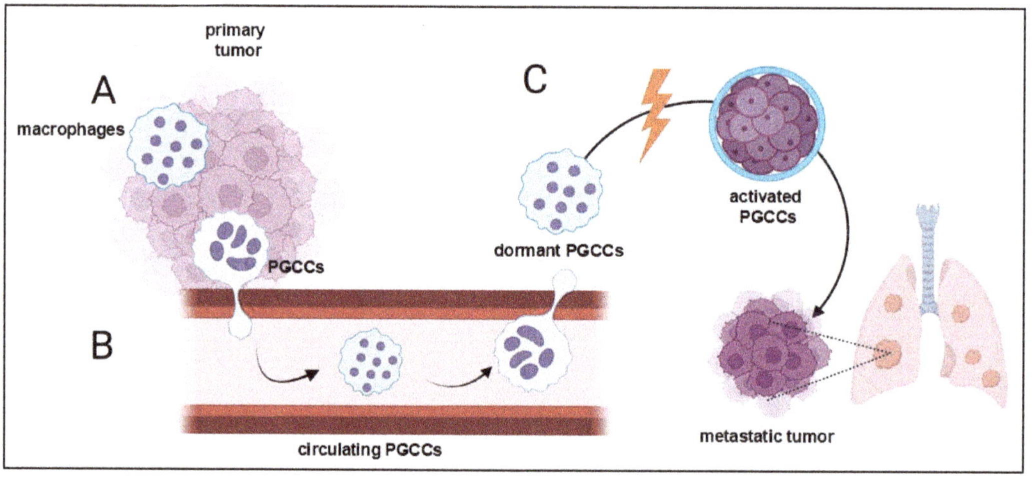

Figure 5. Hypothetical role of polyploid giant cancer cells (PGCCs) in metastasis. (**A**) PGCCs are generated in primary tumors by the fusion of cancer cells or by the fusion of cancer cells with macrophages. (**B**) PGCCs, which are hybrids of macrophages and cancer cells, can enter the circulation and migrate to sites of metastasis. (**C**) PGCCs residing in metastatic niches are activated from a dormant state. Activated PGCCs undergo neosis and release daughter cells, which create metastatic tumors.

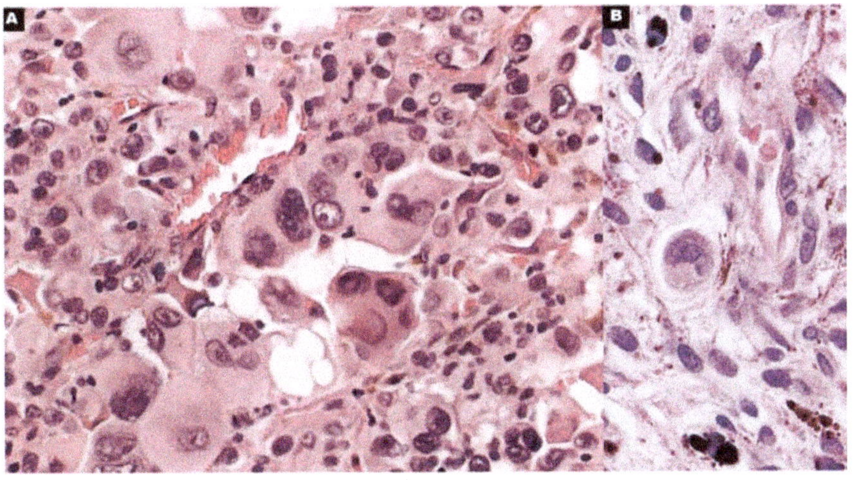

Figure 6. Polyploid giant cancer cells of uveal melanoma in direct contact with small blood vessels ((**A**,**B**); 600×).

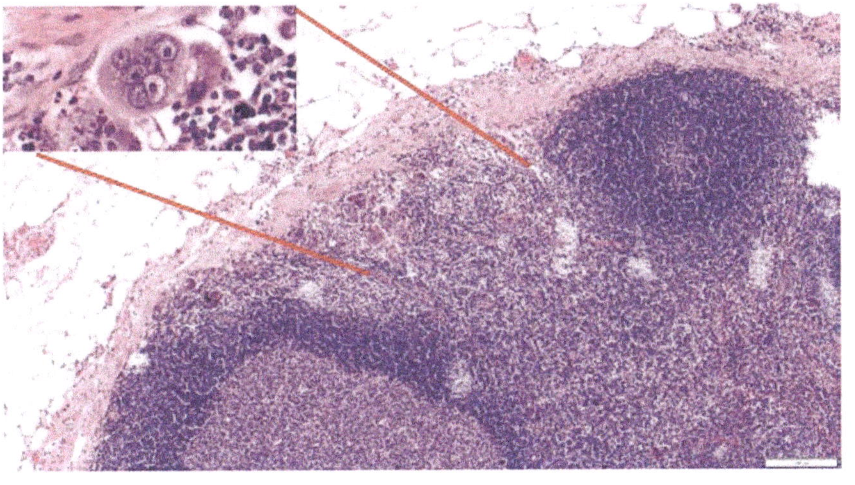

Figure 7. Polyploid giant cancer cells in chemoresistant nodal metastases of poorly differentiated gastric adenocarcinoma (200×); inset: multinucleated giant cell with several highly atypical nuclei (600×).

5. Discussion

Ghost mitochondria (GM), vasculogenic mimicry (VM), and polyploid giant cancer cells (PGCCs) are three underestimated concepts that may help elucidate the process of cancer metastasis. Currently, there are two active clinical trials on VM (NCT03600233) and PGCCs (NCT06389201), discussed hereunder. We also discuss the correlation between PGCCs and VM formation in several neoplasms.

Specific dormant PGCCs have garnered significant attention due to their correlation with the clinical risk of recurrence in nasopharyngeal carcinoma (NPC). In a study by You et al., the authors demonstrate that autophagy serves as a crucial mechanism for the induction of PGCCs. Furthermore, the pharmacological inhibition of autophagy markedly reduced the formation of PGCCs, leading to a significant decrease in metastasis and an improvement in survival rates in a mouse model [98]. Consequently, one clinical trial

(NCT06389201) will evaluate whether an autophagy inhibitor, hydroxychloroquine (HCQ), inhibits the development of therapy-induced dormant PGCCs. This approach has the potential to reduce the risk of recurrence and metastasis in patients with NPC.

CVM-1118 (foslinanib, TRX-818) is a novel small molecule with a high potential for inhibiting VM formation via targeting TNF receptor-associated protein 1 (TRAP1) [99]. The notable occurrence of VM has been documented in neuroendocrine tumors (NETs). Furthermore, VM has been identified as a component of various microvascular changes in mouse models of pancreatic NETs. This compound has demonstrated significant anticancer activity across various human cancer cell lines. The safety profile of CVM-1118 in human subjects is currently being assessed through a phase 1 clinical study. The primary objective of the phase 2 study (NCT03600233) is to further evaluate the efficacy of CVM-1118 in patients diagnosed with advanced NETs [100].

The number of PGCCs is correlated with the formation of VM human glioma. The walls of VM exhibit positive (or negative) staining for PAS and are negative for CD31 staining. Furthermore, the prevalence of VM is significantly greater in high-grade gliomas compared to low-grade gliomas [101]. Similarly, the number of PGCCs is associated with VM formation in human ovarian cancer. In a study by Zhang et al., the metastatic foci of ovarian carcinoma exhibited the highest prevalence of PGCCs and VM. An increase in the number of PGCCs and VM was observed in correlation with the grade of ovarian carcinoma. PGCCs were found to generate erythroid cells through a budding process, and collectively, these cells contributed to the formation of VM [102]. The quantity of PGCCs and VM was also correlated with the differentiation of colorectal cancer. Additionally, daughter cells that originate from PGCCs may facilitate lymph node metastasis through the expression of EMT-related proteins. PGCCs, along with newly generated erythroid cells, contribute to the formation of VM structures in colorectal cancer [103]. In another study, Liu et al. found that primary anorectal malignant melanomas exhibiting lymph node metastasis demonstrate a higher incidence of PGCCs and VM compared to those without metastasis. Authors confirmed that PGCCs and their generated erythroid cells have the ability to form VM [104]. Thus, the number of PGCCs is correlated with disease development, progression, and formation of VM in glioma, ovarian carcinoma, colorectal cancer, and malignant melanoma.

Blocking mitochondrial metabolism in cancer is a relatively new concept that led to clinical trials targeting the mitochondrial energy homeostasis essential for cancer proliferation [105,106]. The new emerging field of cancer therapy could be identifying signaling pathways associated with GM. The GCN2/Akt kinase signaling cascade seems to be one of the most important targets in cancers with GM [43]. Blocking that pathway could be a major component of metastatic cancer therapy that might possibly result in deceleration of tumor progression and extend survival in patients with advanced disease [43]. In vitro studies of available GCN2 inhibitors revealed promising effects on inducing cell death in chronic diseases; however, its value in cancer therapy could be more complex and should be determined in further clinical trials [107]. VM seems to be a promising target in therapy of advanced cancers. Studies of fasudil revealed that the drug inhibited vascular-like structure formation and tumor growth in melanoma mice xenografts [76]. Studies of in vitro glioma models showed that histone deacetylase inhibitors inhibit the VM and could potentially be used in therapy [77]. Achieving a better understanding of VM and determining the effects of blocking VM pathways on overall patient outcome are important directions for further studies. PGCCs are associated with drug resistance in many cancer types. The utilization of drugs targeting PGCC endoreplication provides a possible additional therapy, eliminating drug resistance. Mifepristone promotes apoptosis of endoreplicating PGCCs and blocks their formation. The use of mifepristone with olaparib was more effective in restricting tumor growth than each of the drugs in a monotherapy ovarian cancer model [108]. Targeting GM, VM, and PGCCs is an emerging field for developing new cancer therapies in advanced metastatic cancers.

6. Conclusions

In summary, this article highlights the multifaceted nature of metastatic processes and emphasizes the necessity of incorporating a wide range of scientific viewpoints to enhance our comprehension. Targeting GM, VM, and PGCCs is an emerging field for developing new cancer therapies in advanced metastatic cancers. The future of metastasis research is anchored in the ongoing investigation of these emerging paradigms and their application in clinical settings. These concepts may lead to the development of innovative therapeutic strategies.

Author Contributions: Conceptualization, P.D.; writing—original draft preparation, M.K. and M.T.; writing—review and editing, M.T., J.S. and P.D.; visualization—M.K. and P.D.; supervision, P.D. and J.S.; funding acquisition, P.D. All authors have read and agreed to the published version of the manuscript.

Funding: This research was funded by the Polish Ministry of Science and Higher Education Subvention under number SUBZ.A430.24.049 from the IT simple system of Wroclaw Medical University.

Conflicts of Interest: The authors declare that they have no conflicts of interest.

References

1. Dillekås, H.; Rogers, M.S.; Straume, O. Are 90% of Deaths from Cancer Caused by Metastases? *Cancer Med.* **2019**, *8*, 5574–5576. [CrossRef] [PubMed]
2. Siegel, R.L.; Miller, K.D.; Wagle, N.S.; Jemal, A. Cancer Statistics, 2023. *CA Cancer J. Clin.* **2023**, *73*, 17–48. [CrossRef] [PubMed]
3. Gennari, A.; Conte, P.; Rosso, R.; Orlandini, C.; Bruzzi, P. Survival of Metastatic Breast Carcinoma Patients over a 20-year Period. *Cancer* **2005**, *104*, 1742–1750. [CrossRef]
4. Mani, K.; Deng, D.; Lin, C.; Wang, M.; Hsu, M.L.; Zaorsky, N.G. Causes of Death among People Living with Metastatic Cancer. *Nat. Commun.* **2024**, *15*, 1519. [CrossRef]
5. Paget, S. The distribution of secondary growths in cancer of the breast. *Lancet* **1889**, *133*, 571–573. [CrossRef]
6. Lin, D.; Shen, L.; Luo, M.; Zhang, K.; Li, J.; Yang, Q.; Zhu, F.; Zhou, D.; Zheng, S.; Chen, Y.; et al. Circulating Tumor Cells: Biology and Clinical Significance. *Signal Transduct. Target. Ther.* **2021**, *6*, 404. [CrossRef]
7. Massagué, J.; Obenauf, A.C. Metastatic Colonization by Circulating Tumour Cells. *Nature* **2016**, *529*, 298–306. [CrossRef]
8. Fares, J.; Fares, M.Y.; Khachfe, H.H.; Salhab, H.A.; Fares, Y. Molecular Principles of Metastasis: A Hallmark of Cancer Revisited. *Signal Transduct. Target. Ther.* **2020**, *5*, 28. [CrossRef] [PubMed]
9. Luzzi, K.J.; MacDonald, I.C.; Schmidt, E.E.; Kerkvliet, N.; Morris, V.L.; Chambers, A.F.; Groom, A.C. Multistep Nature of Metastatic Inefficiency: Dormancy of Solitary Cells after Successful Extravasation and Limited Survival of Early Micrometastases. *Am. J. Pathol.* **1998**, *153*, 865–873. [CrossRef]
10. Martin, O.A.; Anderson, R.L.; Narayan, K.; MacManus, M.P. Does the Mobilization of Circulating Tumour Cells during Cancer Therapy Cause Metastasis? *Nat. Rev. Clin. Oncol.* **2017**, *14*, 32–44. [CrossRef]
11. Su, J.-X.; Li, S.-J.; Zhou, X.-F.; Zhang, Z.-J.; Yan, Y.; Liu, S.-L.; Qi, Q. Chemotherapy-Induced Metastasis: Molecular Mechanisms and Clinical Therapies. *Acta Pharmacol. Sin.* **2023**, *44*, 1725–1736. [CrossRef]
12. Warburg, O. The Metabolism of Carcinoma Cells. *J. Cancer Res.* **1925**, *9*, 148–163. [CrossRef]
13. Seyfried, T.N. Cancer as a Mitochondrial Metabolic Disease. *Front. Cell Dev. Biol.* **2015**, *3*, 43. [CrossRef]
14. Kaipparettu, B.A.; Ma, Y.; Park, J.H.; Lee, T.L.; Zhang, Y.; Yotnda, P.; Creighton, C.J.; Chan, W.Y.; Wong, L.J.C. Crosstalk from Non-Cancerous Mitochondria Can Inhibit Tumor Properties of Metastatic Cells by Suppressing Oncogenic Pathways. *PLoS ONE* **2013**, *8*, e61747. [CrossRef]
15. Elliott, R.L.; Jiang, X.P.; Head, J.F. Mitochondria Organelle Transplantation: Introduction of Normal Epithelial Mitochondria into Human Cancer Cells Inhibits Proliferation and Increases Drug Sensitivity. *Breast Cancer Res. Treat.* **2012**, *136*, 347–354. [CrossRef]
16. Shay, J.W. Cytoplasmic Suppression of Tumorigenicity in Reconstructed Mouse Cells. *Cancer Res.* **1988**, *48*, 830–833.
17. Mckinnell, R.G.; Deggins, B.A.; Labat, D.D. Transplantation of Pluripotential Nuclei from Triploid Frog Tumors. *Science* **1969**, *165*, 394–396. [CrossRef]
18. Ishikawa, K.; Takenaga, K.; Akimoto, M.; Koshikawa, N.; Yamaguchi, A.; Imanishi, H.; Nakada, K.; Honma, Y.; Hayashi, J.I. ROS-Generating Mitochondrial DNA Mutations Can Regulate Tumor Cell Metastasis. *Science* **2008**, *320*, 661–664. [CrossRef]
19. Garimella, S.V.; Gampa, S.C.; Chaturvedi, P. Mitochondria in Cancer Stem Cells: From an Innocent Bystander to a Central Player in Therapy Resistance. *Stem Cells Cloning* **2023**, *16*, 19. [CrossRef]
20. Cheng, X.; Henick, B.S.; Cheng, K. Anticancer Therapy Targeting Cancer-Derived Extracellular Vesicles. *ACS Nano* **2024**, *18*, 6748–6765. [CrossRef]
21. Ippolito, L.; Morandi, A.; Taddei, M.L.; Parri, M.; Comito, G.; Iscaro, A.; Raspollini, M.R.; Magherini, F.; Rapizzi, E.; Masquelier, J.; et al. Cancer-Associated Fibroblasts Promote Prostate Cancer Malignancy via Metabolic Rewiring and Mitochondrial Transfer. *Oncogene* **2019**, *38*, 5339–5355. [CrossRef]

22. Zhang, Z.G.; Buller, B.; Chopp, M. Exosomes—Beyond Stem Cells for Restorative Therapy in Stroke and Neurological Injury. *Nat. Rev. Neurol.* **2019**, *15*, 193–203. [CrossRef]
23. Choi, D.; Montermini, L.; Jeong, H.; Sharma, S.; Meehan, B.; Rak, J. Mapping Subpopulations of Cancer Cell-Derived Extracellular Vesicles and Particles by Nano-Flow Cytometry. *ACS Nano* **2019**, *13*, 10499–10511. [CrossRef]
24. Nawaz, M.; Camussi, G.; Valadi, H.; Nazarenko, I.; Ekström, K.; Wang, X.; Principe, S.; Shah, N.; Ashraf, N.M.; Fatima, F.; et al. The Emerging Role of Extracellular Vesicles as Biomarkers for Urogenital Cancers. *Nat. Rev. Urol.* **2014**, *11*, 688–701. [CrossRef]
25. Stier, A. Human Blood Contains Circulating Cell-Free Mitochondria, but Are They Really Functional? *Am. J. Physiol. Endocrinol. Metab.* **2021**, *320*, E859–E863. [CrossRef]
26. Stephens, O.R.; Grant, D.; Frimel, M.; Wanner, N.; Yin, M.; Willard, B.; Erzurum, S.C.; Asosingh, K. Characterization and Origins of Cell-Free Mitochondria in Healthy Murine and Human Blood. *Mitochondrion* **2020**, *54*, 102–112. [CrossRef]
27. Mohd Khair, S.Z.N.; Abd Radzak, S.M.; Mohamed Yusoff, A.A. The Uprising of Mitochondrial DNA Biomarker in Cancer. *Dis. Markers* **2021**, *2021*, 7675269. [CrossRef]
28. Jang, S.C.; Crescitelli, R.; Cvjetkovic, A.; Belgrano, V.; Olofsson Bagge, R.; Sundfeldt, K.; Ochiya, T.; Kalluri, R.; Lötvall, J. Mitochondrial Protein Enriched Extracellular Vesicles Discovered in Human Melanoma Tissues Can Be Detected in Patient Plasma. *J. Extracell. Vesicles* **2019**, *8*, 1635420. [CrossRef]
29. Boudreau, L.H.; Duchez, A.C.; Cloutier, N.; Soulet, D.; Martin, N.; Bollinger, J.; Paré, A.; Rousseau, M.; Naika, G.S.; Lévesque, T.; et al. Platelets Release Mitochondria Serving as Substrate for Bactericidal Group IIA-Secreted Phospholipase a to Promote Inflammation. *Blood* **2014**, *124*, 2173–2183. [CrossRef]
30. Lin, L.; Xu, H.; Bishawi, M.; Feng, F.F.; Samy, K.; Truskey, G.; Barbas, A.S.; Kirk, A.D.; Brennan, T.V. Circulating Mitochondria in Organ Donors Promote Allograft Rejection. *Am. J. Transplant.* **2019**, *19*, 1917–1929. [CrossRef]
31. Sahinbegovic, H.; Jelinek, T.; Hrdinka, M.; Bago, J.R.; Turi, M.; Sevcikova, T.; Kurtovic-Kozaric, A.; Hajek, R.; Simicek, M. Intercellular Mitochondrial Transfer in the Tumor Microenvironment. *Cancers* **2020**, *12*, 1787. [CrossRef]
32. Chang, J.C.; Chang, H.S.; Wu, Y.C.; Cheng, W.L.; Lin, T.T.; Chang, H.J.; Kuo, S.J.; Chen, S.T.; Liu, C.S. Mitochondrial Transplantation Regulates Antitumour Activity, Chemoresistance and Mitochondrial Dynamics in Breast Cancer. *J. Exp. Clin. Cancer Res.* **2019**, *38*, 30. [CrossRef]
33. Zampieri, L.X.; Silva-Almeida, C.; Rondeau, J.D.; Sonveaux, P. Mitochondrial Transfer in Cancer: A Comprehensive Review. *Int. J. Mol. Sci.* **2021**, *22*, 3245. [CrossRef]
34. Guo, X.; Can, C.; Liu, W.; Wei, Y.; Yang, X.; Liu, J.; Jia, H.; Jia, W.; Wu, H.; Ma, D. Mitochondrial Transfer in Hematological Malignancies. *Biomark. Res.* **2023**, *11*, 89. [CrossRef]
35. Hayes, J.D.; Dinkova-Kostova, A.T.; Tew, K.D. Oxidative Stress in Cancer. *Cancer Cell* **2020**, *38*, 167–197. [CrossRef]
36. Sellers, K.; Fox, M.P.; Ii, M.B.; Slone, S.P.; Higashi, R.M.; Miller, D.M.; Wang, Y.; Yan, J.; Yuneva, M.O.; Deshpande, R.; et al. Pyruvate Carboxylase Is Critical for Non-Small-Cell Lung Cancer Proliferation. *J. Clin. Investig.* **2015**, *125*, 687–698. [CrossRef]
37. Anderson, R.G.; Ghiraldeli, L.P.; Pardee, T.S. Mitochondria in Cancer Metabolism, an Organelle Whose Time Has Come? *Biochim. Biophys. Acta Rev. Cancer* **2018**, *1870*, 96–102. [CrossRef]
38. Burt, R.; Dey, A.; Aref, S.; Aguiar, M.; Akarca, A.; Bailey, K.; Day, W.; Hooper, S.; Kirkwood, A.; Kirschner, K.; et al. Activated Stromal Cells Transfer Mitochondria to Rescue Acute Lymphoblastic Leukemia Cells from Oxidative Stress. *Blood* **2019**, *134*, 1415. [CrossRef]
39. Moschoi, R.; Imbert, V.; Nebout, M.; Chiche, J.; Mary, D.; Prebet, T.; Saland, E.; Castellano, R.; Pouyet, L.; Collette, Y.; et al. Protective Mitochondrial Transfer from Bone Marrow Stromal Cells to Acute Myeloid Leukemic Cells during Chemotherapy. *Blood* **2016**, *128*, 253–264. [CrossRef]
40. Saha, T.; Dash, C.; Jayabalan, R.; Khiste, S.; Kulkarni, A.; Kurmi, K.; Mondal, J.; Majumder, P.K.; Bardia, A.; Jang, H.L.; et al. Intercellular Nanotubes Mediate Mitochondrial Trafficking between Cancer and Immune Cells. *Nat. Nanotechnol.* **2022**, *17*, 98. [CrossRef]
41. Levoux, J.; Prola, A.; Lafuste, P.; Gervais, M.; Chevallier, N.; Koumaiha, Z.; Kefi, K.; Braud, L.; Schmitt, A.; Yacia, A.; et al. Platelets Facilitate the Wound-Healing Capability of Mesenchymal Stem Cells by Mitochondrial Transfer and Metabolic Reprogramming. *Cell Metab.* **2021**, *33*, 283–299. [CrossRef]
42. Zhang, W.; Zhou, H.; Li, H.; Mou, H.; Yinwang, E.; Xue, Y.; Wang, S.; Zhang, Y.; Wang, Z.; Chen, T.; et al. Cancer Cells Reprogram to Metastatic State through the Acquisition of Platelet Mitochondria. *Cell Rep.* **2023**, *42*, 113147. [CrossRef]
43. Ghosh, J.C.; Perego, M.; Agarwal, E.; Bertolini, I.; Wang, Y.; Goldman, A.R.; Tang, H.Y.; Kossenkov, A.V.; Libby, C.J.; Languino, L.R.; et al. Ghost Mitochondria Drive Metastasis through Adaptive GCN2/Akt Therapeutic Vulnerability. *Proc. Natl. Acad. Sci. USA* **2022**, *119*, e2115624119. [CrossRef]
44. Altieri, D.C. Mitochondria in Cancer: Clean Windmills or Stressed Tinkerers? *Trends Cell Biol.* **2023**, *33*, 293–299. [CrossRef]
45. Li, H.; Ruan, Y.; Zhang, K.; Jian, F.; Hu, C.; Miao, L.; Gong, L.; Sun, L.; Zhang, X.; Chen, S.; et al. Mic60/Mitofilin Determines MICOS Assembly Essential for Mitochondrial Dynamics and MtDNA Nucleoid Organization. *Cell Death Differ.* **2016**, *23*, 380–392. [CrossRef]
46. Kossenkov, A.V.; Milcarek, A.; Notta, F.; Jang, G.H.; Wilson, J.M.; Gallinger, S.; Zhou, D.C.; Ding, L.; Ghosh, J.C.; Perego, M.; et al. Mitochondrial Fitness and Cancer Risk. *PLoS ONE* **2022**, *17*, e0273520. [CrossRef]

47. Maniotis, A.J.; Folberg, R.; Hess, A.; Seftor, E.A.; Gardner, L.M.G.; Pe'er, J.; Trent, J.M.; Meltzer, P.S.; Hendrix, M.J.C. Vascular Channel Formation by Human Melanoma Cells in Vivo and in Vitro: Vasculogenic Mimicry. *Am. J. Pathol.* **1999**, *155*, 739–752. [CrossRef]
48. Luo, Q.; Wang, J.; Zhao, W.; Peng, Z.; Liu, X.; Li, B.; Zhang, H.; Shan, B.; Zhang, C.; Duan, C. Vasculogenic Mimicry in Carcinogenesis and Clinical Applications. *J. Hematol. Oncol.* **2020**, *13*, 19. [CrossRef]
49. Wei, X.; Chen, Y.; Jiang, X.; Peng, M.; Liu, Y.; Mo, Y.; Ren, D.; Hua, Y.; Yu, B.; Zhou, Y.; et al. Mechanisms of Vasculogenic Mimicry in Hypoxic Tumor Microenvironments. *Mol. Cancer* **2021**, *20*, 19. [CrossRef]
50. Bisht, S.; Nigam, M.; Kunjwal, S.S.; Sergey, P.; Mishra, A.P.; Sharifi-Rad, J. Cancer Stem Cells: From an Insight into the Basics to Recent Advances and Therapeutic Targeting. *Stem Cells Int.* **2022**, *2022*, 9653244. [CrossRef]
51. Sun, B.; Zhang, D.; Zhao, N.; Zhao, X. Epithelial-to-Endothelial Transition and Cancer Stem Cells: Two Cornerstones of Vasculogenic Mimicry in Malignant Tumors. *Oncotarget* **2017**, *8*, 30502–30510. [CrossRef]
52. Lapkina, E.Z.; Esimbekova, A.R.; Ruksha, T.G. Vasculogenic Mimicry. *Arkh Patol.* **2023**, *85*, 508–525. [CrossRef]
53. Andonegui-Elguera, M.A.; Alfaro-Mora, Y.; Cáceres-Gutiérrez, R.; Caro-Sánchez, C.H.S.; Herrera, L.A.; Díaz-Chávez, J. An Overview of Vasculogenic Mimicry in Breast Cancer. *Front. Oncol.* **2020**, *10*, 220. [CrossRef]
54. Saini, G.; Joshi, S.; Garlapati, C.; Li, H.; Kong, J.; Krishnamurthy, J.; Reid, M.D.; Aneja, R. Polyploid Giant Cancer Cell Characterization: New Frontiers in Predicting Response to Chemotherapy in Breast Cancer. *Semin. Cancer Biol.* **2022**, *81*, 220–231. [CrossRef]
55. Liu, P.; Wang, L.; Yu, H. Polyploid Giant Cancer Cells: Origin, Possible Pathways of Formation, Characteristics, and Mechanisms of Regulation. *Front. Cell Dev. Biol.* **2024**, *12*, 1410637. [CrossRef]
56. Liu, Y.; Sun, B.; Lin, Y.; Deng, H.; Wang, X.; Xu, C.; Wang, K.; Yu, N.; Liu, R.; Han, M. Lysyl Oxidase Promotes the Formation of Vasculogenic Mimicry in Gastric Cancer through PDGF-PDGFR Pathway. *J. Cancer* **2024**, *15*, 1816–1825. [CrossRef]
57. Han, C.; Sun, B.; Zhao, X.; Zhang, Y.; Gu, Q.; Liu, F.; Zhao, N.; Wu, L. Phosphorylation of STAT3 Promotes Vasculogenic Mimicry by Inducing Epithelial-to-Mesenchymal Transition in Colorectal Cancer. *Technol. Cancer Res. Treat.* **2017**, *16*, 1209–1219. [CrossRef]
58. Lizárraga-Verdugo, E.; Avendaño-Félix, M.; Bermúdez, M.; Ramos-Payán, R.; Pérez-Plasencia, C.; Aguilar-Medina, M. Cancer Stem Cells and Its Role in Angiogenesis and Vasculogenic Mimicry in Gastrointestinal Cancers. *Front. Oncol.* **2020**, *10*, 4130. [CrossRef]
59. Paulis, Y.W.J.; Soetekouw, P.M.M.B.; Verheul, H.M.W.; Tjan-Heijnen, V.C.G.; Griffioen, A.W. Signalling Pathways in Vasculogenic Mimicry. *Biochim. Biophys. Acta Rev. Cancer* **2010**, *1806*, 18–28. [CrossRef]
60. Pereira-Veiga, T.; Schneegans, S.; Pantel, K.; Wikman, H. Circulating Tumor Cell-Blood Cell Crosstalk: Biology and Clinical Relevance. *Cell Rep.* **2022**, *40*, 111298. [CrossRef]
61. Patel, D.; Thankachan, S.; Sreeram, S.; Kavitha, K.P.; Suresh, P.S. The Role of Tumor-Educated Platelets in Ovarian Cancer: A Comprehensive Review and Update. *Pathol. Res. Pract.* **2023**, *241*, 154267. [CrossRef]
62. Best, M.G.; Sol, N.; In 't Veld, S.G.J.G.; Vancura, A.; Muller, M.; Niemeijer, A.L.N.; Fejes, A.V.; Tjon Kon Fat, L.A.; Huis In 't Veld, A.E.; Leurs, C.; et al. Swarm Intelligence-Enhanced Detection of Non-Small-Cell Lung Cancer Using Tumor-Educated Platelets. *Cancer Cell* **2017**, *32*, 238–252.e9. [CrossRef]
63. Bian, X.; Yin, S.; Yang, S.; Jiang, X.; Wang, J.; Zhang, M.; Zhang, L. Roles of Platelets in Tumor Invasion and Metastasis: A Review. *Heliyon* **2022**, *8*, e12072. [CrossRef]
64. Ruf, W.; Yokota, N.; Schaffner, F. Tissue Factor in Cancer Progression and Angiogenesis. *Thromb. Res.* **2010**, *125*, S36–S38. [CrossRef]
65. Li, S.; Lu, Z.; Wu, S.; Chu, T.; Li, B.; Qi, F.; Zhao, Y.; Nie, G. The Dynamic Role of Platelets in Cancer Progression and Their Therapeutic Implications. *Nat. Rev. Cancer* **2024**, *24*, 72–87. [CrossRef]
66. Cluxton, C.D.; Spillane, C.; O'Toole, S.A.; Sheils, O.; Gardiner, C.M.; O'Leary, J.J. Suppression of Natural Killer Cell NKG2D and CD226 Anti-Tumour Cascades by Platelet Cloaked Cancer Cells: Implications for the Metastatic Cascade. *PLoS ONE* **2019**, *14*, e0211538. [CrossRef]
67. Karsten, E.; Breen, E.; McCracken, S.A.; Clarke, S.; Herbert, B.R. Red Blood Cells Exposed to Cancer Cells in Culture Have Altered Cytokine Profiles and Immune Function. *Sci. Rep.* **2020**, *10*, 7727. [CrossRef]
68. Liang, N.; Jiao, Z.; Zhang, C.; Wu, Y.; Wang, T.; Li, S.; Wang, Y.; Song, T.; Chen, J.Q.; Liang, H.; et al. Mature Red Blood Cells Contain Long DNA Fragments and Could Acquire DNA from Lung Cancer Tissue. *Adv. Sci.* **2023**, *10*, e2206361. [CrossRef]
69. Nilsson, R.J.A.; Balaj, L.; Hulleman, E.; Van Rijn, S.; Pegtel, D.M.; Walraven, M.; Widmark, A.; Gerritsen, W.R.; Verheul, H.M.; Vandertop, W.P.; et al. Blood Platelets Contain Tumor-Derived RNA Biomarkers. *Blood* **2011**, *118*, 3680–3683. [CrossRef]
70. Girardot, M.; Pecquet, C.; Boukour, S.; Knoops, L.; Ferrant, A.; Vainchenker, W.; Giraudier, S.; Constantinescu, S.N. MiR-28 Is a Thrombopoietin Receptor Targeting MicroRNA Detected in a Fraction of Myeloproliferative Neoplasm Patient Platelets. *Blood* **2010**, *116*, 437–445. [CrossRef]
71. Nilsson, J.; Skog, J.; Nordstrand, A.; Baranov, V.; Mincheva-Nilsson, L.; Breakefield, X.O.; Widmark, A. Prostate Cancer-Derived Urine Exosomes: A Novel Approach to Biomarkers for Prostate Cancer. *Br. J. Cancer* **2009**, *100*, 1603–1607. [CrossRef]
72. Ma, J.; Zhang, H.; Tang, K.; Huang, B. Tumor-Derived Microparticles in Tumor Immunology and Immunotherapy. *Eur. J. Immunol.* **2020**, *50*, 1653–1662. [CrossRef]
73. Bagnall, J.S.; Byun, S.; Begum, S.; Miyamoto, D.T.; Hecht, V.C.; Maheswaran, S.; Stott, S.L.; Toner, M.; Hynes, R.O.; Manalis, S.R. Deformability of Tumor Cells versus Blood Cells. *Sci. Rep.* **2015**, *5*, 18542. [CrossRef]

74. Afify, S.M.; Seno, M. Conversion of Stem Cells to Cancer Stem Cells: Undercurrent of Cancer Initiation. *Cancers* **2019**, *11*, 345. [CrossRef]
75. Foster, B.M.; Shi, L.; Harris, K.S.; Patel, C.; Surratt, V.E.; Langsten, K.L.; Kerr, B.A. Bone Marrow-Derived Stem Cell Factor Regulates Prostate Cancer-Induced Shifts in Pre-Metastatic Niche Composition. *Front. Oncol.* **2022**, *12*, 855188. [CrossRef]
76. Xia, Y.; Cai, X.Y.; Fan, J.Q.; Zhang, L.L.; Ren, J.H.; Chen, J.; Li, Z.Y.; Zhang, R.G.; Zhu, F.; Wu, G. Rho Kinase Inhibitor Fasudil Suppresses the Vasculogenic Mimicry of B16 Mouse Melanoma Cells Both in Vitro and in Vivo. *Mol. Cancer Ther.* **2015**, *14*, 1582–1590. [CrossRef]
77. Colucci-D'amato, L.; Pastorino, O.; Teresa Gentile, M.; Mancini, A.; Del Gaudio, N.; Di Costanzo, A.; Bajetto, A.; Franco, P.; Altucci, L.; Florio, T.; et al. Histone Deacetylase Inhibitors Impair Vasculogenic Mimicry from Glioblastoma Cells. *Cancers* **2019**, *11*, 747. [CrossRef]
78. Hazra, S.; Kalyan Dinda, S.; Kumar Mondal, N.; Hossain, S.R.; Datta, P.; Yasmin Mondal, A.; Malakar, P.; Manna, D. Giant Cells: Multiple Cells Unite to Survive. *Front. Cell Infect. Microbiol.* **2023**, *13*, 1220589. [CrossRef]
79. Zhou, X.; Zhou, M.; Zheng, M.; Tian, S.; Yang, X.; Ning, Y.; Li, Y.; Zhang, S. Polyploid Giant Cancer Cells and Cancer Progression. *Front. Cell Dev. Biol.* **2022**, *10*, 1017588. [CrossRef]
80. Jiao, Y.; Yu, Y.; Zheng, M.; Yan, M.; Wang, J.; Zhang, Y.; Zhang, S. Dormant Cancer Cells and Polyploid Giant Cancer Cells: The Roots of Cancer Recurrence and Metastasis. *Clin. Transl. Med.* **2024**, *14*, e1567. [CrossRef]
81. Liu, K.; Zheng, M.; Zhao, Q.; Zhang, K.; Li, Z.; Fu, F.; Zhang, H.; Du, J.; Li, Y.; Zhang, S. Different P53 Genotypes Regulating Different Phosphorylation Sites and Subcellular Location of CDC25C Associated with the Formation of Polyploid Giant Cancer Cells. *J. Exp. Clin. Cancer Res.* **2020**, *39*, 83. [CrossRef]
82. Zhang, S.; Mercado-Uribe, I.; Xing, Z.; Sun, B.; Kuang, J.; Liu, J. Generation of Cancer Stem-like Cells through the Formation of Polyploid Giant Cancer Cells. *Oncogene* **2014**, *33*, 116–128. [CrossRef]
83. Wang, X.; Zheng, M.; Fei, F.; Li, C.; Du, J.; Liu, K.; Li, Y.; Zhang, S. EMT-Related Protein Expression in Polyploid Giant Cancer Cells and Their Daughter Cells with Different Passages after Triptolide Treatment. *Med. Oncol.* **2019**, *36*, 82. [CrossRef]
84. Liu, Y.; Shi, Y.; Wu, M.; Liu, J.; Wu, H.; Xu, C.; Chen, L. Hypoxia-Induced Polypoid Giant Cancer Cells in Glioma Promote the Transformation of Tumor-Associated Macrophages to a Tumor-Supportive Phenotype. *CNS Neurosci. Ther.* **2022**, *28*, 1326–1338. [CrossRef]
85. Lopez-Sánchez, L.M.; Jimenez, C.; Valverde, A.; Hernandez, V.; Peñarando, J.; Martinez, A.; Lopez-Pedrera, C.; Muñoz-Castañeda, J.R.; De La Haba-Rodríguez, J.R.; Aranda, E.; et al. $CoCl_2$, a Mimic of Hypoxia, Induces Formation of Polyploid Giant Cells with Stem Characteristics in Colon Cancer. *PLoS ONE* **2014**, *9*, e99143. [CrossRef]
86. Fei, F.; Qu, J.; Liu, K.; Li, C.; Wang, X.; Li, Y.; Zhang, S. The Subcellular Location of Cyclin B1 and CDC25 Associated with the Formation of Polyploid Giant Cancer Cells and Their Clinicopathological Significance. *Lab. Investig.* **2019**, *99*, 483–498. [CrossRef]
87. Lu, P.; White-Gilbertson, S.; Beeson, G.; Beeson, C.; Ogretmen, B.; Norris, J.; Voelkel-Johnson, C. Ceramide Synthase 6 Maximizes P53 Function to Prevent Progeny Formation from Polyploid Giant Cancer Cells. *Cancers* **2021**, *13*, 2212. [CrossRef]
88. Li, X.; Zhong, Y.; Zhang, X.; Sood, A.K.; Liu, J. Spatiotemporal View of Malignant Histogenesis and Macroevolution via Formation of Polyploid Giant Cancer Cells. *Oncogene* **2023**, *42*, 665–678. [CrossRef]
89. Zhao, S.; Xing, S.; Wang, L.; Ouyang, M.; Liu, S.; Sun, L.; Yu, H. IL-1β Is Involved in Docetaxel Chemoresistance by Regulating the Formation of Polyploid Giant Cancer Cells in Non-Small Cell Lung Cancer. *Sci. Rep.* **2023**, *13*, 12763. [CrossRef]
90. Niu, N.; Mercado-Uribe, I.; Liu, J. Dedifferentiation into Blastomere-like Cancer Stem Cells via Formation of Polyploid Giant Cancer Cells. *Oncogene* **2017**, *36*, 4887–4900. [CrossRef]
91. Erenpreisa, J.A.; Cragg, M.S.; Fringes, B.; Sharakhov, I.; Illidge, T.M. Release of Mitotic Descendants by Giant Cells from Irradiated Burkitt's Lymphoma Cell Lines. *Cell Biol. Int.* **2000**, *24*, 635–648. [CrossRef]
92. Sundaram, M.; Guernsey, D.L.; Rajaraman, M.M.; Rajaraman, R. Neosis: A Novel Type of Cell Division in Cancer. *Cancer Biol. Ther.* **2004**, *3*, 207–218. [CrossRef]
93. Zhang, Z.; Feng, X.; Deng, Z.; Cheng, J.; Wang, Y.; Zhao, M.; Zhao, Y.; He, S.; Huang, Q. Irradiation-Induced Polyploid Giant Cancer Cells Are Involved in Tumor Cell Repopulation via Neosis. *Mol. Oncol.* **2021**, *15*, 2219–2234. [CrossRef]
94. Liu, J. The "Life Code": A Theory That Unifies the Human Life Cycle and the Origin of Human Tumors. *Semin. Cancer Biol.* **2020**, *60*, 380–397. [CrossRef]
95. Adams, D.L.; Martin, S.S.; Alpaugh, R.K.; Charpentier, M.; Tsai, S.; Bergan, R.C.; Ogden, I.M.; Catalona, W.; Chumsri, S.; Tang, C.M.; et al. Circulating Giant Macrophages as a Potential Biomarker of Solid Tumors. *Proc. Natl. Acad. Sci. USA* **2014**, *111*, 3514–3519. [CrossRef]
96. Mantovani, A.; Allavena, P.; Marchesi, F.; Garlanda, C. Macrophages as Tools and Targets in Cancer Therapy. *Nat. Rev. Drug Discov.* **2022**, *21*, 799–820. [CrossRef]
97. Sutton, T.L.; Patel, R.K.; Anderson, A.N.; Bowden, S.G.; Whalen, R.; Giske, N.R.; Wong, M.H. Circulating Cells with Macrophage-like Characteristics in Cancer: The Importance of Circulating Neoplastic-Immune Hybrid Cells in Cancer. *Cancers* **2022**, *14*, 3871. [CrossRef]
98. You, B.; Xia, T.; Gu, M.; Zhang, Z.; Zhang, Q.; Shen, J.; Fan, Y.; Yao, H.; Pan, S.; Lu, Y.; et al. AMPK-MTOR-Mediated Activation of Autophagy Promotes Formation of Dormant Polyploid Giant Cancer Cells. *Cancer Res.* **2022**, *82*, 846–858. [CrossRef]

99. Shen, L.; Chen, Y.L.; Huang, C.C.; Shyu, Y.C.; Seftor, R.E.B.; Seftor, E.A.; Hendrix, M.J.C.; Chien, D.S.; Chu, Y.W. CVM-1118 (Foslinanib), a 2-Phenyl-4-Quinolone Derivative, Promotes Apoptosis and Inhibits Vasculogenic Mimicry via Targeting TRAP1. *Pathol. Oncol. Res.* **2023**, *29*, 1611038. [CrossRef]
100. Su, W.-C.; Chen, M.H.; Bai, L.-Y.; Chen, J.-S.; Chen, Y.-Y.; Shih, Y.-H.; Wu, I.-C.; Gutheil, J.; Melink, T.J.; Chu, Y.-W.; et al. CVM-1118: A Potent Oral Anti-Vasculogenic Mimicry (VM) Agent in Patients with Advanced Neuroendocrine Tumors (NETs)-A Phase IIa Study. *Am. Soc. Clin. Oncol.* **2023**, *41*, e16235. [CrossRef]
101. Qu, Y.; Zhang, L.; Rong, Z.; He, T.; Zhang, S. Number of Glioma Polyploid Giant Cancer Cells (PGCCs) Associated with Vasculogenic Mimicry Formation and Tumor Grade in Human Glioma. *J. Exp. Clin. Cancer Res.* **2013**, *32*, 75. [CrossRef]
102. Zhang, L.; Ding, P.; Lv, H.; Zhang, D.; Liu, G.; Yang, Z.; Li, Y.; Liu, J.; Zhang, S. Number of Polyploid Giant Cancer Cells and Expression of EZH2 Are Associated with VM Formation and Tumor Grade in Human Ovarian Tumor. *Biomed. Res. Int.* **2014**, *2014*, 903542. [CrossRef]
103. Zhang, D.; Yang, X.; Yang, Z.; Fei, F.; Li, S.; Qu, J.; Zhang, M.; Li, Y.; Zhang, X.; Zhang, S. Daughter Cells and Erythroid Cells Budding from PGCCs and Their Clinicopathological Significances in Colorectal Cancer. *J. Cancer* **2017**, *8*, 469–478. [CrossRef]
104. Liu, G.; Wang, Y.; Fei, F.; Wang, X.; Li, C.; Liu, K.; Du, J.; Cao, Y.; Zhang, S. Clinical Characteristics and Preliminary Morphological Observation of 47 Cases of Primary Anorectal Malignant Melanomas. *Melanoma Res.* **2018**, *28*, 592–599. [CrossRef]
105. Sainero-Alcolado, L.; Liaño-Pons, J.; Ruiz-Pérez, M.V.; Arsenian-Henriksson, M. Targeting Mitochondrial Metabolism for Precision Medicine in Cancer. *Cell Death Differ.* **2022**, *29*, 1304–1317. [CrossRef]
106. Vasan, K.; Werner, M.; Chandel, N.S. Mitochondrial Metabolism as a Target for Cancer Therapy. *Cell Metab.* **2020**, *32*, 341–352. [CrossRef]
107. Piecyk, M.; Ferraro-Peyret, C.; Laville, D.; Perros, F.; Chaveroux, C. Novel Insights into the GCN2 Pathway and Its Targeting. Therapeutic Value in Cancer and Lessons from Lung Fibrosis Development. *FEBS J.* **2024**. [CrossRef]
108. Zhang, X.; Yao, J.; Li, X.; Niu, N.; Liu, Y.; Hajek, R.A.; Peng, G.; Westin, S.; Sood, A.K.; Liu, J. Targeting Polyploid Giant Cancer Cells Potentiates a Therapeutic Response and Overcomes Resistance to PARP Inhibitors in Ovarian Cancer. *Sci. Adv.* **2023**, *9*, eadf7195. [CrossRef]

cancers

MDPI

Review

Components of the Endosome-Lysosome Vesicular Machinery as Drivers of the Metastatic Cascade in Prostate Cancer

Bukuru Dieu-Donne Nturubika [1,*], Jessica Logan [1], Ian R. D. Johnson [1], Courtney Moore [1], Ka Lok Li [1], Jingying Tang [1], Giang Lam [2], Emma Parkinson-Lawrence [1], Desmond B. Williams [1], James Chakiris [1], Madison Hindes [1], Robert D. Brooks [1], Mark A. Miles [3], Stavros Selemidis [3], Philip Gregory [2], Roberto Weigert [4], Lisa Butler [5,6], Mark P. Ward [7], David J. J. Waugh [2], John J. O'Leary [8] and Douglas A. Brooks [1,8,*]

1 Clinical and Health Sciences, University of South Australia, Adelaide, SA 5000, Australia; jessica.logan@unisa.edu.au (J.L.); courtney.moore@unisa.edu.au (C.M.); ka_lok.li@mymail.unisa.edu.au (K.L.L.); jingying.tang@unisa.edu.au (J.T.); emma.parkinson-lawrence@unisa.edu.au (E.P.-L.); des.williams@unisa.edu.au (D.B.W.); james.chakiris@mymail.unisa.edu.au (J.C.); madison.hindes@mymail.unisa.edu.au (M.H.); rob.brooks@unisa.edu.au (R.D.B.)
2 Centre for Cancer Biology, University of South Australia, Adelaide, SA 5000, Australia; giang.lam@unisa.edu.au (G.L.); philip.gregory@unisa.edu.au (P.G.); david.waugh@unisa.edu.au (D.J.J.W.)
3 Centre for Respiratory Science and Health, School of Health and Biomedical Sciences, RMIT University, Bundoora, VIC 3083, Australia; mark.miles@rmit.edu.au (M.A.M.); stavros.selemidis@rmit.edu.au (S.S.)
4 Laboratory of Cellular and Molecular Biology, Center for Cancer Research, National Cancer Institute, National Institutes of Health, Bethesda, MD 20892, USA; weigertr@mail.nih.gov
5 South Australian ImmunoGENomics Cancer Institute, Freemasons Centre for Male Health and Wellbeing, University of Adelaide, Adelaide, SA 5000, Australia; lisa.butler@adelaide.edu.au
6 Solid Tumour Program, Precision Cancer Medicine Theme, South Australian Health and Medical Research Institute, Adelaide, SA 5000, Australia
7 Department of Pathology, The Coombe Women and Infants University Hospital, Trinity College Dublin, D08 XW7X Dublin, Ireland; wardm6@tcd.ie
8 Department of Histopathology, Trinity College Dublin, D08 XW7X Dublin, Ireland; olearyjj@tcd.ie
* Correspondence: bukuru.nturubika@mymail.unisa.edu.au (B.D.-D.N.); doug.brooks@unisa.edu.au (D.A.B.)

Simple Summary: Prostate cancer is a major global health issue, with over 1.4 million new cases and more than 330,000 deaths annually. The main challenge in treating prostate cancer is predicting and managing metastasis, which is currently incurable. This review highlights that metastasis may be driven by disruptions in the vesicular trafficking machinery that controls endosome-lysosome biology. Understanding how changes in the transport of endosomes and lysosomes contribute to cancer progression could lead to new ways to detect and treat metastatic prostate cancer, improving patient outcomes.

Abstract: Prostate cancer remains a significant global health concern, with over 1.4 million new cases diagnosed and more than 330,000 deaths each year. The primary clinical challenge that contributes to poor patient outcomes involves the failure to accurately predict and treat at the onset of metastasis, which remains an incurable stage of the disease. This review discusses the emerging paradigm that prostate cancer metastasis is driven by a dysregulation of critical molecular machinery that regulates endosome-lysosome homeostasis. Endosome and lysosome compartments have crucial roles in maintaining normal cellular function but are also involved in many hallmarks of cancer pathogenesis, including inflammation, immune response, nutrient sensing, metabolism, proliferation, signalling, and migration. Here we discuss new insight into how alterations in the complex network of trafficking machinery, responsible for the microtubule-based transport of endosomes and lysosomes, may be involved in prostate cancer progression. A better understanding of endosome-lysosome dynamics may facilitate the discovery of novel strategies to detect and manage prostate cancer metastasis and improve patient outcomes.

 https://doi.org/10.3390/cancers17010043

Keywords: prostate cancer; metastasis; prognosis; biomarkers; endosome-lysosome dynamics; Rabs; molecular motors; vesicular trafficking machinery

1. Background

Prostate cancer is one of the most prevalent malignancies affecting men, and globally this involves the diagnosis of over 1.4 million new cases and >330,000 patient deaths each year [1–3]. Metastases and castration resistance are directly associated with the significant mortality from prostate cancer and represent critical turning points in the disease course that impact adversely on patient outcomes [4]. Treatment intervention with androgen receptor-targeted agents, taxanes, immunotherapy, targeted radiotherapy, and bone-targeted therapy in most cases only marginally improves survival for patients with advanced disease, highlighting a need to further identify mechanisms underlying prostate cancer metastasis, which can then be targeted. However, current risk stratification strategies have significant problems with accuracy for identifying those patients who will progress to have metastatic disease [4–6].

Alterations to the endosome-lysosome system have been reported to drive key biological features of prostate disease progression (Figure 1). This is presumably due to the central role of these organelles in immunity/inflammation, cell growth, nutrient sensing/recycling, intracellular signalling, intercellular communication, and cell migration [7–9]. Recognising the critical contribution of endosome-lysosome function to all the classical hallmarks of cancer biology was the underlying premise for the breakthrough discovery of endosomal biomarkers that define the primary pathogenesis in prostate cancer [8,10–12]. A set of endosome-lysosome biomarkers was identified that enable more accurate grading of prostate cancer tissue and prediction of patient biochemical and clinical recurrence/metastasis [8,11–16]. The high specificity of these tissue diagnostics suggests that endosomal-lysosomal biology is directly involved in both prostate cancer onset and progression. In this review, we take this a further step and discuss how alterations to vesicular trafficking machinery may regulate endosome-lysosome function and dynamics during prostate cancer progression and metastasis. We also summarise the potential to target this vesicular machinery for biomarker discovery and drug target identification to help address key clinical management problems for patients with prostate cancer metastasis.

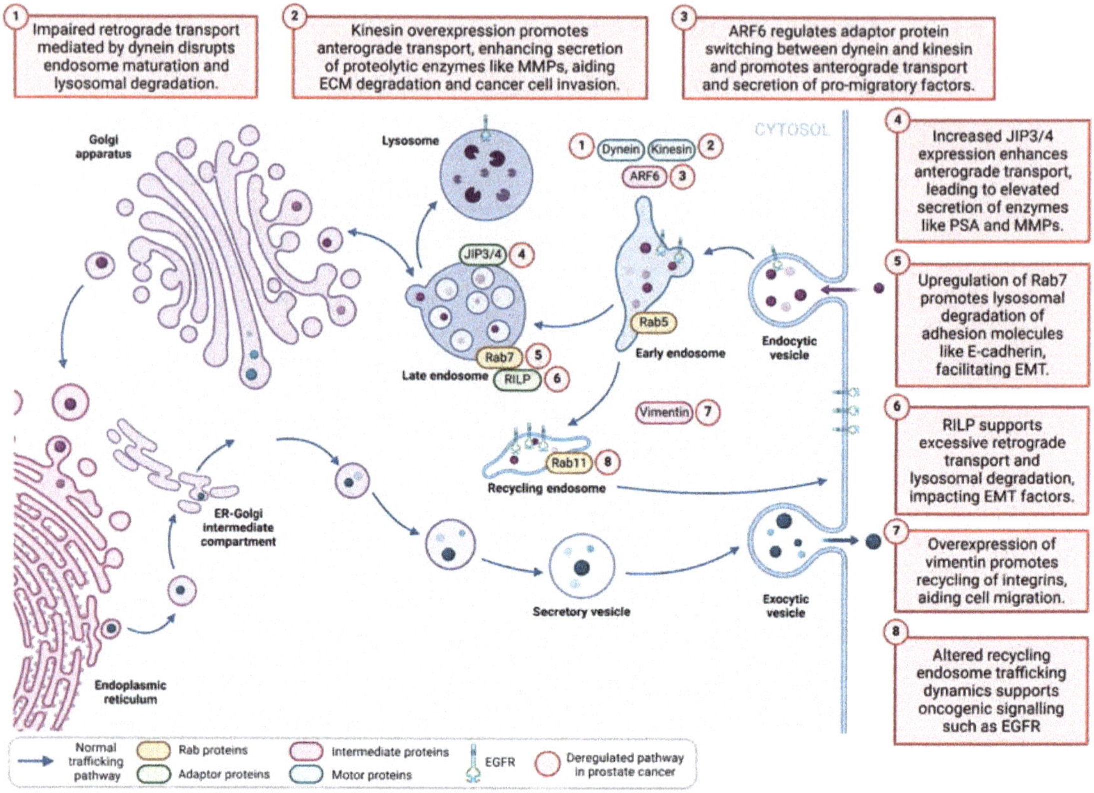

Figure 1. Alterations in the endosome-lysosome system may be key drivers of prostate cancer metastasis.

2. Dysregulation of Endosome-Lysosome Homoeostasis May Promote Prostate Cancer Metastasis

Two main hypotheses are generally accepted to describe the metastatic process, both of which converge on the acquisition of a metastatic phenotype (see [17,18] for a review of both hypotheses). Briefly, the phenotypic plasticity model hypothesises that cancer cells undergo molecular rearrangements that are essential for these cancer cells to leave the primary tumour site, and this results in epithelial-to-mesenchymal transition (EMT) and a more migratory and invasive phenotype. The second clonal model of metastasis hypothesises that a subpopulation of cancer cells is genetically predisposed to metastasis and that driver or acquired mutations endow cancer cells with metastatic potential. These attributes are not mutually exclusive, and it has become increasingly evident that both genetic and non-genetic alterations cooperatively mediate metastatic progression [19]. In addition, alterations to the endosome-lysosome system have been shown to facilitate cancer cell plasticity and activate signalling cascades that promote proliferation, migration, and invasion of metastatic prostate cancer cells [20]. However, the contribution of the endosome-lysosome system to prostate cancer metastasis is less well-defined [21].

Functional alterations of endosomes and lysosomes in prostate cancer cells often feature concomitant disruption to organelle positioning, suggesting that the vesicular trafficking machinery transporting these compartments may be utilised during prostate cancer metastasis [8,22]. For instance, changes in the spatial-temporal dynamics of endosome and lysosome trafficking could lead to increased secretion of proteases that degrade the extracellular matrix, facilitating cancer cell invasion (Figure 1) [23]. Additionally, altered endosome and lysosome function could impact the recycling of cell surface receptors, influencing cell signalling and promoting cell survival, particularly in foreign microenvironments [24]. Thus, the role of vesicular trafficking machinery and how altered organelle dynamics might contribute to prostate cancer progression and metastasis is an attractive concept.

A multitude of adaptor and scaffolding protein complexes are involved in recruiting specific vesicular trafficking machinery to endosome-lysosome compartments to enable effective organelle transport and positioning (reviewed in [25–29]). For example, the assembly of molecular complexes containing the motor proteins dynein or kinesin enables the bidirectional movement (respectively, retrograde vs. anterograde transport) of endosomes and lysosomes along microtubules, utilising ATP hydrolysis to fuel the transport (Figure 2). The molecular motor dynein is a large multiprotein complex that requires stabilisation from its co-factors to engage endosome-lysosome adaptor proteins to mediate retrograde transport and maturation of endosomes to lysosomes [28,30]. Conversely, there are 15 families of the kinesin molecular motor proteins that are selectively recruited by specific adaptor proteins to mediate anterograde vesicular transport, although several classes of these motor proteins can also mediate retrograde transport [31]. In prostate cancer, a mispositioning of endosome compartments has been observed, which is indicative of altered endosomal biogenesis and maturation and may culminate from dysregulated dynein- or kinesin-mediated retrograde or anterograde transport [8]. Despite this circumstantial evidence for the altered spatial-temporal organisation of endosomes and lysosomes in prostate cancer, the altered regulation, bidirectional transport, and functional impact of this dynamic process are yet to be fully investigated in the context of prostate cancer metastasis.

Adaptor protein complexes can be shared between dynein and kinesin and require regulatory proteins to facilitate the switch between dynein-mediated retrograde and kinesin-mediated anterograde transport [32]. Dysfunction of regulatory proteins and adaptor protein recruitment may lead to impaired endosome trafficking events that promote the metastasis of cancer cells. For instance, the increased expression of the regulatory small GTPase (ADP-ribosylation factor 6) ARF6 in cancer cells leads to the disengagement of the dynein adaptor complex, which is comprised of JNK-interacting protein (JIP)-3 and -4 adaptor proteins. This disengagement of dynein promotes the recruitment of kinesin to mediate anterograde transport of late endosomes and subsequent secretion of metalloproteinases that degrade the extracellular matrix during cancer metastasis [32, 33]. While the endosomal release of proteolytic factors such as prostate-specific antigen (PSA) is supported by changes in myosin motor protein expression, which regulates cargo transport on actin cytoskeletal filaments proximal to the plasma membrane, it may also be modulated by microtubule-based anterograde transport, as both systems operate on Rab27 endosomes that transport PSA [34–36]. Moreover, changes in expression of components of the microtubule vesicular trafficking system, which control the switch between anterograde and retrograde transport, such as increased ARF6 and JIP4 (encoded by SPAG9), have been reported to support the proliferation and invasion of prostate cancer cells, possibly by promoting endosome anterograde transport, which supports the release of proteolytic factors like PSA [34,35,37,38]. Therefore, an aberrant shift in the dynamic balance between retrograde and anterograde vesicular transport is a potential contributor to prostate cancer metastasis. The known and theoretical components of microtubule vesicular trafficking machinery that may be involved in prostate cancer metastasis are summarised in Table 1.

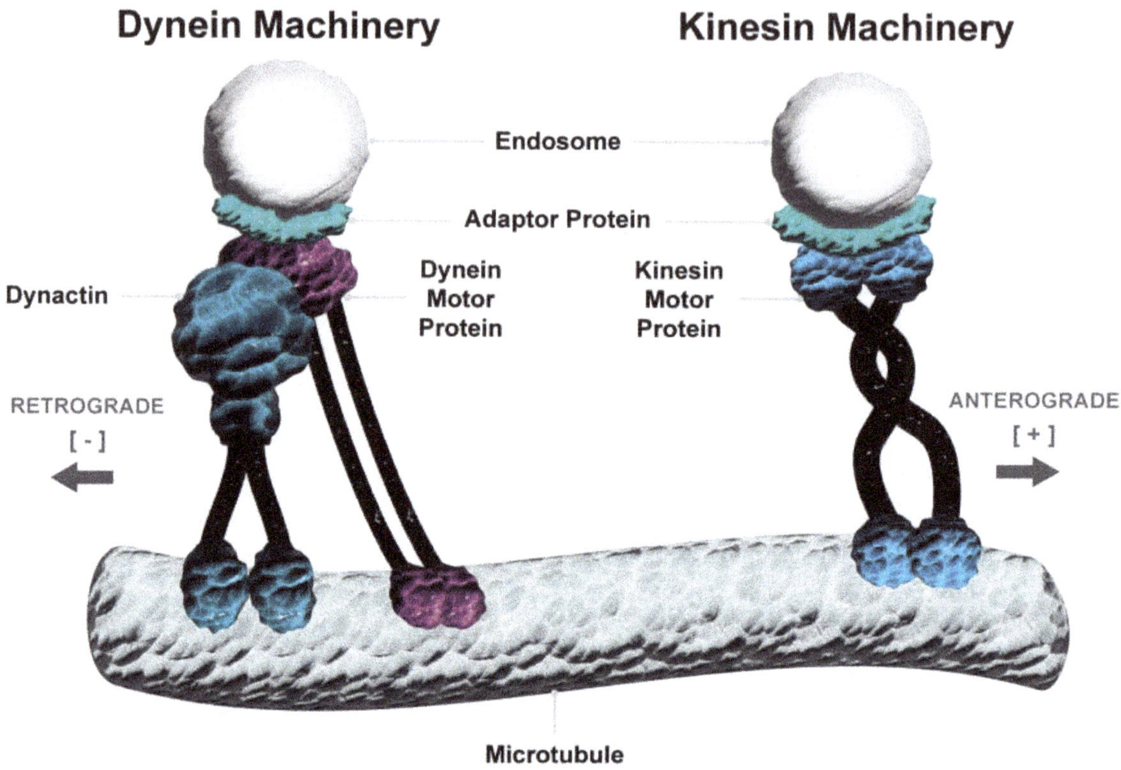

Figure 2. The movement of endosomes and lysosomes on microtubules requires dynein and kinesin motor proteins. Specific adaptor proteins recruit vesicular compartments to dynein and kinesin motor complexes to mediate retrograde movement towards microtubule minus and plus ends, respectively, using ATP hydrolysis as an energy source to drive the process. The dynein motor complex requires stabilisation from the co-factor dynactin.

Table 1. Microtubule vesicular trafficking machinery in prostate cancer metastasis: highlight of known and theoretical components.

Mode of Transport	Vesicular Trafficking Machinery	Component	Endosome Compartment	Role in PCa Metastasis	Refs.
Anterograde	Regulatory protein	ARF6	Lysosomes & Multivesicular bodies	Promotes cancer cell proliferation and invasion & May regulate extracellular vesicle release	[37–40]
	Motor protein	KIFC2	Multivesicular bodies	May regulate extracellular vesicle release	[41,42]
	Motor protein	KIF3A	Recycling endosomes	May regulate invasion and migration	[43,44]
	Motor protein	KIF3C	Recycling endosomes	May regulate treatment response	[45,46]

Table 1. *Cont.*

Mode of Transport	Vesicular Trafficking Machinery	Component	Endosome Compartment	Role in PCa Metastasis	Refs.
Anterograde	Motor protein	KIF5B	Lysosomes	Promotes autophagy	[47]
	Motor protein	KIF5B	Early endosomes	May modulate immune responses	[48]
	Motor protein	KIF13A	Recycling endosomes	May regulate EMT	[43,49,50]
	Motor protein	KIF16B	Recycling endosomes	May regulate migration and invasion	[51,52]
	Motor protein	KLC1/2	Endosomes/lysosomes	May regulate EMT and treatment response	[53–55]
	Adaptor protein	ARL8B	Lysosomes	May modulate treatment response	[53,56]
Retrograde	Adaptor protein	JIP3/4	Lysosomes	Promotes cancer cell migration and invasion	[38,53]
	Motor protein	DYNC1I1	Endosomes/lysosomes	Modulates cancer cell migration and regulates EMT May modulate treatment response	[53,57,58]
	Adaptor protein	Hook3	Early endosomes	Associated with poor prognosis and may promote cancer cell proliferation	[59]
	Regulatory protein	NDEL1	Late endosomes	May regulate EMT	[60]
	Adaptor protein	RILP	Late endosomes/lysosomes	Modulates cancer cell proliferation, migration, and invasion	[61]
	Adaptor protein	TMEM55B	Lysosomes	Promotes autophagy	[62]
	Adaptor protein	Neuropilin-2	Early endosomes	May modulate autophagy and immune response	[63–65]
	Regulatory protein	NDE1	Endosomes/lysosomes	May regulate migration and invasion	[66]

3. Migration and Tissue Invasion by Cancer Cells from the Primary Tumour Within the Prostate

Activation of EMT by cancer cells during the initiation of metastasis results in a shift towards a mesenchymal phenotype. This EMT shift is characterised by the selective, temporal, and reversible dysregulation of adhesion receptors, cytokeratins, and mesenchymal intermediate filaments, which facilitate detachment from the primary tumour and enhance cell migration [67–70]. Partial activation of EMT enhances the metastatic potential of prostate cancer cells and is influenced by transcription factors, cytokines, hypoxia, and androgen signalling. This suggests that an intricate level of plasticity may be operating to exploit endosome and lysosome trafficking and that this is dynamically regulated together with EMT [67,71–76]. For example, the observation of high protein expression of the epithelial adhesion receptor E-cadherin in the presence of low mRNA in prostate cancer cells indicates post-translational regulation that may involve a dysfunction of the endosome machinery that regulates the turn-over of these proteins, as observed in other cancers such as colorectal cancer [77,78]. Moreover, analysis of high-grade prostate cancer tissue showed a redistribution of E-cadherin expression from the plasma membrane to the cytosol when compared to low-grade prostate cancers, which is consistent with an altered regulation of E-cadherin recycling or endosome delivery to the plasma membrane (Figure 3A) [79]. This concept is further supported by the observation that the overexpression of Rab11, which regulates slow endosome recycling kinetics, results in increased sequestration of E-cadherin in recycling endosomes and reduces cell surface delivery in HeLa, HT29, and MDCK cells [77,80]. This process may involve alterations in anterograde transport of recycling endosomes, as the depletion of kinesin-3 motor protein impedes the formation of Rab11 recycling endosomes and promotes the generation of enlarged endosomes that sequester cargo [49,50]. Overexpression of wildtype and constitutively active Rab11 promotes the retrograde transport and perinuclear positioning of recycling endosomes [81]. However, the overexpression of a dominant negative mutant promotes anterograde transport and peripheral accumulation of recycling endosomes [81]. This suggests that recycling endosome function, which is critical for the regulation of epithelial adhesion proteins, is modulated by the bidirectional transport of these organelles. Further exemplifying the importance of endosome trafficking kinetics in modulating EMT factors is the observation that the induction of EMT in MDCK cells resulted in E-cadherin accumulation in enlarged vesicles wherein Rab7 activation promoted E-cadherin lysosomal degradation [82,83]. Rab7 activation is known to be sustained by the Rab-interacting lysosomal protein RILP, which concomitantly recruits the dynein motor protein to promote maturation of late endosomes to lysosomes. This suggests that alterations to retrograde transport that mediate endosome maturation may also contribute to E-cadherin dysregulation [84]. Indeed, the inhibition of Rab7 results in a redistribution of E-cadherin-containing endosomes to the cell periphery and promotes E-cadherin recycling, an event that is consistent with attenuated retrograde endosome transport [83]. We have recently reported the upregulation of EMT proteins, including E-cadherin, and decreased cell migration of prostate cancer cells upon knockdown of dynein light intermediate light chain, possibly due to altered lysosome degradation of EMT factors, which may be modulated by dynein-driven retrograde lysosome trafficking [53,85]. These data also suggest that alterations to the trafficking of E-cadherin may involve disrupted activation of Rab proteins, leading to dysregulation of endosome bidirectional transport. This may have consequences for prostate cancer cells because they have increased levels of Rab GDP dissociation inhibitor, which affects the activation of Rab proteins and would therefore influence the endosome trafficking of epithelial adhesion receptors [86]. These changes in endosome kinetics may also be exploited by cancer cells to modulate the expression of mesenchymal adhesion receptors. For example, the inhibition of N-cadherin recycling can also be induced by reducing recycling endosome populations [87,88]. Interestingly, N-cadherin overexpression is increased with androgen receptor inhibition in prostate cancer cells, with no accompanying augmentation of mRNA levels for the EMT transcription factors that regulate mesenchymal protein expression [89]. Moreover, exogenous androgens increase the expression of dynein, while pharmacological inhibition of dynein in castration-resistant prostate cancer cells promotes increased

N-cadherin expression. This suggests that increased N-cadherin expression is possibly supported by a downregulation of retrograde trafficking machinery (Figure 3A) [89]. Thus, dysregulation of endosome trafficking appears to mediate altered expression and positioning of epithelial and mesenchymal proteins in prostate cancer to facilitate a partial EMT that promotes cancer metastasis.

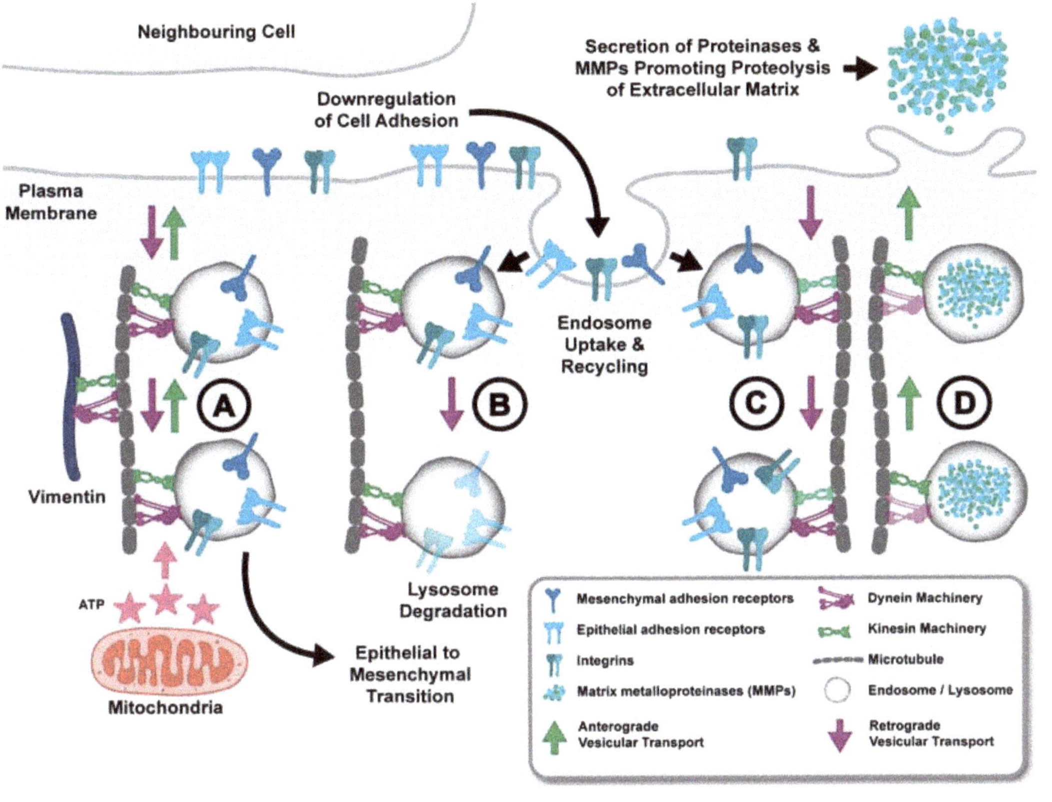

Figure 3. Abnormal localisation and kinetics of endosomes and lysosomes mediated by defective bidirectional transport may aid prostate cancer metastasis via several mechanisms. (A) The dysregulated anterograde and retrograde transport of endosome-lysosome compartments may promote dynamic changes to EMT proteins that facilitate cancer cell detachment and invasion. The metabolic status of the cell may alter the concentration of ATP, which may affect motor protein recruitment to endosomes and impact on ATP-hydrolysis driven processivity to alter endosome bidirectional transport. (B) (degradation) and (C) (internalisation/intracellular traffic). Retrograde transport by dynein trafficking machinery may promote downregulation of integrin proteins to support prostate cancer cell migration. (D) Anterograde transport supported by kinesin trafficking machinery may promote lysosome and endosome secretion of proteolytic factors such as metalloproteinases (MMP) to modify the extracellular matrix (ECM) and support detachment of prostate cancer cells.

Altered expression of integrin adhesion receptors and mesenchymal intermediate filaments, such as vimentin, are also important features of prostate cancer cells undergoing EMT and can potentiate cell migration to facilitate metastasis [90–92]. Notably, vimentin is increased in metastases and circulating tumour cells relative to primary well-differentiated tumours, suggesting that exploring the mechanistic interplay with the endosome-lysosome system could reveal selective therapeutic targets [91,93,94]. Overexpression of vimentin has been shown to promote the recycling of integrins and cell migration, suggesting that overexpression of vimentin in cancer cells may increase the endosomal recycling of migratory factors [95]. This concept is supported by the observation that decreased vimentin interferes with endosome localisation and the dynamic maturation and acidification of early endosomes, leading to endosomal enlargement and cargo arrest [96]. Vimentin interacts with and is phosphorylated by active Rab7, a mechanism that potentiates the rearrangement of the vimentin network [97,98]. Notably, this interaction happens during a Rab7 GDP to GTP switch, which recruits RILP and dynein trafficking machinery to promote Rab7 late endosome maturation to lysosomes [99,100]. Furthermore, vimentin also undergoes bi-directional movement on microtubules and appears to require dynein and kinesin regulation for precise cellular localisation [60,101,102]. This therefore acts as a potential mechanism for regulating cell migration, which requires the transport of vimentin to cell protrusions (Figure 3A). Vimentin positioning mediated by vesicular trafficking machinery, which in turn affects key endosome functions that regulate migratory factors, appears to be a critical determinant for cancer cell migration. A further understanding of the interplay between vimentin, vesicular trafficking machinery, and endosome transport could provide valuable insights into potential therapeutic targets that inhibit prostate cancer progression.

Prostate cancer cells exploit aberrant expression of integrin adhesion receptors, which mediate cell adherence to the ECM and generate tractional forces for cell migration, to support increased tumour cell detachment and directed migration during metastasis [103–106]. While most integrins exhibit decreased expression in prostate cancer compared to normal prostate tissue, some have increased expression, and others can change during cancer metastasis [107,108]. This plasticity in integrin expression may be achieved by dynamic modulation of endosomes and lysosomes and may play a role in promoting a shift from an inefficient random to a highly efficient persistent directional cancer cell migration, which requires organelle polarisation at cell protrusions [109]. Indeed, endosomal regulation of integrin cell surface expression requires both Rab4- and Rab11-positive endosomes [104,110]. Increased Rab4 endosomal recycling of the integrin $\alpha v\beta 3$ and attenuated Rab11 endosomal recycling of $\alpha 5\beta 1$ at migratory cell projections promotes persistent cell migration [104]. Conversely, increased Rab11 endosomal recycling of $\alpha 5\beta 1$ and attenuated Rab4 endosomal recycling of $\alpha v\beta 3$ promotes random migration, suggesting that differential activity and kinetics of particular recycling endosomes carrying specific integrins may affect the migration potential of cells [104]. The ability to selectively downregulate and upregulate Rab4 and Rab11 endosome recycling kinetics is likely mediated by trafficking machinery. The selective shift in balance between respectively fast and slow recycling depends on the expression of specific trafficking motor proteins. For example, reduced kinesin motor protein KIF16B promotes transferrin recycling, whilst concomitantly promoting EGFR degradation whereas overexpression of KIF16B has the opposite effect [51]. Therefore, plasticity to engage specific recycling endosome kinetics to modulate integrin-mediated cell adhesion may play a critical role in prostate cancer cell migration (Figure 3C).

Proteolysis of the extracellular matrix is required for prostate cancer cell migration and involves the secretion of proteolytic enzymes such as the matrix metalloproteinases (MMPs) MMP-2, 3, 7, 9, and cathepsins from late endosomes and lysosomes [8,111–113]. The increased activity of these proteases in prostate cancer cells may reflect concomitant upregulation of endosome anterograde trafficking, which promotes delivery of MMP-containing endosomes-lysosomes to the cell periphery, resulting in secretion or localisation on the cell surface (Figure 3D) [33]. Indeed, in prostate cancer cells, MMP secretion can be stimulated by hepatocyte growth factor, which has been shown to recruit kinesin machinery to promote endosome anterograde transport [22,112,114]. This may further be supported by inefficient recruitment of anterograde transport machinery and increased engagement of retrograde transport machinery, involving expression changes in adaptor proteins or endosome transmembrane proteins. This has been observed in invasive breast cancer cells and leads to the perinuclear arrest of late endosomes and decreased delivery of MMPs to invadopodia [33,115]. Similarly, increased secretion of proteinases may be mediated by mispositioning of lysosomes to the periphery of cancer cells, which can be induced by factors in the tumour microenvironment such as growth factors, acidic pH, and chemotherapeutic agents [116–118]. This plasticity of endosome-lysosome positioning in response to external factors may be mediated by vesicular trafficking machinery adaptor proteins with specific sensor capacity that recruit endosomes-lysosomes for microtubule-based transport depending on cellular context such as nutrient availability [119]. Of particular note is the lysosome transmembrane protein TMEM55B, which is upregulated by the transcription factor TFEB in response to cell starvation to promote dynein-mediated perinuclear clustering of lysosomes that in turn mediate autophagy [62]. TFEB is upregulated in prostate cancer and is thought to promote cancer progression by regulating lysosome biogenesis and plays a role in the resistance of cancer cells to the taxane docetaxel [118,120]. Lysosome trafficking machinery engaged downstream of TFEB, to mediate lysosome mispositioning, may therefore be critical targets for modulating the pro-metastatic effect of altered TFEB expression [118]. Thus, mispositioning of endosomes and lysosomes, mediated by altered trafficking machinery, may therefore enable cancer cells to proteolytically modify the ECM and respond to factors in the tumour microenvironment to support their metastatic dissemination.

During the initiation of metastasis, prostate cancer cells respond to a hypoxic microenvironment, which can be overcome by several mechanisms, including metabolic adaptations that may be mediated by alterations to endosome trafficking [121]. Indeed, prostate cancer cells induce hypoxia response transcription factors, such as HIF-1α, that upregulate endosome recycling of factors, such as glucose transporters that are crucial for metabolic adaptation [122,123]. Prostate cancer cells have increased levels of ATP relative to benign cells, which may affect the processivity of dynein and kinesin motors to support endosome recycling kinetics that are imperative for metabolic adaptation (Figure 3A) [124]. Indeed, kinesin and dynein processivity is increased by higher ATP concentrations, depending on the density of each motor protein recruited to cargo on microtubules [125–128]. Consequently, it is plausible that a metabolic switch from an efficient ATP-yielding glucose to an inefficient ATP-yielding lipid metabolism in prostate cancer, or vice versa, may involve alterations to endosome trafficking events that are mediated by kinesin and dynein. Consistent with this idea is the finding that prostate cancer cells that utilise glucose metabolism have an increased plasma membrane expression of glucose transporters, but in prostate cancer cells that utilise lipid metabolism, the glucose transporters are sequestered into endosomal pools [14]. This observation aligns with the possibility of increased kinesin-mediated endosome recycling of glucose transporters in an efficient ATP production metabolic state and increased dynein-mediated lysosome degradation in an inefficient ATP production metabolic state. However, a significant gap in our understanding is whether the copy number of dynein or kinesin subunits affects the density of the motor proteins for cargo that is recruited to microtubules or which adaptor proteins can alter the density to modulate the direction of bidirectional endosome transport in different metabolic states of the cell.

4. Tumour Cells Hijack Endosome/Cell-Surface Receptors and Signalling Cascades to Access and Migrate Through the Vasculature

Dysregulation of endosome machinery can facilitate the modification of the local microenvironment during prostate cancer invasion and migration and may also be exploited for successful intravasation of prostate cancer cells during their metastatic dissemination. This may involve a capacity to directly interact with endothelial factors to promote angiogenesis and endothelial permeability. For instance, increased endosomal secretion of vascular endothelial growth factor (VEGF) by prostate cancer cells not only promotes angiogenesis but also increases endothelial permeability to enable cancer cell intravasation (Figure 4A,B) [129]. Interestingly, increased VEGF secretion from prostate cancer cells was mediated by increased exocytosis from Rab11 endosomes, which was supported by increased expression of the myosin VI motor protein that promotes exocytosis by tethering secretory cargo to the actin network [34,130]. Studies on hippocampal neurons have shown that the transport of VEGF-containing endosomes is predominantly anterograde and is disrupted by the knockdown of kinesin-1B [131]. Therefore, dysregulated anterograde endosome trafficking machinery may account for the increased secretion of VEGF in prostate cancer cells, in combination with factors such as myosin-mediated endosome tethering (Figure 4A). The increased expression of chemokines that are regulated by endosomes and lysosomes, such as CXCL8, CXCL12, CCL2, and CXCR4, in prostate cancer cells enables them to directly interact with endothelial cells to support intravasation [132,133]. It has been shown that the chemokines are mainly sorted for lysosomal degradation upon endocytosis, suggesting that their upregulation in prostate cancer may also reflect a down-regulation of lysosome degradation kinetics (Figure 4C) [134]. Thus, the dysregulation of endosome trafficking machinery appears to be imperative for the endosomal mechanisms that enable the intravasation of prostate cancer cells. Further exploration of the specific adaptor proteins and motor proteins involved in this process is an important focus for further research on cancer metastasis.

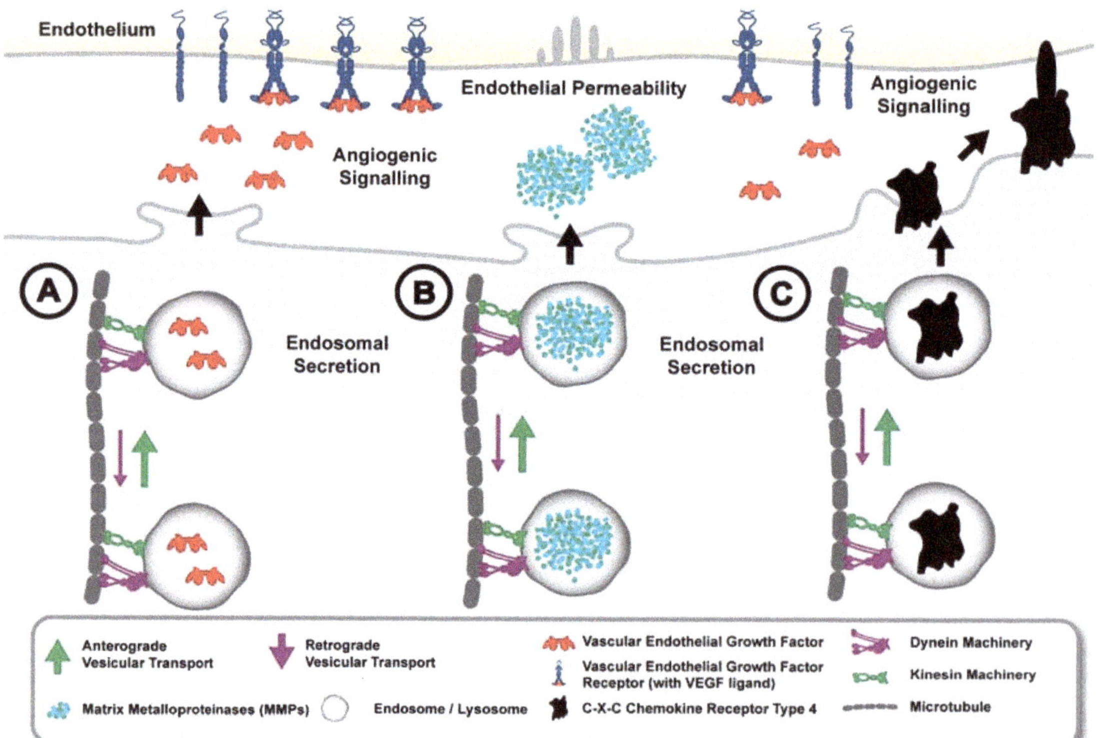

Figure 4. Dysregulated anterograde vesicular transport in prostate cancer cells may support secretion of factors and expression of receptors that promote interaction with the endothelium and the secretion of proteolytic factors that promote endothelial permeability. For example, (A), increased endosome secretion of proteins such as vascular endothelial growth factor, (B), secretion of proteinases, and (C), increased

expression of receptors such as CXCR4 may be mediated by dysregulated anterograde trafficking machinery to support angiogenic signalling and endothelial permeability, which enables prostate cancer cells to intravasate and migrate into the circulation.

5. Survival of Cancer Cells in the Circulation Is Dependent on the Evasion of the Immune System and Interactions with Endothelial Factors

Immune evasion and cytoprotective interactions with endothelial factors are mechanisms regulated by endosomes and lysosomes and may be utilised by prostate circulating tumour cells (CTCs) to support their survival in the circulation [135,136]. The downregulation of major histocompatibility complex (MHC) class I molecules by lysosome degradation to facilitate immune evasion may involve the recruitment of specific trafficking machinery that mediates endosome maturation to lysosomes (Figure 5C) [135,137]. It has been reported that late endosomes carrying MHC class II are retained in a perinuclear region when their dynein-mediated motility overpowers kinesin-mediated motility, and this attenuates MHC class II antigen presentation, implicating a requirement for vesicular trafficking machinery activity for the endosome regulation of MHC proteins [138]. MHC class I colocalises with MHC class II endosomes, suggesting that alterations to the transport of MHC class II endosomes may also affect MHC class I trafficking [139]. Indeed, KIF13A and KIF5B kinesin motors are key mediators of the endosome traffic that may promote recycling of MHC class I molecules, highlighting a role for kinesins in modulating antigen presentation in the immune system [48,140,141]. KIF5B is increased in prostate cancer cells, but whether it mediates the ability of prostate cancer CTCs to downregulate MHC class I as a mechanism for immune evasion has not been explored [47,48]. Prostate cancer cells also overexpress endosome-regulated ligands such as programmed death-ligand 1 (PD-L1), which enable direct interaction with T lymphocytes to suppress the immune system, suggesting that prostate cancer cells harbour the capacity for endosomal immunomodulatory plasticity to downregulate and upregulate pathways that mediate a net effect of immune suppression when interacting with lymphocytes (Figure 5B) [142]. Anti-tumour activity of immune checkpoint inhibitors, such as those directed against CTLA4, has demonstrated that enhancing the interactions of cancer cells with immune cells is a targetable mechanism for prostate cancer immunotherapy [143]. Thus, the potential role of endosome trafficking machinery to regulate endosomal immunomodulatory plasticity in prostate cancer cells presents an attractive target for intervention.

Impaired regulation of cell surface receptors by endosomes and lysosomes can lead to abnormal receptor expression and dysregulated endosome secretory function, which may support the survival of CTCs in the vasculature by facilitating interactions with endothelial factors and platelets. For example, prostate cancer cells derived from bone metastases overexpress protease-activated receptor 1 (PAR1), a thrombin receptor that enables CTCs to interact with platelets and to form protective heterotypic clusters (Figure 5A) [144,145]. Normally, PAR1 is quickly endocytosed and targeted for endosomal-lysosomal degradation upon activation by thrombin [146,147]. However, Rab11 endosomal recycling can maintain cell surface expression and signalling duration for both inactive and active PAR1 [146,147]. Interestingly, invasive breast cancer cells overexpress and maintain the activation of PAR1 even without thrombin, possibly due to increased recycling of PAR1 and failure to sort it into degradative endosomes-lysosomes [146]. This upregulation of PAR1 in cancer cells is typically seen in advanced stages of metastasis, where endosomes and lysosomes also show altered localisation and function, suggesting dysfunction in vesicular trafficking machinery may be involved in mediating this endosome positioning and function (Figure 5) [117,144,146]. The dysregulation of endosome secretion of metalloproteinases due to altered trafficking machinery may also function in CTCs to support their interactions with platelets. Consequently, upregulated MMP-2 secretion by metastatic prostate cancer cells can promote platelet aggregation [148]. Moreover, the secretion of growth factors by platelets, such as platelet-derived growth factor, may help promote the survival of prostate cancer circulating tumour cells. This may operate by sustained endosomal recycling of receptors that are receptive to these growth factors [149,150]. In this context, the activation of platelet-derived growth factor receptor is known to mediate the interaction with phosphatidylinositol 3-kinase, which promotes kinesin-mediated anterograde transport of recycling endosomes [150,151]. This suggests a role for endosome trafficking machinery in promoting processes that are imperative for the survival of CTCs. Alterations to the vesicular trafficking machinery that mediates the endosomal upregulation of factors, which facilitate cancer cell interactions with endothelial factors, may therefore be a critical mechanism that supports the survival and metastasis of cancer cells.

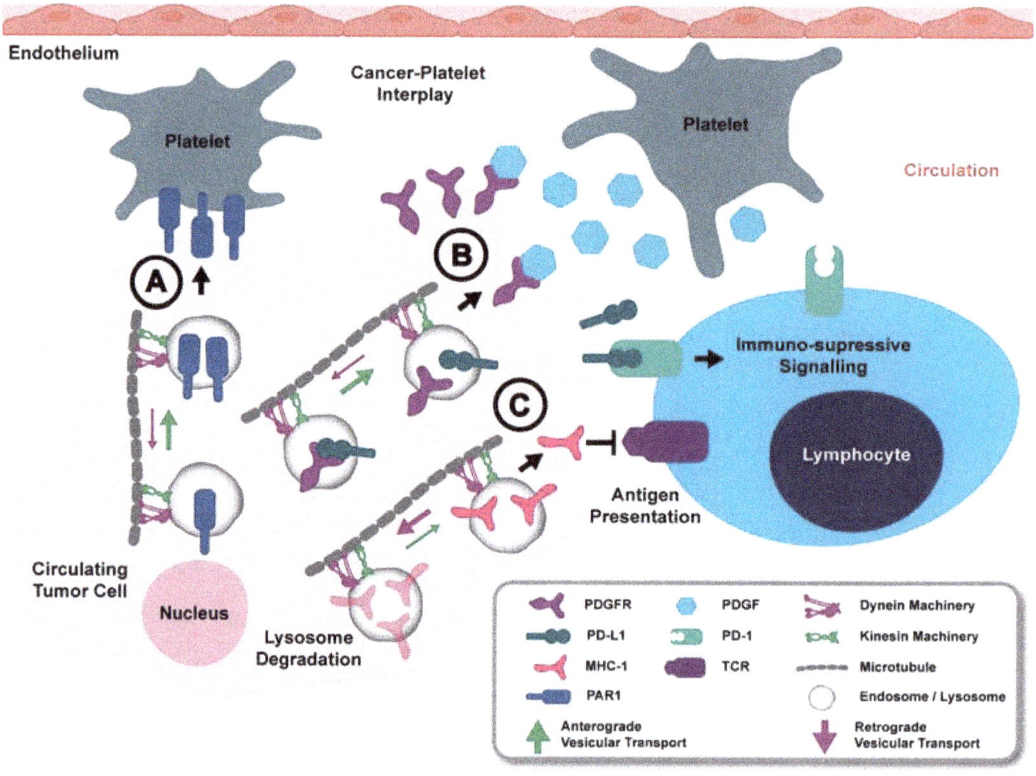

Figure 5. Altered endosome trafficking may support survival of prostate cancer circulating tumour cells. (A) Anterograde trafficking machinery may promote the trafficking of factors such as Protease-activated receptors 1 (PAR1) that enable interaction with platelets. (B) Altered endosome trafficking machinery may promote endosome recycling of receptors such as platelet-derived growth factor receptor (PDGFR), which promotes survival of circulating tumour cells by responding to growth factors secreted by platelets, such as platelet-derived growth factor (PDGF). This altered endosome recycling dynamics may also support the expression of programmed death-ligand 1 (PD-L1), which interacts with the programmed death 1 (PD-1) receptor on lymphocytes to propagate immune suppression signalling. (C) Retrograde trafficking of immune presentation factors such as major histocompatibility complex class I (MHC-I) molecules may contribute to cyto-protection of cancer cells against immune attack by lymphocytes potentially inhibiting binding to T-cell receptor (TCR).

6. Invadopodium Formation, Aberrant Receptor Expression, and Protein Secretion Support the Extravasation, Dormancy, and Colonisation of Metastatic Cancer Cells

Prostate cancer CTCs form invadopodium projections, which are enriched with endosome-regulated proteins such as the metalloproteinase MT1-MMP and tyrosine kinase substrate 5 (Tks5), and they can utilise these to proteolytically modify the endothelium and gain access to distant sites (Figure 6) [152–155]. Tks5 is an important scaffolding protein that requires regulation by the serine/threonine-protein kinase 3 (TAO3) and endosome transport for incorporation in invadopodia [155–157]. Interestingly, TAO3 promoted invadopodium formation by promoting Tks5 trafficking in recycling endosomes, involving inhibiting the activation of the dynein light chain LIC2, suggesting that recycling endosome anterograde kinetics may supersede retrograde kinetics to mediate invadopodium formation and to potentiate extravasation [157]. Moreover, Rab40b is an important Tks5 binding partner that localises to recycling endosomes, and its upregulation in cancer cells to promote MMP secretion and invadopodium-mediated proteolysis of the ECM [158]. Rab40b also localises to the leading edge of migrating cells, where it interacts with Rap2 GTPase, a small GTPase required for cell migration, preventing its lysosomal degradation and recycling back to migratory projections. This further highlights the exploitation of endosome trafficking kinetics in pathways that regulate invadopodium formation to mediate cell migration [159]. Altered endosome spatial and temporal kinetics may play a critical role in invadopodium formation, proteolysis of the ECM, and cancer cell migration, which are required for the successful extravasation of prostate cancer cells.

Tumour cells disseminating to a distant site may attenuate their response to therapy and evade clinical detection by remaining dormant until conditions favour proliferation and formation of a metastatic tumour [160]. Indeed, disseminated prostate cancer cells (DTCs) that have gained access to the bone marrow can remain dormant by exploiting quiescent cell signalling, which is usually reserved for the regulation of haemopoietic stem cells [160,161]. DTCs may exhibit dysregulated receptor expression that enables them to colonise the haematopoietic stem cell niche and respond to ligands secreted by resident osteoblasts and endothelial cells that can induce DTC dormancy [162]. For example, AXL is an important receptor tyrosine kinase that is expressed by prostate cancer DTCs and responds to growth arrest-specific 6 (Gas6) protein secreted by osteoblasts and promotes dormancy by limiting DTC proliferation (Figure 6) [163]. Regulation of AXL by the endosome-lysosome system is thought to be initiated when AXL is endocytosed upon Gas6 stimulation, whereby it is either recycled back to the plasma membrane or targeted to endosomes-lysosomes for degradation [164]. Under hypoxic conditions, increased AXL expression and signalling is independent of Gas6, probably due to its delayed endosomal-lysosomal degradation, and this may also promote prolonged signalling (Figure 6) [164,165]. Interestingly, overexpression of the kinesin motor protein KIF16B can inhibit the endosomal-lysosomal degradation of EGFR, highlighting that the degradation kinetics are directly affected by alterations to specific vesicular trafficking machinery [51]. Moreover, altered expression of several kinesins has been described to play a role in prostate cancer metastasis by promoting aberrant AR signalling and inhibiting apoptosis, but their role in regulating endosome-lysosome degradation dynamics remain relatively unexplored [166,167]. The attenuated endosomal-lysosomal degradation of aberrantly expressed receptors may be a key mechanism used by DTCs to prolong signalling cascades. This would support their dormancy at a secondary site, and the potential role of vesicular trafficking machinery in this process warrants further investigation.

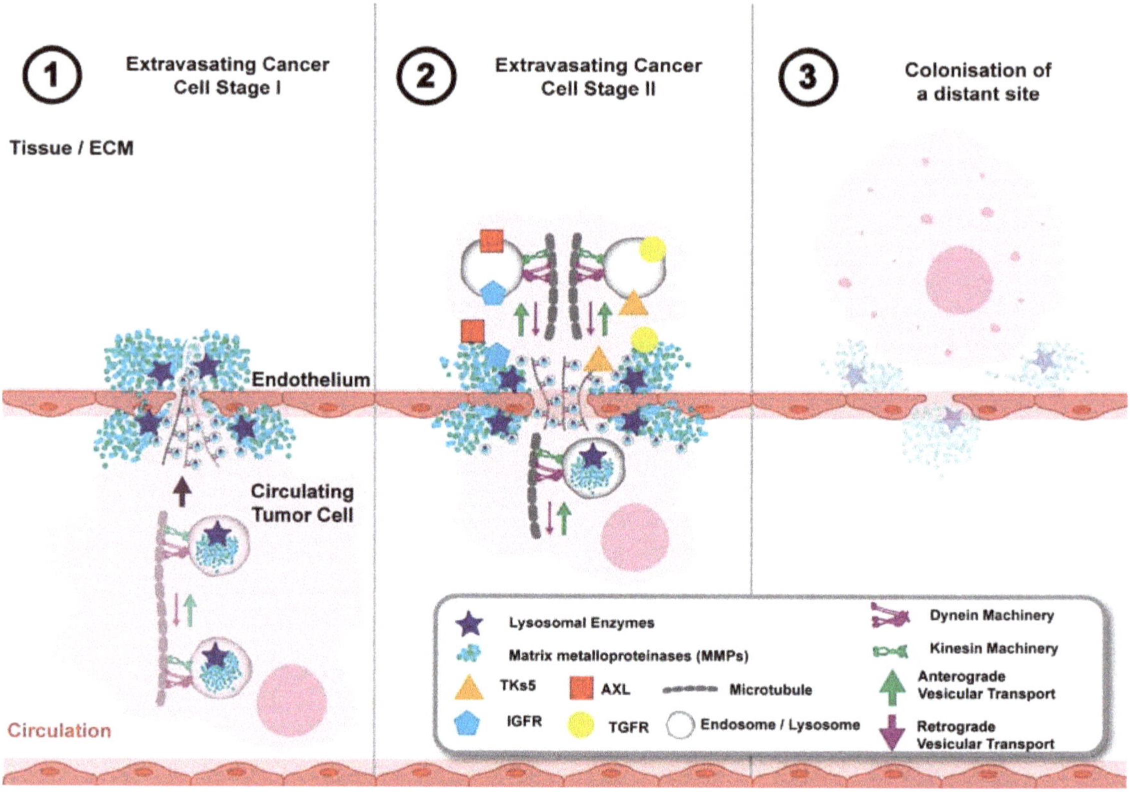

Figure 6. Trafficking of endosomes and lysosomes to the cell periphery may support the stages of extravasation and survival of prostate cancer cells at distant sites. During stages 1–2, dysregulated anterograde and retrograde trafficking mediated by kinesin and dynein trafficking machinery may be involved in the recruitment of endosomes carrying proteins such as tyrosine kinase substrate 5 (Tks5) and serine/threonine-protein kinase 3 (TAO3) to invadopodia projections to support the secretion of metalloproteinases (MMPs) to proteolytically modify the endothelium at sites of metastasis. During stages 2–3, dysregulation of trafficking machinery that results in anterograde trafficking of endosomes may contribute to mispositioning that supports secretion of proteins such as bone morphogenic proteins (BMP), parathyroid hormone-related protein (PTHrP), and expression of receptors such as AXL, transforming growth factor receptor (TGFR), and insulin growth factor receptor (IGFR) that integrate signalling cascades to enable interactions with resident cells in the site of metastasis and enable proteolytic modification of the extracellular matrix (ECM) at the site of metastasis.

Prostate cancer DTCs utilise signalling from growth factors such as transforming growth factor beta (TGF-β) and insulin-like growth factor 1 (IGF-1), which are liberated by osteoblastic and osteoclastic remodelling of the bone, to support their proliferation, survival, and subsequent colonisation [162,168,169]. To induce osteoclast and osteoblast bone remodelling, prostate cancer cells secrete endosome-regulated factors such as parathyroid hormone-related protein (PTHrP) and bone morphogenic proteins (BMPs) [170]. Both BMP and PTHrP secretion require trafficking by Rab11 endosomes and altered fusion with the plasma membrane; hence the upregulation of Rab11 in prostate cancer may facilitate increased secretion of these factors for bone remodelling [171,172]. The sorting of cargo to Rab11 recycling endosomes is dependent on kinesin motors, which promotes increased peripheral trafficking and exocytosis of cargo from recycling endosomes [50,173]. Indeed, kinesin-1 and kinesin-3 trafficking machinery have been shown to potentiate endosome exocytosis, while increased Rab11 endosomes in prostate cancer provide circumstantial evidence of potential alterations to recycling endosome dynamics [174–179]. Further characterisation of the role of vesicular trafficking machinery is required to determine if this is critical for endosomal secretion of bone resorption factors to support the colonisation and survival of prostate cancer cells.

Extracellular vesicles (EVs) are crucial for intercellular communication and are released by prostate cancer cells to aid in interactions with the extracellular matrix and to modulate the immune response. EVs are therefore able to influence therapeutic outcomes and to support cancer cell dissemination [180,181]. Emerging evidence suggests that the contents of prostate cancer-derived EVs, such as miRNAs, proteins, and lipids, may also serve as biomarkers for disease diagnosis and prognosis, highlighting important clinical potential [182–185]. The fusion of multivesicular bodies with the plasma membrane is an important mechanism of EV release and may be affected by changes in endosome trafficking dynamics [186]. Indeed, the upregulation of the kinesin motor KIFC2 in prostate cancer is known to regulate the transport of multivesicular bodies [41,42]. Rab27 is one of the critical Rab GTPases that regulates the release of EVs and is also upregulated in prostate cancer [187,188]. Interestingly, the ability of Rab27 to promote exosome release is regulated by the scaffolding protein KIBRA, which is also upregulated in prostate cancer and requires interaction with dynein for its co-activator functions [189–191]. Consequently, the altered endosome biology that promotes increased EV release by prostate cancer cells involves utilising specific components of the vesicular trafficking machinery.

7. The Potential for Altered Vesicular Trafficking Machinery to Inform on Prostate Cancer Diagnosis and Prognosis

There is an urgent need for more effective biomarkers to improve risk stratification in the management of patients with prostate cancer and particularly to accurately assign patients to either active surveillance or specific interventions [192]. The kallikrein panel, consisting of the endosomal secreted proteins total PSA, free PSA, intact PSA, and human kallikrein-2, is the current gold standard screening approach for prostate cancer but was originally only proposed to be used for monitoring [193]. PSA colocalises with motor protein myosin VI in transferrin-positive recycling endosomes in prostate cancer cells, and its secretion was reduced when myosin VI was knocked down, highlighting a critical role for endosome transport in modulating PSA secretion [34]. Interestingly, myosin VI is overexpressed in prostate cancer, with the highest expression in high-grade prostate cancers, suggesting that the motor proteins regulating endosome transport and secretion of kallikrein biomarkers may help inform on advanced prostate cancer [194]. It is likely that bidirectional endosome transport on microtubules may also impact the secretion of kallikreins, as kinesin and dynein motor proteins are required for the transport of transferrin-positive recycling endosomes, and changes in their expression are sufficient to modulate endosome trafficking (Figure 7A) [51,195]. Proteomic analysis of microvesicles secreted by prostate cancer cells revealed a high content of cytoskeletal proteins, including the heavy chain of dynein, suggesting a potential avenue for the clinical detection of aberrantly expressed vesicular trafficking machinery that may inform on prostate cancer prognosis [196].

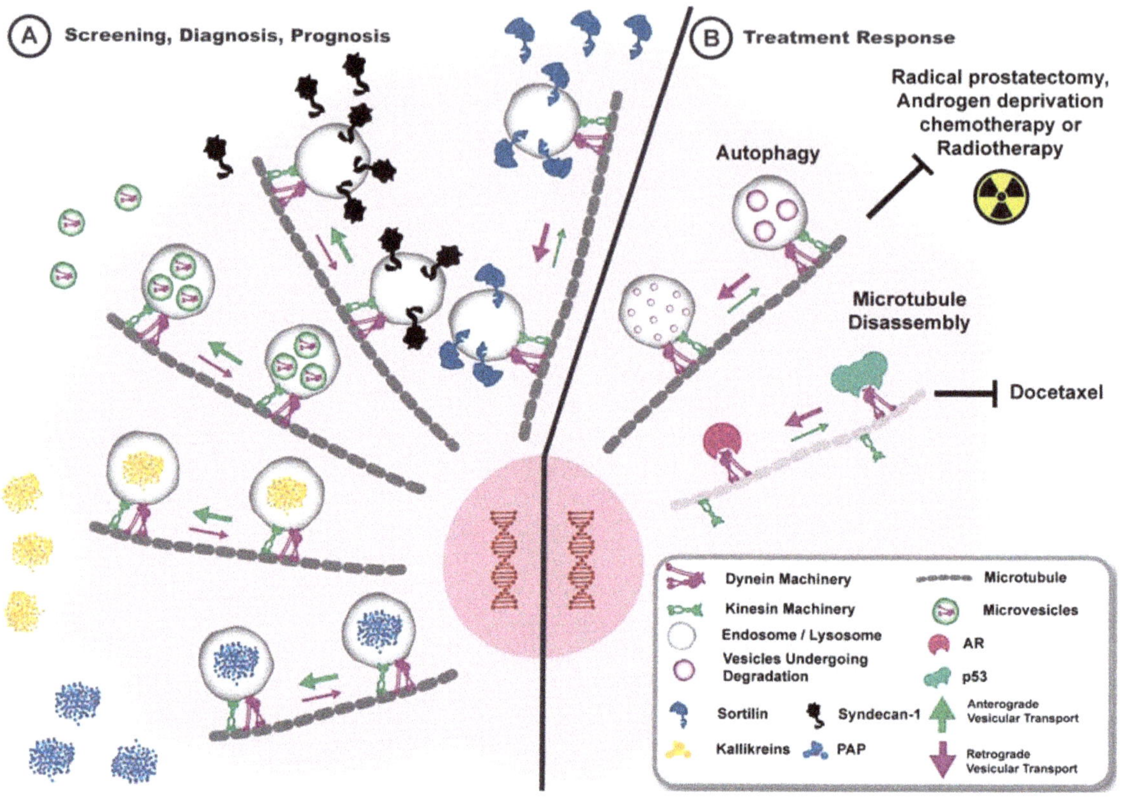

Figure 7. Alterations to vesicular trafficking machinery may have implications for the screening, diagnosis, prognosis, and treatment response for prostate cancer. (A) Anterograde vesicular trafficking machinery may operate to promote endosomal secretion of biomarkers such as kallikreins, prostatic acid phosphatase, and secretion of microvesicles enriched with vesicular trafficking machinery proteins. Upregulation of syndecan-1 and reduction in sortilin in advanced prostate cancer is associated with defects in endosome trafficking that may involve vesicular trafficking machinery. (B) Germline and somatic mutations may directly affect vesicular trafficking machinery or indirectly through promoting vesicular trafficking machinery-regulated endosomal pathways such as autophagy that facilitate poor treatment response for prostate cancer. Microtubule disassembly mediated by kinesin motor proteins may operate as a mechanism of resistance for agents such as docetaxel. Dynein motor proteins may be involved in aberrant translocation transcription factors such as androgen receptor (AR) and tumour suppressors such as p53 to the nucleus.

Assessment of prostate histopathology using H&E staining, which is often complemented by immunohistochemistry, remains the gold standard for prostate cancer diagnosis [197]. Upregulation of biomarkers, including PSA, prostatic acid phosphatase, and prostein, in prostate cancer may involve altered endosome trafficking kinetics that regulate protein localisation, degradation, and secretion, and is likely controlled by changes in vesicular trafficking machinery (Figure 7A) [34,35,197]. Direct examination of the endosome-lysosome system revealed that there is increased syndecan-1 and reduced sortilin in high-grade prostate cancer, and mechanistic studies further linked the activity of these proteins to dysregulated endosome trafficking and a lipogenic phenotype associated with a higher metastatic propensity [8,14]. Altered vesicular trafficking machinery may play a critical role in the aberrant protein expression, location, and function of these endosome-regulated biomarkers. For instance, sortilin interacts with early and recycling endosomes (positive for Rabs 4, 10, 11, and 14) to regulate glucose transporter 1 transport in androgen-responsive prostate cancer cells [14,198–200]. Syndecan-1, when overexpressed in advanced prostate cancer cells, promotes an invasive lipogenic phenotype by interaction with β3 integrin, facilitated by concurrent expression of MT1-MMP [14]. Additionally, prostatic acid phosphatase relies on trafficking within Rab27 endosomes for secretion [35]. Earlier we described a model for altered vesicular trafficking machinery that contributes to dysfunction in endosomal recycling and secretion during prostate cancer metastasis and which impacts the function of proteins implicated in prostate cancer metastasis. This highlights the potential opportunity to explore vesicular trafficking machinery biology further for biomarker discovery in prostate cancer.

8. Altered Vesicular Trafficking Machinery May Correlate with Poorer Outcomes for Patients with Prostate Cancer

Vesicular trafficking machinery requires microtubules to transport endosomal cargo and would therefore be affected by chemotherapeutic agents that stabilise these microtubules. Indeed, prostate cancer chemotherapeutics such as docetaxel function by stabilising microtubules, leading to mitotic arrest and apoptosis of cancer cells, in addition to inhibiting translocation of the oncogenic androgen receptor (AR) to the nucleus [201,202]. It has been postulated that overexpression of kinesins in the kinesin 1, 5, 12, and 14 families may facilitate docetaxel resistance by binding to and altering microtubule stabilisation (Figure 6B) [203–206]. Moreover, activation of EMT and autophagy, which we highlighted as being regulated by vesicular transport machinery, are recognised as mechanisms for docetaxel resistance that complicate treatment outcomes [207,208]. Several other lines of evidence point to a critical role of trafficking machinery in modulating treatment response. For instance, breast cancer cells that were resistant to ionising radiation engaged the kinesin adaptor protein Arl8b to recruit lysosomes for anterograde transport and subsequent proteinase secretion promoted cancer cell invasion into the ECM [56]. This suggests that aberrant trafficking of endosomes and lysosomes may contribute to radiotherapy resistance, a concept that needs further investigation in prostate cancer [209,210]. The ability of dynein and kinesin motor proteins to modulate the stability of microtubules by mediating the assembly and disassembly of tubulin proteins may also have direct consequences for treatment response [211]. Indeed, decreased dynein expression in patient tissues with colorectal cancer was found to be associated with poor outcome and was postulated to relate to microtubule dynamics, possibly by enabling kinesin-mediated microtubule disassembly to induce a phenotype that is resistant to therapy [212]. Furthermore, whilst colorectal cancer cells are sensitised to therapeutic agents such as oxaliplatin in the absence of dynein, taxanes that stabilise the microtubule are ineffective, possibly due to increased kinesin-promoted disassembly (Figure 7B) [57,213]. Thus, altered trafficking machinery-mediated transport of endosomes and lysosomes may persist in cancer cells and dictate response to therapy. Proof of concept for the therapeutic targeting of trafficking machinery motor proteins in cancer has been demonstrated with the discovery of the aminothiazole dynarrestin, a novel inhibitor of dynein that inhibited the proliferation of oesophageal squamous cell carcinoma cells [214]. Furthermore, the potential for targeting kinesins in cancer has also been reviewed, though the current compounds developed target kinesins predominantly involved in regulating mitosis and cytokinesis, and not those involved in regulating endosome dynamics during cancer metastasis [215–217].As dysfunction of vesicular trafficking machinery affects specific endosome pathways in the metastatic cascade, targeting the machinery that is recruited to mediate specific endosome functions may be attractive in precision medicine applications for the clinical management of prostate cancer.

Chimeric antigen receptor T-cell (CAR-T) therapy and immune checkpoint inhibitors are emerging as promising therapeutic approaches for prostate cancer and rely on targeting increased expression of antigens on prostate cancer cells, such as prostate-specific membrane antigen (PSMA), or modulating the expression of increased immunosuppressive receptors, such as PD-L1, respectively [142,218,219]. Challenges with clinical efficacy of these approaches include the immunosuppressive microenvironment, which involves vesicular trafficking to dynamically modulate the surface expression of receptors and antigens to facilitate prostate cancer cell immune escape during metastasis [220,221]. Targeting the vesicular trafficking machinery to control expression of immunomodulatory factors for therapeutic benefit is compelling and is supported by the observation that inhibition of the endosome trafficking regulator neuropilin-2 reduces PD-L1 expression and enables antitumor immune cell activation in prostate cancer cells [63,64].

DNA sequencing for oncogenic mutations that are frequently detected in prostate cancer metastasis is an important approach for understanding prostate cancer disease predisposition, progression, and, more increasingly, for monitoring clinical outcomes. This is supported by recent advances in sequencing technology and knowledge of specific germline and somatic mutations involved in prostate cancer metastasis [222,223]. Mutations in vesicular trafficking machinery genes or known oncogenes could potentially disrupt endosome-lysosome function, which may disrupt or promote a metastatic phenotype in prostate cancer cells. In the former case, mutations in motor proteins may impair their processivity and ability to engage adaptor proteins and recruit cargo for microtubule-based transport [224,225]. In the latter case, germline and somatic mutations in known oncogenes may promote cell adaptation mechanisms that involve engagement and disruption of vesicular trafficking systems. Analysis of the prostate cancer exome revealed that the kinesin heavy chain gene KIF5A was mutated in some prostate cancers, and amplification of KIF18B was significantly higher in tumours with PTEN mutations, supporting the hypotheses that vesicular trafficking machinery mutations may be oncogenic and that oncogenic mutations may involve vesicular trafficking machinery (Figure 7B) [166,226]. Specific mutations have significant implications for treatment response in prostate cancer. For instance, germline mutations such as BRAC1/BRAC2 and somatic mutations such as tumour protein p53 (p53) are notably associated with increased risk of prostate cancer progression and poor survival outcomes following treatments such as radical prostatectomy, androgen deprivation chemotherapy, or radiotherapy [223,227–231]. Inducing autophagy, a process regulated by endosomes and lysosomes, is one of the mechanisms for treatment resistance, and this is affected by these mutations. This suggests that engagement of the endosome-lysosome system downstream of the oncogenic mutations could be a potential mechanism for how these mutations potentiate poor treatment response in prostate cancer (Figure 7B) [232–234]. However, it is important to acknowledge that while we highlight specific mutations, there are other mutations in prostate cancer that also impact treatment response [222,227]. These mutations in prostate cancer warrant further investigation to understand how they may modulate spatial-temporal dynamics in the endosome-lysosome system as a potential mechanism of therapy resistance. p53 mutations are particularly interesting as p53 is transported to the nucleus by dynein motor proteins, and it would be interesting to explore how the germline mutations in p53 change this interaction [235]. Moreover, p53 mutants have been shown to promote amplification of the endosome-related protein dynamin, which is involved in the recruitment of APPL1 early endosome recycling machinery and increased recycling of EGFR and β-integrin, which promotes migration and invasion of non-small cell lung cancer cells [236]. In prostate cancer cells with p53 mutations, a similar mechanism may operate, as these cells exhibit high expression and pro-migratory signalling of EGFR and altered expression of the APPL1 protein, consistent with increased endosomal recycling [8,237,238]. This may have significant clinical significance, as inhibiting EGFR signalling can sensitise prostate cancer cells to chemotherapeutic agents such as Adriamycin [239]. Furthermore, mutant p53 promotes the activation of ARF6, which in turn functions to modulate endosome recycling kinetics by facilitating the switch between dynein and kinesin motor proteins [33,240]. We have described the role of dysfunctional trafficking machinery in mediating changes in vesicular transport that may be required for autophagy and endosome recycling to promote prostate cancer metastasis. It is therefore conceivable that specific combinations of germline and somatic mutations may promote prostate cancer metastasis and poor treatment response by manipulating this vesicular trafficking machinery to control the function of the endosome-lysosome system (Figure 7).

Clinical trials have demonstrated that benefits in overall survival can be derived from the combination therapy of docetaxel and next-generation anti-androgens such as enzalutamide [241]. Although there is little or no clinical data on correcting endosome dysfunction by directly targeting vesicular trafficking machinery in metastatic prostate cancer, current therapeutics that target endosome function show the potential of synergising with androgen deprivation compounds. For instance, inhibiting endosome recycling by the anti-malarial small molecule primaquine in prostate cancer cells reduced the expression of androgen receptor, modulated lysosome degradation kinetics, and resulted in decreased cell survival independently. This effect was greater when used in combination with enzalutamide, demonstrating the viability of targeting recycling endosome function, which we contend contributes to prostate cancer metastasis and is modulated by vesicular trafficking machinery [242]. Drug discovery efforts focusing on modifying the vesicular trafficking machinery may provide a novel therapeutic strategy for metastatic prostate cancer to improve patient outcomes.

9. Conclusions

An increased understanding of the clinical significance of the altered expression of specific vesicular trafficking machinery during metastasis could potentially identify novel markers for prostate cancer management and intervention. Circumstantial evidence points to the potential role of adaptor proteins in prostate cancer progression. However, a comprehensive investigation into endosome-lysosome recruitment and interaction with specific motor proteins, plus the involvement in mediating particular events in the prostate cancer metastatic cascade, is yet to be completed. For example, the adaptor protein JIP4, which regulates the endosomal trafficking of metalloproteinases during cancer cell migration, is increased in prostate cancer tissue and is associated with poor prognosis for prostate cancer patients [38]. Moreover, the adaptor protein Hook3, which interacts with both the dynein and kinesin motor proteins to recruit early endosomes for retrograde transport, is overexpressed in prostate cancer and has been associated with poor prognosis [59,243]. The observed alterations in early endosome biogenesis and trafficking in prostate cancer, along with the potential involvement of alterations in the vesicular trafficking machinery, underscore the need for further research into the diagnostic and prognostic potential of this cellular machinery. Our discussion also highlights the potential for targeting the dynein and kinesin trafficking machinery for therapeutic intervention in patients with prostate cancer. This could be a promising strategy for treating metastatic prostate cancer, especially when resistance is encountered with androgen-related therapies that target microtubules and AR nuclear translocation and poor efficacy of novel immunomodulatory approaches such as CAR-T cells and immune checkpoint inhibitors. Endosome-lysosome proteins can report on the primary pathogenesis in prostate cancer, and the vesicular trafficking machinery is involved in regulating critical events like microtubule polymerisation. This machinery can also alter the sensitivity of cancer cells to specific drugs based on motor protein expression. Therefore, a deeper understanding together with the potential to manipulate these mechanisms could pave the way for more effective diagnostics and treatments for patients with prostate cancer.

Author Contributions: B.D.-D.N., I.R.D.J. and D.A.B. conceived the original concept for the manuscript; B.D.-D.N., D.A.B., C.M., J.L. and R.D.B. drafted the manuscript; B.D.-D.N., J.L., I.R.D.J., C.M., K.L.L., J.T., G.L., E.P.-L., D.B.W., J.C., M.H., R.D.B., M.A.M., S.S., P.G., R.W., L.B., M.P.W., D.J.J.W., J.J.O. and D.A.B. revised the manuscript and contributed to specific sections; R.D.B. generated figures for the manuscript under guidance from the other authors. All authors have read and agreed to the published version of the manuscript.

Funding: This research received no external funding.

Conflicts of Interest: D.A.B. and J.J.O. are shareholders for Envision Sciences Pty Ltd. and benefit from this company's research funding. D.A.B., I.R.D.J. and L.B. hold patent WO2014197937A1 that has been licensed by UniSA Ventures to Envision Sciences Pty Ltd. for commercialisation. J.L., C.M. and I.R.D.J. were employed by the University of South Australia using funding from Envision Sciences Pty Ltd.

Abbreviations

ARF6	ADP-ribosylation factor 6
BMP	Bone morphogenic protein
CTC	Circulating tumour cell
DCT	Disseminated tumour cell
ECM	Extracellular matrix
EGFR	Epidermal growth factor receptor
EMT	Epithelial to mesenchymal transition
Gas6	Growth arrest-specific 6
IGF-1	Insulin-like growth factor 1
JIP3	JNK-interacting protein 3
JIP4	JNK-interacting protein 4
MHC	Major histocompatibility complex
MMP	Matrix metalloproteinase
PAR1	Protease-activated receptor 1
PD-L1	Programmed death-ligand 1
PSA	Prostate specific antigen
PTHrP	Parathyroid hormone-related protein
RILP	Rab interacting lysosomal protein
TAO3	serine/threonine-protein kinase 3
TGF-β	Transforming growth factor beta
Tks5	Tyrosine kinase substrate 5
VEGF	Vascular endothelial growth factor
VEGFR	Vascular endothelial growth factor receptor

References

1. Culp, M.B.; Soerjomataram, I.; Efstathiou, J.A.; Bray, F.; Jemal, A. Recent global patterns in prostate cancer incidence and mortality rates. *Eur. Urol.* **2020**, *77*, 38–52. [CrossRef] [PubMed]
2. Gandaglia, G.; Leni, R.; Bray, F.; Fleshner, N.; Freedland, S.J.; Kibel, A.; Stattin, P.; Van Poppel, H.; La Vecchia, C. Epidemiology and prevention of prostate cancer. *Eur. Urol. Oncol.* **2021**, *4*, 877–892. [CrossRef] [PubMed]
3. Sung, H.; Ferlay, J.; Siegel, R.L.; Laversanne, M.; Soerjomataram, I.; Jemal, A.; Bray, F. Global Cancer Statistics 2020: GLOBOCAN Estimates of Incidence and Mortality Worldwide for 36 Cancers in 185 Countries. *CA Cancer J. Clin.* **2021**, *71*, 209–249. [CrossRef] [PubMed]

4. Elmehrath, A.O.; Afifi, A.M.; Al-Husseini, M.J.; Saad, A.M.; Wilson, N.; Shohdy, K.S.; Pilie, P.; Sonbol, M.B.; Alhalabi, O. Causes of Death Among Patients with Metastatic Prostate Cancer in the US From 2000 to 2016. *JAMA Netw. Open* **2021**, *4*, e2119568. [CrossRef]
5. Al Hussein Al Awamlh, B.; Wu, X.; Barocas, D.A.; Moses, K.A.; Hoffman, R.M.; Basourakos, S.P.; Lewicki, P.; Smelser, W.W.; Arenas-Gallo, C.; Shoag, J.E. Intensity of observation with active surveillance or watchful waiting in men with prostate cancer in the United States. *Prostate Cancer Prostatic Dis.* **2023**, *26*, 395–402. [CrossRef]
6. Srigley, J.R.; Delahunt, B.; Samaratunga, H.; Billis, A.; Cheng, L.; Clouston, D.; Evans, A.; Furusato, B.; Kench, J.; Leite, K.; et al. Controversial issues in Gleason and International Society of Urological Pathology (ISUP) prostate cancer grading: Proposed recommendations for international implementation. *Pathology* **2019**, *51*, 463–473. [CrossRef]
7. Banushi, B.; Joseph, S.R.; Lum, B.; Lee, J.J.; Simpson, F. Endocytosis in cancer and cancer therapy. *Nat. Rev. Cancer* **2023**, *23*, 450–473. [CrossRef]
8. Johnson, I.R.D.; Parkinson-Lawrence, E.J.; Shandala, T.; Weigert, R.; Butler, L.M.; Brooks, D.A. Altered endosome biogenesis in prostate cancer has biomarker potential. *Mol. Cancer Res.* **2014**, *12*, 1851–1862. [CrossRef]
9. Pathak, C.; Vaidya, F.U.; Waghela, B.N.; Jaiswara, P.K.; Gupta, V.K.; Kumar, A.; Rajendran, B.K.; Ranjan, K. Insights of endocytosis signaling in health and disease. *Int. J. Mol. Sci.* **2023**, *24*, 2971. [CrossRef]
10. Porzycki, P.; Ciszkowicz, E. Modern biomarkers in prostate cancer diagnosis. *Cent. Eur. J. Urol.* **2020**, *73*, 300.
11. Martini, C.; Logan, J.M.; Sorvina, A.; Gordon, C.; Beck, A.R.; Ung, B.S.-Y.; Caruso, M.C.; Moore, C.; Hocking, A.; Johnson, I.R.; et al. Aberrant protein expression of Appl1, Sortilin and Syndecan-1 during the biological progression of prostate cancer. *Pathology* **2023**, *55*, 40–51. [CrossRef] [PubMed]
12. Sorvina, A.; Martini, C.; Prabhakaran, S.; Logan, J.M.; Ung, B.S.; Moore, C.; Johnson, I.R.; Lazniewska, J.; Tewari, P.; Malone, V.; et al. Appl1, Sortilin and Syndecan-1 immunohistochemistry on intraductal carcinoma of the prostate provides evidence of retrograde spread. *Pathology* **2023**, *55*, 792–799. [CrossRef] [PubMed]
13. Logan, J.M.; Hopkins, A.M.; Martini, C.; Sorvina, A.; Tewari, P.; Prabhakaran, S.; Huzzell, C.; Johnson, I.R.; Hickey, S.M.; Ung, B.S.Y.; et al. Prediction of prostate cancer biochemical and clinical recurrence is improved by IHC-assisted grading using Appl1, Sortilin and Syndecan-1. *Cancers* **2023**, *15*, 3215. [CrossRef] [PubMed]
14. Lazniewska, J.; Li, K.L.; Johnson, I.R.D.; Sorvina, A.; Logan, J.M.; Martini, C.; Moore, C.; Ung, B.S.-Y.; Karageorgos, L.; Hickey, S.M.; et al. Dynamic interplay between sortilin and syndecan-1 contributes to prostate cancer progression. *Sci. Rep.* **2023**, *13*, 13489. [CrossRef]
15. Johnson, I.R.; Parkinson-Lawrence, E.J.; Butler, L.M.; Brooks, D.A. Prostate cell lines as models for biomarker discovery: Performance of current markers and the search for new biomarkers. *Prostate* **2014**, *74*, 547–560. [CrossRef]
16. Johnson, I.R.; Parkinson-Lawrence, E.J.; Keegan, H.; Spillane, C.D.; Barry-O'Crowley, J.; Watson, W.R.; Selemidis, S.; Butler, L.M.; O'Leary, J.J.; Brooks, D.A. Endosomal gene expression: A new indicator for prostate cancer patient prognosis? *Oncotarget* **2015**, *6*, 37919–37929. [CrossRef]
17. Vanharanta, S.; Massagué, J. Origins of metastatic traits. *Cancer Cell* **2013**, *24*, 410–421. [CrossRef]
18. Hanahan, D. Hallmarks of cancer: New dimensions. *Cancer Discov.* **2022**, *12*, 31–46. [CrossRef]
19. Karras, P.; Black, J.R.M.; McGranahan, N.; Marine, J.-C. Decoding the interplay between genetic and non-genetic drivers of metastasis. *Nature* **2024**, *629*, 543–554. [CrossRef]
20. Evergren, E.; Mills, I.G.; Kennedy, G. Adaptations of membrane trafficking in cancer and tumorigenesis. *J. Cell Sci.* **2024**, *137*, jcs260943. [CrossRef]
21. Manna, F.L.; Karkampouna, S.; Zoni, E.; De Menna, M.; Hensel, J.; Thalmann, G.N.; Kruithof-de Julio, M. Metastases in Prostate Cancer. *Cold Spring Harb. Perspect. Med.* **2019**, *9*, a033688. [CrossRef] [PubMed]
22. Steffan, J.J.; Williams, B.C.; Welbourne, T.; Cardelli, J.A. HGF-induced invasion by prostate tumor cells requires anterograde lysosome trafficking and activity of Na^+-H^+ exchangers. *J. Cell Sci.* **2010**, *123*, 1151–1159. [CrossRef] [PubMed]
23. Wenzel, E.M.; Pedersen, N.M.; Elfmark, L.A.; Wang, L.; Kjos, I.; Stang, E.; Malerød, L.; Brech, A.; Stenmark, H.; Raiborg, C. Intercellular transfer of cancer cell invasiveness via endosome-mediated protease shedding. *Nat. Commun.* **2024**, *15*, 1277. [CrossRef] [PubMed]
24. Hu, C.-T.; Wu, J.-R.; Wu, W.-S. The role of endosomal signaling triggered by metastatic growth factors in tumor progression. *Cell. Signal.* **2013**, *25*, 1539–1545. [CrossRef]
25. Horgan, C.P.; McCaffrey, M.W. Rab GTPases and microtubule motors. *Biochem. Soc. Trans.* **2011**, *39*, 1202–1206. [CrossRef]
26. Reck-Peterson, S.L.; Redwine, W.B.; Vale, R.D.; Carter, A.P. The cytoplasmic dynein transport machinery and its many cargoes. *Nat. Rev. Mol. Cell Biol.* **2018**, *19*, 382–398. [CrossRef]
27. Hirokawa, N.; Takemura, R. Kinesin superfamily proteins and their various functions and dynamics. *Exp. Cell Res.* **2004**, *301*, 50–59. [CrossRef]
28. Olenick, M.A.; Holzbaur, E.L.F. Dynein activators and adaptors at a glance. *J. Cell Sci.* **2019**, *132*, jcs227132. [CrossRef]

29. Ballabio, A.; Bonifacino, J.S. Lysosomes as dynamic regulators of cell and organismal homeostasis. *Nat. Rev. Mol. Cell Biol.* **2020**, *21*, 101–118. [CrossRef]
30. York, H.M.; Patil, A.; Moorthi, U.K.; Kaur, A.; Bhowmik, A.; Hyde, G.J.; Gandhi, H.; Fulcher, A.; Gaus, K.; Arumugam, S. Rapid whole cell imaging reveals a calcium-APPL1-dynein nexus that regulates cohort trafficking of stimulated EGF receptors. *Commun. Biol.* **2021**, *4*, 224. [CrossRef]
31. Miki, H.; Okada, Y.; Hirokawa, N. Analysis of the kinesin superfamily: Insights into structure and function. *Trends Cell Biol.* **2005**, *15*, 467–476. [CrossRef] [PubMed]
32. Montagnac, G.; Sibarita, J.-B.; Loubéry, S.; Daviet, L.; Romao, M.; Raposo, G.; Chavrier, P. ARF6 Interacts with JIP4 to control a motor switch mechanism regulating endosome traffic in cytokinesis. *Curr. Biol.* **2009**, *19*, 184–195. [CrossRef] [PubMed]
33. Marchesin, V.; Castro-Castro, A.; Lodillinsky, C.; Castagnino, A.; Cyrta, J.; Bonsang-Kitzis, H.; Fuhrmann, L.; Irondelle, M.; Infante, E.; Montagnac, G.; et al. ARF6–JIP3/4 regulate endosomal tubules for MT1-MMP exocytosis in cancer invasion. *J. Cell Biol.* **2015**, *211*, 339–358. [CrossRef]
34. Puri, C.; Chibalina, M.V.; Arden, S.D.; Kruppa, A.J.; Kendrick-Jones, J.; Buss, F. Overexpression of myosin VI in prostate cancer cells enhances PSA and VEGF secretion, but has no effect on endocytosis. *Oncogene* **2010**, *29*, 188–200. [CrossRef]
35. Johnson, J.L.; Ellis, B.A.; Noack, D.; Seabra, M.C.; Catz, S.D. The Rab27a-binding protein, JFC1, regulates androgen-dependent secretion of prostate-specific antigen and prostatic-specific acid phosphatase. *Biochem. J.* **2005**, *391*, 699–710. [CrossRef]
36. Kurowska, M.; Goudin, N.; Nehme, N.T.; Court, M.; Garin, J.; Fischer, A.; de Saint Basile, G.; Ménasché, G. Terminal transport of lytic granules to the immune synapse is mediated by the kinesin-1/Slp3/Rab27a complex. *Blood* **2012**, *119*, 3879–3889. [CrossRef]
37. Lei, H.; Ma, F.; Jia, R.; Tan, B. Effects of Arf6 downregulation on biological characteristics of human prostate cancer cells. *Int. Braz. J. Urol.* **2020**, *46*, 950–961. [CrossRef]
38. Sun, H.-F.; Wang, W.-D.; Feng, L. Effect of SPAG9 on migration, invasion and prognosis of prostate cancer. *Int. J. Clin. Exp. Pathol.* **2017**, *10*, 9468–9474.
39. Morgan, C.; Lewis, P.D.; Hopkins, L.; Burnell, S.; Kynaston, H.; Doak, S.H. Increased expression of ARF GTPases in prostate cancer tissue. *Springerplus* **2015**, *4*, 342. [CrossRef]
40. Ghossoub, R.; Lembo, F.; Rubio, A.; Gaillard, C.B.; Bouchet, J.; Vitale, N.; Slavík, J.; Machala, M.; Zimmermann, P. Syntenin-ALIX exosome biogenesis and budding into multivesicular bodies are controlled by ARF6 and PLD2. *Nat. Commun.* **2014**, *5*, 3477. [CrossRef]
41. Saito, N.; Okada, Y.; Noda, Y.; Kinoshita, Y.; Kondo, S.; Hirokawa, N. KIFC2 is a novel neuron-specific C-terminal type kinesin superfamily motor for dendritic transport of multivesicular body-like organelles. *Neuron* **1997**, *18*, 425–438. [CrossRef] [PubMed]
42. Liu, X.; Lin, Y.; Long, W.; Yi, R.; Zhang, X.; Xie, C.; Jin, N.; Qiu, Z.; Liu, X. The kinesin-14 family motor protein KIFC2 promotes prostate cancer progression by regulating p65. *J. Biol. Chem.* **2023**, *299*, 105253. [CrossRef] [PubMed]
43. Gifford, V.; Woskowicz, A.; Ito, N.; Balint, S.; Lagerholm, B.C.; Dustin, M.L.; Itoh, Y. Coordination of two kinesin superfamily motor proteins, KIF3A and KIF13A, is essential for pericellular matrix degradation by membrane-type 1 matrix metalloproteinase (MT1-MMP) in cancer cells. *Matrix Biol.* **2022**, *107*, 1–23. [CrossRef] [PubMed]
44. Liu, Z.; Rebowe, R.E.; Wang, Z.; Li, Y.; Wang, Z.; DePaolo, J.S.; Guo, J.; Qian, C.; Liu, W. KIF3a promotes proliferation and invasion via Wnt signaling in advanced prostate cancer. *Mol. Cancer Res.* **2014**, *12*, 491–503. [CrossRef]
45. Wang, J.; Liu, P.; Zhang, R.; Xing, B.; Chen, G.; Han, L.; Yu, J. VASH2 enhances KIF3C-mediated EGFR-endosomal recycling to promote aggression and chemoresistance of lung squamous cell carcinoma by increasing tubulin detyrosination. *Cell Death Dis.* **2024**, *15*, 772. [CrossRef]
46. Ma, H.; Zhang, F.; Zhong, Q.; Hou, J. METTL3-mediated m6A modification of KIF3C-mRNA promotes prostate cancer progression and is negatively regulated by miR-320d. *Aging* **2021**, *13*, 22332–22344. [CrossRef]
47. Turner, L.S.; Cheng, J.C.; Beckham, T.H.; Keane, T.E. Autophagy is increased in prostate cancer cells overexpressing acid ceramidase and enhances resistance to C6 ceramide. *Prostate Cancer Prostatic Dis.* **2011**, *14*, 30–37. [CrossRef]
48. Belabed, M.; Mauvais, F.-X.; Maschalidi, S.; Kurowska, M.; Goudin, N.; Huang, J.-D.; Fischer, A.; Basile, G.d.S.; van Endert, P.; Sepulveda, F.E.; et al. Kinesin-1 regulates antigen cross-presentation through the scission of tubulations from early endosomes in dendritic cells. *Nat. Commun.* **2020**, *11*, 1817. [CrossRef]
49. Thankachan, J.M.; Setty, S.R.G. KIF13A-A Key Regulator of Recycling Endosome Dynamics. *Front. Cell Dev. Biol.* **2022**, *10*, 877532. [CrossRef]
50. Delevoye, C.; Miserey-Lenkei, S.; Montagnac, G.; Gilles-Marsens, F.; Paul-Gilloteaux, P.; Giordano, F.; Waharte, F.; Marks, M.S.; Goud, B.; Raposo, G. Recycling endosome tubule morphogenesis from sorting endosomes requires the kinesin motor KIF13A. *Cell Rep.* **2014**, *6*, 445–454. [CrossRef]
51. Hoepfner, S.; Severin, F.; Cabezas, A.; Habermann, B.; Runge, A.; Gillooly, D.; Stenmark, H.; Zerial, M. Modulation of receptor recycling and degradation by the endosomal kinesin KIF16B. *Cell* **2005**, *121*, 437–450. [CrossRef] [PubMed]
52. Hey, S.; Wiesner, C.; Barcelona, B.; Linder, S. KIF16B drives MT1-MMP recycling in macrophages and promotes co-invasion of cancer cells. *Life Sci. Alliance* **2023**, *6*, e202302158. [CrossRef]

53. Nturubika, B.D.; Guardia, C.M.; Gershlick, D.C.; Logan, J.M.; Martini, C.; Heatlie, J.K.; Lazniewska, J.; Moore, C.; Lam, G.T.; Li, K.L.; et al. Altered expression of vesicular trafficking machinery in prostate cancer affects lysosomal dynamics and provides insight into the underlying biology and disease progression. *Br. J. Cancer* **2024**, *131*, 1263–1278. [CrossRef] [PubMed]
54. Li, S.; Zhu, X. The influence of kinesin light chain-2 on the radiosensitivity of non-small cell lung cancer cells and the underlying mechanisms. *J. Clin. Oncol.* **2019**, *37*, e14512. [CrossRef]
55. Moamer, A.; Hachim, I.Y.; Binothman, N.; Wang, N.; Lebrun, J.-J.; Ali, S. A role for kinesin-1 subunits KIF5B/KLC1 in regulating epithelial mesenchymal plasticity in breast tumorigenesis. *eBiomedicine* **2019**, *45*, 92–107. [CrossRef] [PubMed]
56. Wu, P.-H.; Onodera, Y.; Giaccia, A.J.; Le, Q.-T.; Shimizu, S.; Shirato, H.; Nam, J.-M. Lysosomal trafficking mediated by Arl8b and BORC promotes invasion of cancer cells that survive radiation. *Commun. Biol.* **2020**, *3*, 620. [CrossRef]
57. Chang, C.-C.; Chao, K.-C.; Huang, C.-J.; Hung, C.-S.; Wang, Y.-C. Association between aberrant dynein cytoplasmic 1 light intermediate chain 1 expression levels, mucins and chemosensitivity in colorectal cancer. *Mol. Med. Rep.* **2020**, *22*, 185–192. [CrossRef]
58. Gong, L.-B.; Wen, T.; Li, Z.; Xin, X.; Che, X.-F.; Wang, J.; Liu, Y.-P.; Qu, X.-J. Corrigendum: DYNC1I1 Promotes the Proliferation and Migration of Gastric Cancer by Up-Regulating IL-6 Expression. *Front. Oncol.* **2022**, *12*, 819244. [CrossRef]
59. Melling, N.; Harutyunyan, L.; Hube-Magg, C.; Kluth, M.; Simon, R.; Lebok, P.; Minner, S.; Tsourlakis, M.C.; Koop, C.; Graefen, M.; et al. High-Level HOOK3 Expression Is an Independent Predictor of Poor Prognosis Associated with Genomic Instability in Prostate Cancer. *PLoS ONE* **2015**, *10*, e0134614. [CrossRef]
60. Shim, S.Y.; Samuels, B.A.; Wang, J.; Neumayer, G.; Belzil, C.; Ayala, R.; Shi, Y.; Shi, Y.; Tsai, L.-H.; Nguyen, M.D. Ndel1 controls the dynein-mediated transport of vimentin during neurite outgrowth. *J. Biol. Chem.* **2008**, *283*, 12232–12240. [CrossRef]
61. Wang, Z.; Zhou, Y.; Nie, D.; Tan, Y.; Zhao, S.; Wang, G.; Wang, T. RILP inhibits proliferation, migration, and invasion of PC3 prostate cancer cells. *Acta Histochem.* **2022**, *124*, 151938. [CrossRef] [PubMed]
62. Willett, R.; Martina, J.A.; Zewe, J.P.; Wills, R.; Hammond, G.R.V.; Puertollano, R. TFEB regulates lysosomal positioning by modulating TMEM55B expression and JIP4 recruitment to lysosomes. *Nat. Commun.* **2017**, *8*, 1580. [CrossRef] [PubMed]
63. Wang, M.; Wisniewski, C.A.; Xiong, C.; Chhoy, P.; Goel, H.L.; Kumar, A.; Zhu, L.J.; Li, R.; Louis, P.A.S.; Ferreira, L.M.; et al. Therapeutic blocking of VEGF binding to neuropilin-2 diminishes PD-L1 expression to activate antitumor immunity in prostate cancer. *Sci. Transl. Med.* **2023**, *15*, eade5855. [CrossRef] [PubMed]
64. Dutta, S.; Roy, S.; Polavaram, N.S.; Stanton, M.J.; Zhang, H.; Bhola, T.; Hönscheid, P.; Donohue, T.M.; Band, H.; Batra, S.K.; et al. Neuropilin-2 regulates endosome maturation and EGFR trafficking to support cancer cell pathobiology. *Cancer Res.* **2016**, *76*, 418–428. [CrossRef]
65. Benwell, C.J.; Johnson, R.T.; Taylor, J.A.G.E.; Lambert, J.; Robinson, S.D. A proteomics approach to isolating neuropilin-dependent α5 integrin trafficking pathways: Neuropilin 1 and 2 co-traffic α5 integrin through endosomal p120RasGAP to promote polarised fibronectin fibrillogenesis in endothelial cells. *Commun. Biol.* **2024**, *7*, 629. [CrossRef]
66. Wang, P.; Ning, J.; Chen, W.; Zou, F.; Yu, W.; Rao, T.; Cheng, F. Comprehensive analysis indicated that NDE1 is a potential biomarker for pan-cancer and promotes bladder cancer progression. *Cancer Med.* **2024**, *13*, e6931. [CrossRef]
67. Odero-Marah, V.; Hawsawi, O.; Henderson, V.; Sweeney, J. Epithelial-Mesenchymal Transition (EMT) and Prostate Cancer. *Adv. Exp. Med. Biol.* **2018**, *1095*, 101–110.
68. Ruscetti, M.; Quach, B.; Dadashian, E.L.; Mulholland, D.J.; Wu, H. Tracking and functional characterization of epithelial–mesenchymal transition and mesenchymal tumor cells during prostate cancer metastasis. *Cancer Res.* **2015**, *75*, 2749–2759. [CrossRef]
69. Chao, Y.; Wu, Q.; Acquafondata, M.; Dhir, R.; Wells, A. Partial mesenchymal to epithelial reverting transition in breast and prostate cancer metastases. *Cancer Microenviron.* **2012**, *5*, 19–28. [CrossRef]
70. Saxena, K.; Jolly, M.K.; Balamurugan, K. Hypoxia, partial EMT and collective migration: Emerging culprits in metastasis. *Transl. Oncol.* **2020**, *13*, 100845. [CrossRef]
71. Kitz, J.; Lefebvre, C.; Carlos, J.; Lowes, L.E.; Allan, A.L. Reduced Zeb1 Expression in Prostate Cancer Cells Leads to an Aggressive Partial-EMT Phenotype Associated with Altered Global Methylation Patterns. *Int. J. Mol. Sci.* **2021**, *22*, 12840. [CrossRef] [PubMed]
72. Jolly, M.K.; Somarelli, J.A.; Sheth, M.; Biddle, A.; Tripathi, S.C.; Armstrong, A.J.; Hanash, S.M.; Bapat, S.A.; Rangarajan, A.; Levine, H. Hybrid epithelial/mesenchymal phenotypes promote metastasis and therapy resistance across carcinomas. *Pharmacol. Ther.* **2019**, *194*, 161–184. [CrossRef] [PubMed]
73. Loh, C.-Y.; Chai, J.Y.; Tang, T.F.; Wong, W.F.; Sethi, G.; Shanmugam, M.K.; Chong, P.P.; Looi, C.Y. The E-Cadherin and N-Cadherin Switch in Epithelial-to-Mesenchymal Transition: Signaling, Therapeutic Implications, and Challenges. *Cells* **2019**, *8*, 1118. [CrossRef] [PubMed]
74. Rojas, A.; Liu, G.; Coleman, I.; Nelson, P.S.; Zhang, M.; Dash, R.; Fisher, P.B.; Plymate, S.R.; Wu, J.D. IL-6 promotes prostate tumorigenesis and progression through autocrine cross-activation of IGF-IR. *Oncogene* **2011**, *30*, 2345–2355. [CrossRef]

75. Li, M.; Wang, Y.X.; Luo, Y.; Zhao, J.; Li, Q.; Zhang, J.; Jiang, Y. Hypoxia inducible factor-1α-dependent epithelial to mesenchymal transition under hypoxic conditions in prostate cancer cells. *Oncol. Rep.* **2016**, *36*, 521–527. [CrossRef]
76. Sun, Y.; Wang, B.-E.; Leong, K.G.; Yue, P.; Li, L.; Jhunjhunwala, S.; Chen, D.; Seo, K.; Modrusan, Z.; Gao, W.-Q.; et al. Androgen deprivation causes epithelial-mesenchymal transition in the prostate: Implications for androgen-deprivation therapy. *Cancer Res.* **2012**, *72*, 527–536. [CrossRef]
77. Chung, Y.-C.; Wei, W.-C.; Huang, S.-H.; Shih, C.-M.; Hsu, C.-P.; Chang, K.-J.; Chao, W.-T. Rab11 regulates E-cadherin expression and induces cell transformation in colorectal carcinoma. *BMC Cancer* **2014**, *14*, 587. [CrossRef]
78. Putzke, A.P.; Ventura, A.P.; Bailey, A.M.; Akture, C.; Opoku-Ansah, J.; Çeliktaş, M.; Hwang, M.S.; Darling, D.S.; Coleman, I.M.; Nelson, P.S.; et al. Metastatic progression of prostate cancer and e-cadherin regulation by zeb1 and SRC family kinases. *Am. J. Pathol.* **2011**, *179*, 400–410. [CrossRef]
79. Yuen, H.-F.; Chua, C.-W.; Chan, Y.-P.; Wong, Y.-C.; Wang, X.; Chan, K.-W. Significance of TWIST and E-cadherin expression in the metastatic progression of prostatic cancer. *Histopathology* **2007**, *50*, 648–658. [CrossRef]
80. Lock, J.G.; Stow, J.L. Rab11 in recycling endosomes regulates the sorting and basolateral transport of E-cadherin. *Mol. Biol. Cell* **2005**, *16*, 1744–1755. [CrossRef]
81. Pasqualato, S.; Senic-Matuglia, F.; Renault, L.; Goud, B.; Salamero, J.; Cherfils, J. The structural GDP/GTP cycle of Rab11 reveals a novel interface involved in the dynamics of recycling endosomes. *J. Biol. Chem.* **2004**, *279*, 11480–11488. [CrossRef] [PubMed]
82. Palacios, F.; Tushir, J.S.; Fujita, Y.; D'Souza-Schorey, C. Lysosomal targeting of E-cadherin: A unique mechanism for the down-regulation of cell-cell adhesion during epithelial to mesenchymal transitions. *Mol. Cell. Biol.* **2005**, *25*, 389–402. [CrossRef] [PubMed]
83. Frasa, M.A.; Maximiano, F.C.; Smolarczyk, K.; Francis, R.E.; Betson, M.E.; Lozano, E.; Goldenring, J.; Seabra, M.C.; Rak, A.; Ahmadian, M.R.; et al. Armus is a Rac1 effector that inactivates Rab7 and regulates E-cadherin degradation. *Curr. Biol.* **2010**, *20*, 198–208. [CrossRef] [PubMed]
84. Jordens, I.; Fernandez-Borja, M.; Marsman, M.; Dusseljee, S.; Janssen, L.; Calafat, J.; Janssen, H.; Wubbolts, R.; Neefjes, J. The Rab7 effector protein RILP controls lysosomal transport by inducing the recruitment of dynein-dynactin motors. *Curr. Biol.* **2001**, *11*, 1680–1685. [CrossRef]
85. Sun, H.; Weidner, J.; Allamargot, C.; Piper, R.C.; Misurac, J.; Nester, C. Dynein-Mediated Trafficking: A New Mechanism of Diabetic Podocytopathy. *Kidney360* **2023**, *4*, 162–176. [CrossRef]
86. Xu, D.D.; Xu, C.B.; Lam, H.M.; Wong, F.-L.; Leung, A.W.N.; Leong, M.M.L.; Cho, W.C.S.; Hoeven, R.; Lv, Q.; Rong, R. Proteomic analysis reveals that pheophorbide a-mediated photodynamic treatment inhibits prostate cancer growth by hampering GDP-GTP exchange of ras-family proteins. *Photodiagnosis Photodyn. Ther.* **2018**, *23*, 35–39. [CrossRef]
87. Miao, L.; Li, J.; Li, J.; Lu, Y.; Shieh, D.; Mazurkiewicz, J.E.; Barroso, M.; Schwarz, J.J.; Xin, H.-B.; Singer, H.A.; et al. Cardiomyocyte orientation modulated by the Numb family proteins-N-cadherin axis is essential for ventricular wall morphogenesis. *Proc. Natl. Acad. Sci. USA* **2019**, *116*, 15560–15569. [CrossRef]
88. Gil, D.; Zarzycka, M.; Ciołczyk-Wierzbicka, D.; Laidler, P. Integrin linked kinase regulates endosomal recycling of N-cadherin in melanoma cells. *Cell. Signal.* **2020**, *72*, 109642. [CrossRef]
89. Martin, S.K.; Banuelos, C.A.; Sadar, M.D.; Kyprianou, N. N-terminal targeting of androgen receptor variant enhances response of castration resistant prostate cancer to taxane chemotherapy. *Mol. Oncol.* **2014**, *9*, 628–639. [CrossRef]
90. Kuburich, N.A.; den Hollander, P.; Pietz, J.T.; Mani, S.A. Vimentin and cytokeratin: Good alone, bad together. *Semin. Cancer Biol.* **2021**, *86*, 816–826. [CrossRef]
91. Armstrong, A.J.; Marengo, M.S.; Oltean, S.; Kemeny, G.; Bitting, R.L.; Turnbull, J.D.; Herold, C.I.; Marcom, P.K.; George, D.J.; Garcia-Blanco, M.A. Circulating tumor cells from patients with advanced prostate and breast cancer display both epithelial and mesenchymal markers. *Mol. Cancer Res.* **2011**, *9*, 997–1007. [CrossRef] [PubMed]
92. Satelli, A.; Li, S. Vimentin in cancer and its potential as a molecular target for cancer therapy. *Cell. Mol. Life Sci.* **2011**, *68*, 3033–3046. [CrossRef] [PubMed]
93. Vyas, A.R.; Singh, S.V. Functional relevance of D,L-sulforaphane-mediated induction of vimentin and plasminogen activator inhibitor-1 in human prostate cancer cells. *Eur. J. Nutr.* **2014**, *53*, 843–852. [CrossRef]
94. Jones, M.L.; Siddiqui, J.; Pienta, K.J.; Getzenberg, R.H. Circulating fibroblast-like cells in men with metastatic prostate cancer. *Prostate* **2013**, *73*, 176–181. [CrossRef]
95. Ivaska, J.; Vuoriluoto, K.; Huovinen, T.; Izawa, I.; Inagaki, M.; Parker, P.J. PKCepsilon-mediated phosphorylation of vimentin controls integrin recycling and motility. *EMBO J.* **2005**, *24*, 3834–3845. [CrossRef]
96. Wu, W.; Panté, N. Vimentin plays a role in the release of the influenza A viral genome from endosomes. *Virology* **2016**, *497*, 41–52. [CrossRef]
97. Romano, R.; Calcagnile, M.; Margiotta, A.; Franci, L.; Chiariello, M.; Alifano, P.; Bucci, C. RAB7A Regulates Vimentin Phosphorylation through AKT and PAK. *Cancers* **2021**, *13*, 2220. [CrossRef]

98. Cogli, L.; Progida, C.; Bramato, R.; Bucci, C. Vimentin phosphorylation and assembly are regulated by the small GTPase Rab7a. *Biochim. Biophys. Acta Mol. Cell Res.* **2013**, *1833*, 1283–1293. [CrossRef]
99. Hanafusa, H.; Yagi, T.; Ikeda, H.; Hisamoto, N.; Nishioka, T.; Kaibuchi, K.; Shirakabe, K.; Matsumoto, K. LRRK1 phosphorylation of Rab7 at S72 links trafficking of EGFR-containing endosomes to its effector RILP. *J. Cell Sci.* **2019**, *132*, jcs228809. [CrossRef]
100. Yap, C.C.; Digilio, L.; McMahon, L.P.; Wang, T.; Winckler, B. Dynein is required for Rab7-dependent endosome maturation, retrograde dendritic transport, and degradation. *J. Neurosci.* **2022**, *42*, 4415–4434. [CrossRef]
101. Clarke, E.J.; Allan, V. Intermediate filaments: Vimentin moves in. *Curr. Biol.* **2002**, *12*, R596–R598. [CrossRef] [PubMed]
102. Lois-Bermejo, I.; González-Jiménez, P.; Duarte, S.; Pajares, M.A.; Pérez-Sala, D. Vimentin Tail Segments Are Differentially Exposed at Distinct Cellular Locations and in Response to Stress. *Front. Cell Dev. Biol.* **2022**, *10*, 908263. [CrossRef] [PubMed]
103. Sun, Y.-X.; Fang, M.; Wang, J.; Cooper, C.R.; Pienta, K.J.; Taichman, R.S. Expression and activation of alpha v beta 3 integrins by SDF-1/CXC12 increases the aggressiveness of prostate cancer cells. *Prostate* **2007**, *67*, 61–73. [CrossRef] [PubMed]
104. White, D.P.; Caswell, P.T.; Norman, J.C. alpha v beta3 and alpha5beta1 integrin recycling pathways dictate downstream Rho kinase signaling to regulate persistent cell migration. *J. Cell Biol.* **2007**, *177*, 515–525. [CrossRef]
105. Gardel, M.L.; Sabass, B.; Ji, L.; Danuser, G.; Schwarz, U.S.; Waterman, C.M. Traction stress in focal adhesions correlates biphasically with actin retrograde flow speed. *J. Cell Biol.* **2008**, *183*, 999–1005. [CrossRef]
106. Krishn, S.R.; Singh, A.; Bowler, N.; Duffy, A.N.; Friedman, A.; Fedele, C.; Kurtoglu, S.; Tripathi, S.K.; Wang, K.; Hawkins, A.; et al. Prostate cancer sheds the $\alpha v\beta 3$ integrin in vivo through exosomes. *Matrix Biol.* **2019**, *77*, 41–57. [CrossRef]
107. Goel, H.L.; Alam, N.; Johnson, I.N.S.; Languino, L.R. Integrin signaling aberrations in prostate cancer. *Am. J. Transl. Res.* **2009**, *1*, 211–220.
108. Goel, H.L.; Li, J.; Kogan, S.; Languino, L.R. Integrins in prostate cancer progression. *Endocr.-Relat. Cancer* **2008**, *15*, 657–664. [CrossRef]
109. Vaidžiulytė, K.; Macé, A.-S.; Battistella, A.; Beng, W.; Schauer, K.; Coppey, M. Persistent cell migration emerges from a coupling between protrusion dynamics and polarized trafficking. *eLife* **2022**, *11*, e69229. [CrossRef]
110. Linford, A.; Yoshimura, S.-I.; Bastos, R.N.; Langemeyer, L.; Gerondopoulos, A.; Rigden, D.J.; Barr, F.A. Rab14 and its exchange factor FAM116 link endocytic recycling and adherens junction stability in migrating cells. *Dev. Cell* **2012**, *22*, 952–966. [CrossRef]
111. Morgia, G.; Falsaperla, M.; Malaponte, G.; Madonia, M.; Indelicato, M.; Travali, S.; Mazzarino, M.C. Matrix metalloproteinases as diagnostic (MMP-13) and prognostic (MMP-2, MMP-9) markers of prostate cancer. *Urol. Res.* **2005**, *33*, 44–50. [CrossRef] [PubMed]
112. Nagakawa, O.; Murakami, K.; Yamaura, T.; Fujiuchi, Y.; Murata, J.; Fuse, H.; Saiki, I. Expression of membrane-type 1 matrix metalloproteinase (MT1-MMP) on prostate cancer cell lines. *Cancer Lett.* **2000**, *155*, 173–179. [CrossRef] [PubMed]
113. Koistinen, H.; Kovanen, R.; Hollenberg, M.D.; Dufour, A.; Radisky, E.S.; Stenman, U.; Batra, J.; Clements, J.; Hooper, J.D.; Diamandis, E.; et al. The roles of proteases in prostate cancer. *IUBMB Life* **2023**, *75*, 493–513. [CrossRef] [PubMed]
114. Wiesner, C.; Faix, J.; Himmel, M.; Bentzien, F.; Linder, S. KIF5B and KIF3A/KIF3B kinesins drive MT1-MMP surface exposure, CD44 shedding, and extracellular matrix degradation in primary macrophages. *Blood* **2010**, *116*, 1559–1569. [CrossRef]
115. Kajiho, H.; Kajiho, Y.; Frittoli, E.; Confalonieri, S.; Bertalot, G.; Viale, G.; Di Fiore, P.P.; Oldani, A.; Garre, M.; Beznoussenko, G.V.; et al. RAB2A controls MT1-MMP endocytic and E-cadherin polarized Golgi trafficking to promote invasive breast cancer programs. *EMBO Rep.* **2016**, *17*, 1061–1080. [CrossRef]
116. Steffan, J.J.; Snider, J.L.; Skalli, O.; Welbourne, T. Na^+/H^+ Exchangers and RhoA Regulate Acidic Extracellular pH-Induced Lysosome Trafficking in Prostate Cancer Cells. *Traffic* **2009**, *10*, 737–753. [CrossRef]
117. Dykes, S.S.; Steffan, J.J.; Cardelli, J.A. Lysosome trafficking is necessary for EGF-driven invasion and is regulated by p38 MAPK and Na^+/H^+ exchangers. *BMC Cancer* **2017**, *17*, 672. [CrossRef]
118. Zhang, J.; Wang, J.; Wong, Y.K.; Sun, X.; Chen, Y.; Wang, L.; Yang, L.; Lu, L.; Shen, H.-M.; Huang, D. Docetaxel enhances lysosomal function through TFEB activation. *Cell Death Dis.* **2018**, *9*, 614. [CrossRef]
119. Raiborg, C. How Nutrients Orchestrate Lysosome Positioning. *Contact* **2018**, *1*, 2515256418756111. [CrossRef]
120. Zhu, X.; Zhuo, Y.; Wu, S.; Chen, Y.; Ye, J.; Deng, Y.; Feng, Y.; Liu, R.; Cai, S.; Zou, Z.; et al. TFEB Promotes Prostate Cancer Progression via Regulating ABCA2-Dependent Lysosomal Biogenesis. *Front. Oncol.* **2021**, *11*, 632524.
121. Mohamed, O.A.A.; Tesen, H.S.; Hany, M.; Sherif, A.; Abdelwahab, M.M.; Elnaggar, M.H. The role of hypoxia on prostate cancer progression and metastasis. *Mol. Biol. Rep.* **2023**, *50*, 3873–3884. [CrossRef] [PubMed]
122. Raskov, H.; Gaggar, S.; Tajik, A.; Orhan, A.; Gögenur, I. Metabolic switch in cancer—Survival of the fittest. *Eur. J. Cancer* **2023**, *180*, 30–51. [CrossRef] [PubMed]
123. Pei, Y.; Lv, S.; Shi, Y.; Jia, J.; Ma, M.; Han, H.; Zhang, R.; Tan, J.; Zhang, X. RAB21 controls autophagy and cellular energy homeostasis by regulating retromer-mediated recycling of SLC2A1/GLUT1. *Autophagy* **2023**, *19*, 1070–1086. [CrossRef] [PubMed]
124. Li, C.; He, C.; Xu, Y.; Xu, H.; Tang, Y.; Chavan, H.; Duan, S.; Artigues, A.; Forrest, M.L.; Krishnamurthy, P.; et al. Alternol eliminates excessive ATP production by disturbing Krebs cycle in prostate cancer. *Prostate* **2019**, *79*, 628–639. [CrossRef]

125. Toprak, E.; Yildiz, A.; Hoffman, M.T.; Rosenfeld, S.S.; Selvin, P.R. Why kinesin is so processive. *Proc. Natl. Acad. Sci. USA* **2009**, *106*, 12717–12722. [CrossRef]
126. Hirakawa, E.; Higuchi, H.; Toyoshima, Y.Y. Processive movement of single 22S dynein molecules occurs only at low ATP concentrations. *Proc. Natl. Acad. Sci. USA* **2000**, *97*, 2533–2537. [CrossRef]
127. Kinoshita, Y.; Kambara, T.; Nishikawa, K.; Kaya, M.; Higuchi, H. Step Sizes and Rate Constants of Single-headed Cytoplasmic Dynein Measured with Optical Tweezers. *Sci. Rep.* **2018**, *8*, 16333. [CrossRef]
128. Monzon, G.A.; Scharrel, L.; DSouza, A.; Henrichs, V.; Santen, L.; Diez, S. Stable tug-of-war between kinesin-1 and cytoplasmic dynein upon different ATP and roadblock concentrations. *J. Cell Sci.* **2020**, *133*, jcs249938. [CrossRef]
129. Kempers, L.; Wakayama, Y.; van der Bijl, I.; Furumaya, C.; De Cuyper, I.M.; Jongejan, A.; Kat, M.; van Stalborch, A.M.D.; van Boxtel, A.L.; Hubert, M.; et al. The endosomal RIN2/Rab5C machinery prevents VEGFR2 degradation to control gene expression and tip cell identity during angiogenesis. *Angiogenesis* **2021**, *24*, 695–714. [CrossRef]
130. Tomatis, V.M.; Papadopulos, A.; Malintan, N.T.; Martin, S.; Wallis, T.; Gormal, R.S.; Kendrick-Jones, J.; Buss, F.; Meunier, F.A. Myosin VI small insert isoform maintains exocytosis by tethering secretory granules to the cortical actin. *J. Cell Biol.* **2013**, *200*, 301–320. [CrossRef]
131. Yang, P.; Sun, X.; Kou, Z.-W.; Wu, K.-W.; Huang, Y.-L.; Sun, F.-Y. VEGF Axonal Transport Dependent on Kinesin-1B and Microtubules Dynamics. *Front. Mol. Neurosci.* **2017**, *10*, 424. [CrossRef] [PubMed]
132. Kukreja, P.; Abdel-Mageed, A.B.; Mondal, D.; Liu, K. Up-regulation of CXCR4 expression in PC-3 cells by stromal-derived factor-1α (CXCL12) increases endothelial adhesion and transendothelial migration: Role of MEK/ERK Signaling Pathway–Dependent NF-κB Activation. *Cancer Res.* **2005**, *65*, 9891–9898. [CrossRef] [PubMed]
133. Mughees, M.; Kaushal, J.B.; Sharma, G.; Wajid, S.; Batra, S.K.; Siddiqui, J.A. Chemokines and cytokines: Axis and allies in prostate cancer pathogenesis. *Semin. Cancer Biol.* **2022**, *86*, 497–512. [CrossRef]
134. Marchese, A. Endocytic trafficking of chemokine receptors. *Curr. Opin. Cell Biol.* **2014**, *27*, 72–77. [CrossRef]
135. Mohme, M.; Riethdorf, S.; Pantel, K. Circulating and disseminated tumour cells—Mechanisms of immune surveillance and escape. *Nat. Rev. Clin. Oncol.* **2017**, *14*, 155–167. [CrossRef]
136. Gakhar, G.; Navarro, V.N.; Jurish, M.; Lee, G.Y.; Tagawa, S.T.; Akhtar, N.H.; Seandel, M.; Geng, Y.; Liu, H.; Bander, N.H.; et al. Circulating tumor cells from prostate cancer patients interact with E-selectin under physiologic blood flow. *PLoS ONE* **2013**, *8*, e85143. [CrossRef]
137. Qiao, X.; Cheng, Z.; Xue, K.; Xiong, C.; Zheng, Z.; Jin, X.; Li, J. Tumor-associated macrophage-derived exosomes LINC01592 induce the immune escape of esophageal cancer by decreasing MHC-I surface expression. *J. Exp. Clin. Cancer Res.* **2023**, *42*, 289. [CrossRef]
138. Wubbolts, R.; Fernandez-Borja, M.; Jordens, I.; Reits, E.; Dusseljee, S.; Echeverri, C.; Vallee, R.B.; Neefjes, J. Opposing motor activities of dynein and kinesin determine retention and transport of MHC class II-containing compartments. *J. Cell Sci.* **1999**, *112 Pt 6*, 785–795. [CrossRef]
139. Grommé, M.; Uytdehaag, F.G.; Janssen, H.; Calafat, J.; van Binnendijk, R.S.; Kenter, M.J.H.; Tulp, A.; Verwoerd, D.; Neefjes, J. Recycling MHC class I molecules and endosomal peptide loading. *Proc. Natl. Acad. Sci. USA* **1999**, *96*, 10326–10331. [CrossRef]
140. Weigert, R.; Yeung, A.C.; Li, J.; Donaldson, J.G. Rab22a regulates the recycling of membrane proteins internalized independently of clathrin. *Mol. Biol. Cell* **2004**, *15*, 3758–3770. [CrossRef]
141. Shakya, S.; Sharma, P.; Bhatt, A.M.; Jani, R.A.; Delevoye, C.; Gangi Setty, S.R. Rab22A recruits BLOC-1 and BLOC-2 to promote the biogenesis of recycling endosomes. *EMBO Rep.* **2018**, *19*, e45918. [CrossRef] [PubMed]
142. Gevensleben, H.; Dietrich, D.; Golletz, C.; Steiner, S.; Jung, M.; Thiesler, T.; Majores, M.; Stein, J.; Uhl, B.; Müller, S.; et al. The immune checkpoint regulator PD-L1 is highly expressed in aggressive primary prostate cancer. *Clin. Cancer Res.* **2016**, *22*, 1969–1977. [CrossRef] [PubMed]
143. Kwek, S.S.; Cha, E.; Fong, L. Unmasking the immune recognition of prostate cancer with CTLA4 blockade. *Nat. Rev. Cancer* **2012**, *12*, 289–297. [CrossRef] [PubMed]
144. Chay, C.H.; Cooper, C.R.; Gendernalik, J.D.; Dhanasekaran, S.M.; Chinnaiyan, A.M.; A Rubin, M.; Schmaier, A.H.; Pienta, K.J. A functional thrombin receptor (PAR1) is expressed on bone-derived prostate cancer cell lines. *Urology* **2002**, *60*, 760–765. [CrossRef] [PubMed]
145. Nierodzik, M.L.; Chen, K.; Takeshita, K.; Li, J.J.; Huang, Y.Q.; Feng, X.S.; D'Andrea, M.R.; Andrade-Gordon, P.; Karpatkin, S. Protease-activated receptor 1 (PAR-1) is required and rate-limiting for thrombin-enhanced experimental pulmonary metastasis. *Blood* **1998**, *92*, 3694–3700. [CrossRef]
146. Booden, M.A.; Eckert, L.B.; Der, C.J.; Trejo, J. Persistent signaling by dysregulated thrombin receptor trafficking promotes breast carcinoma cell invasion. *Mol. Cell. Biol.* **2004**, *24*, 1990–1999. [CrossRef]
147. Grimsey, N.J.; Coronel, L.J.; Cordova, I.C.; Trejo, J. Recycling and Endosomal Sorting of Protease-activated Receptor-1 Is Distinctly Regulated by Rab11A and Rab11B Proteins. *J. Biol. Chem.* **2016**, *291*, 2223–2236. [CrossRef]
148. Jurasz, P.; North, S.; Venner, P.; Radomski, M.W. Matrix metalloproteinase-2 contributes to increased platelet reactivity in patients with metastatic prostate cancer: A preliminary study. *Thromb. Res.* **2003**, *112*, 59–64. [CrossRef]

149. Ustach, C.V.; Taube, M.E.; Hurst, N.J., Jr.; Bhagat, S.; Bonfil, R.D.; Cher, M.L.; Schuger, L.; Kim, H.R.C. A potential oncogenic activity of platelet-derived growth factor d in prostate cancer progression. *Cancer Res.* **2004**, *64*, 1722–1729. [CrossRef]
150. Wang, Y.; Pennock, S.D.; Chen, X.; Kazlauskas, A.; Wang, Z. Platelet-derived growth factor receptor-mediated signal transduction from endosomes. *J. Biol. Chem.* **2004**, *279*, 8038–8046. [CrossRef]
151. Imamura, T.; Huang, J.; Usui, I.; Satoh, H.; Bever, J.; Olefsky, J.M. Insulin-induced GLUT4 translocation involves protein kinase C-lambda-mediated functional coupling between Rab4 and the motor protein kinesin. *Mol. Cell Biol.* **2003**, *23*, 4892–4900. [CrossRef] [PubMed]
152. Leong, H.S.; Robertson, A.E.; Stoletov, K.; Leith, S.J.; Chin, C.A.; Chien, A.E.; Hague, M.N.; Ablack, A.; Carmine-Simmen, K.; McPherson, V.A.; et al. Invadopodia are required for cancer cell extravasation and are a therapeutic target for metastasis. *Cell Rep.* **2014**, *8*, 1558–1570. [CrossRef] [PubMed]
153. Huang, S.-S.; Liao, W.-Y.; Hsu, C.-C.; Chan, T.-S.; Liao, T.-Y.; Yang, P.-M.; Chen, L.-T.; Sung, S.-Y.; Tsai, K.K. A novel invadopodia-specific marker for invasive and pro-metastatic cancer stem cells. *Front. Oncol.* **2021**, *11*, 638311. [CrossRef] [PubMed]
154. Manuelli, V.; Cahill, F.; Wylie, H.; Gillett, C.; Correa, I.; Heck, S.; Rimmer, A.; Haire, A.; Van Hemelrijck, M.; Rudman, S.; et al. Invadopodia play a role in prostate cancer progression. *BMC Cancer* **2022**, *22*, 386. [CrossRef]
155. Eckert, M.A.; Lwin, T.M.; Chang, A.T.; Kim, J.; Danis, E.; Ohno-Machado, L.; Yang, J. Twist1-induced invadopodia formation promotes tumor metastasis. *Cancer Cell* **2011**, *19*, 372–386. [CrossRef]
156. Daly, C.; Logan, B.; Breeyear, J.; Whitaker, K.; Ahmed, M.; Seals, D.F. Tks5 SH3 domains exhibit differential effects on invadopodia development. *PLoS ONE* **2020**, *15*, e0227855. [CrossRef]
157. Iizuka, S.; Quintavalle, M.; Navarro, J.C.; Gribbin, K.P.; Ardecky, R.J.; Abelman, M.M.; Ma, C.-T.; Sergienko, E.; Zeng, F.-Y.; Pass, I.; et al. Serine-Threonine Kinase TAO3-Mediated Trafficking of Endosomes Containing the Invadopodia Scaffold TKS5α Promotes Cancer Invasion and Tumor Growth. *Cancer Res.* **2021**, *81*, 1472–1485. [CrossRef]
158. Jacob, A.; Linklater, E.; Bayless, B.A.; Lyons, T.; Prekeris, R. The role and regulation of Rab40b-Tks5 complex during invadopodia formation and cancer cell invasion. *J. Cell Sci.* **2016**, *129*, 4341–4353. [CrossRef]
159. Duncan, E.D.; Han, K.-J.; Trout, M.A.; Prekeris, R. Ubiquitylation by Rab40b/Cul5 regulates Rap2 localization and activity during cell migration. *J. Cell Biol.* **2022**, *221*, e202107114. [CrossRef]
160. Neophytou, C.M.; Kyriakou, T.-C.; Papageorgis, P. Mechanisms of Metastatic Tumor Dormancy and Implications for Cancer Therapy. *Int. J. Mol. Sci.* **2019**, *20*, 6158. [CrossRef]
161. Ruppender, N.; Larson, S.; Lakely, B.; Kollath, L.; Brown, L.; Coleman, I.; Coleman, R.; Nguyen, H.; Nelson, P.S.; Corey, E.; et al. Cellular Adhesion Promotes Prostate Cancer Cells Escape from Dormancy. *PLoS ONE* **2015**, *10*, e0130565. [CrossRef] [PubMed]
162. Shiozawa, Y.; Pedersen, E.A.; Havens, A.M.; Jung, Y.; Mishra, A.; Joseph, J.; Kim, J.K.; Patel, L.R.; Ying, C.; Ziegler, A.M.; et al. Human prostate cancer metastases target the hematopoietic stem cell niche to establish footholds in mouse bone marrow. *J. Clin. Investig.* **2011**, *121*, 1298–1312. [CrossRef] [PubMed]
163. Axelrod, H.D.; Valkenburg, K.C.; Amend, S.R.; Hicks, J.L.; Parsana, P.; Torga, G.; DeMarzo, A.M.; Pienta, K.J. AXL Is a Putative Tumor Suppressor and Dormancy Regulator in Prostate Cancer. *Mol. Cancer Res.* **2019**, *17*, 356–369. [CrossRef] [PubMed]
164. Poświata, A.; Kozik, K.; Miączyńska, M.; Zdżalik-Bielecka, D. Endocytic trafficking of GAS6–AXL complexes is associated with sustained AKT activation. *Cell. Mol. Life Sci.* **2022**, *79*, 316. [CrossRef] [PubMed]
165. Mishra, A.; Wang, J.; Shiozawa, Y.; McGee, S.; Kim, J.; Jung, Y.; Joseph, J.; Berry, J.E.; Havens, A.; Pienta, K.J.; et al. Hypoxia Stabilizes GAS6/Axl Signaling in Metastatic Prostate Cancer. *Mol. Cancer Res.* **2012**, *10*, 703–712. [CrossRef]
166. Wu, Y.-P.; Ke, Z.-B.; Zheng, W.-C.; Chen, Y.-H.; Zhu, J.-M.; Lin, F.; Li, X.-D.; Chen, S.-H.; Cai, H.; Zheng, Q.-S.; et al. Kinesin family member 18B regulates the proliferation and invasion of human prostate cancer cells. *Cell Death Dis.* **2021**, *12*, 302. [CrossRef]
167. Copello, V.A.; Burnstein, K.L. The kinesin KIF20A promotes progression to castration-resistant prostate cancer through autocrine activation of the androgen receptor. *Oncogene* **2022**, *41*, 2824–2832. [CrossRef]
168. Al Nakouzi, N.; Bawa, O.; Le Pape, A.; Lerondel, S.; Gaudin, C.; Opolon, P.; Gonin, P.; Fizazi, K.; Chauchereau, A. The IGR-CaP1 xenograft model recapitulates mixed osteolytic/blastic bone lesions observed in metastatic prostate cancer. *Neoplasia* **2012**, *14*, 376–387. [CrossRef]
169. Mishra, S.; Tang, Y.; Wang, L.; Degraffenried, L.; Yeh, I.T.; Werner, S.; Troyer, D.; Copland, J.A.; Sun, L.Z. Blockade of transforming growth factor-beta (TGFβ) signaling inhibits osteoblastic tumorigenesis by a novel human prostate cancer cell line: TGFβ Promotes Osteoblastic Tumor Formation. *Prostate* **2011**, *71*, 1441–1454. [CrossRef]
170. Darby, S.; Cross, S.S.; Brown, N.J.; Hamdy, F.C.; Robson, C.N. BMP-6 over-expression in prostate cancer is associated with increased Id-1 protein and a more invasive phenotype. *J. Pathol.* **2008**, *214*, 394–404. [CrossRef]
171. Reyes-Ibarra, A.P.; García-Regalado, A.; Ramírez-Rangel, I.; Esparza-Silva, A.L.; Valadez-Sánchez, M.; Vázquez-Prado, J.; Reyes-Cruz, G. Calcium-sensing receptor endocytosis links extracellular calcium signaling to parathyroid hormone-related peptide secretion via a Rab11a-dependent and AMSH-sensitive mechanism. *Mol. Endocrinol.* **2007**, *21*, 1394–1407. [CrossRef] [PubMed]
172. Michel, M.; Raabe, I.; Kupinski, A.P.; Pérez-Palencia, R.; Bökel, C. Local BMP receptor activation at adherens junctions in the Drosophila germline stem cell niche. *Nat. Commun.* **2011**, *2*, 415. [CrossRef] [PubMed]

173. Skjeldal, F.M.; Strunze, S.; Bergeland, T.; Walseng, E.; Gregers, T.F.; Bakke, O. The fusion of early endosomes induces molecular-motor-driven tubule formation and fission. *J. Cell Sci.* **2012**, *125*, 1910–1919. [CrossRef]
174. Astanina, K.; Jacob, R. KIF5C, a kinesin motor involved in apical trafficking of MDCK cells. *Cell. Mol. Life Sci.* **2010**, *67*, 1331–1342. [CrossRef]
175. Burgo, A.; Proux-Gillardeaux, V.; Sotirakis, E.; Bun, P.; Casano, A.; Verraes, A.; Liem, R.K.; Formstecher, E.; Coppey-Moisan, M.; Galli, T. A molecular network for the transport of the TI-VAMP/VAMP7 vesicles from cell center to periphery. *Dev. Cell* **2012**, *23*, 166–180. [CrossRef]
176. Jaulin, F.; Xue, X.; Rodriguez-Boulan, E.; Kreitzer, G. Polarization-dependent selective transport to the apical membrane by KIF5B in MDCK cells. *Dev. Cell* **2007**, *13*, 511–522. [CrossRef]
177. Nakagawa, T.; Setou, M.; Seog, D.H.; Ogasawara, K.; Dohmae, N.; Takio, K.; Hirokawa, N. A novel motor, KIF13A, transports mannose-6-phosphate receptor to plasma membrane through direct interaction with AP-1 complex. *Cell* **2000**, *103*, 569–581. [CrossRef]
178. Yamada, K.H.; Nakajima, Y.; Geyer, M.; Wary, K.K.; Ushio-Fukai, M.; Komarova, Y.; Malik, A.B. KIF13B regulates angiogenesis through Golgi to plasma membrane trafficking of VEGFR2. *J. Cell Sci.* **2014**, *127*, 4518–4530. [CrossRef]
179. Serra-Marques, A.; Martin, M.; Katrukha, E.A.; Grigoriev, I. Concerted action of kinesins KIF5B and KIF13B promotes efficient secretory vesicle transport to microtubule plus ends. *eLife* **2020**, *9*, e61302. [CrossRef]
180. Heatlie, J.K.; Lazniewska, J.; Moore, C.R.; Johnson, I.R.; Nturubika, B.D.; Williams, R.; Ward, M.P.; O'Leary, J.J.; Butler, L.M.; Brooks, D.A. Prostate Cancer Intercellular Communication. *Preprints* **2024**, 2024110469.
181. Mezzasoma, L.; Costanzi, E.; Scarpelli, P.; Talesa, V.N.; Bellezza, I. Extracellular vesicles from human advanced-stage prostate cancer cells modify the inflammatory response of microenvironment-residing cells. *Cancers* **2019**, *11*, 1276. [CrossRef] [PubMed]
182. Fairey, A.; Paproski, R.J.; Pink, D.; Sosnowski, D.L.; Vasquez, C.; Donnelly, B.; Hyndman, E.; Aprikian, A.; Kinnaird, A.; Beatty, P.H.; et al. Clinical analysis of EV-Fingerprint to predict grade group 3 and above prostate cancer and avoid prostate biopsy. *Cancer Med.* **2023**, *12*, 15797–15808. [CrossRef] [PubMed]
183. Wang, Y.-T.; Shi, T.; Srivastava, S.; Kagan, J.; Liu, T.; Rodland, K.D. Proteomic analysis of exosomes for discovery of protein biomarkers for prostate and bladder cancer. *Cancers* **2020**, *12*, 2335. [CrossRef] [PubMed]
184. Zenner, M.L.; Kirkpatrick, B.; Leonardo, T.R.; Schlicht, M.J.; Saldana, A.C.; Loitz, C.; Valyi-Nagy, K.; Maienschein-Cline, M.; Gann, P.H.; Abern, M.; et al. Prostate-derived circulating microRNAs add prognostic value to prostate cancer risk calculators. *J. Extracell. Biol.* **2023**, *2*, e122. [CrossRef]
185. Wang, J.; Ni, J.; Beretov, J.; Thompson, J.; Graham, P.; Li, Y. Exosomal microRNAs as liquid biopsy biomarkers in prostate cancer. *Crit. Rev. Oncol. Hematol.* **2020**, *145*, 102860. [CrossRef]
186. Minciacchi, V.R.; Freeman, M.R.; Di Vizio, D. Extracellular vesicles in cancer: Exosomes, microvesicles and the emerging role of large oncosomes. *Semin. Cell Dev. Biol.* **2015**, *40*, 41–51. [CrossRef]
187. Li, Z.; Fang, R.; Fang, J.; He, S.; Liu, T. Functional implications of Rab27 GTPases in Cancer. *Cell Commun. Signal.* **2018**, *16*, 44. [CrossRef]
188. Worst, T.S.; Meyer, Y.; Gottschalt, M.; Weis, C.-A.; Von Hardenberg, J.; Frank, C.; Steidler, A.; Michel, M.S.; Erben, P. RAB27A, RAB27B and VPS36 are downregulated in advanced prostate cancer and show functional relevance in prostate cancer cells. *Int. J. Oncol.* **2017**, *50*, 920–932. [CrossRef]
189. Song, L.; Tang, S.; Han, X.; Jiang, Z.; Dong, L.; Liu, C.; Liang, X.; Dong, J.; Qiu, C.; Wang, Y.; et al. KIBRA controls exosome secretion via inhibiting the proteasomal degradation of Rab27a. *Nat. Commun.* **2019**, *10*, 1639. [CrossRef]
190. Rayala, S.K.; Den Hollander, P.; Manavathi, B.; Talukder, A.H.; Song, C.; Peng, S.; Barnekow, A.; Kremerskothen, J.; Kumar, R. Essential role of KIBRA in co-activator function of dynein light chain 1 in mammalian cells. *J. Biol. Chem.* **2006**, *281*, 19092–19099. [CrossRef]
191. Arivazhagan, L.; Surabhi, R.P.; Kanakarajan, A.; Sundaram, S.; Pitani, R.S.; Mudduwa, L.; Kremerskothen, J.; Venkatraman, G.; Rayala, S.K. KIBRA attains oncogenic activity by repressing RASSF1A. *Br. J. Cancer* **2017**, *117*, 553–562.
192. Ventimiglia, E.; Gedeborg, R.; Styrke, J.; Robinson, D.; Stattin, P.; Garmo, H. Natural history of nonmetastatic prostate cancer managed with watchful waiting. *JAMA Netw. Open* **2024**, *7*, e2414599. [CrossRef] [PubMed]
193. Assel, M.J.; Ulmert, H.D.; Karnes, R.J.; Boorjian, S.A.; Hillman, D.W.; Vickers, A.J.; Klee, G.G.; Lilja, H. Kallikrein markers performance in pretreatment blood to predict early prostate cancer recurrence and metastasis after radical prostatectomy among very high-risk men. *Prostate* **2020**, *80*, 51–56. [CrossRef] [PubMed]
194. Dunn, T.A.; Chen, S.; Faith, D.A.; Hicks, J.L.; Platz, E.A.; Chen, Y.; Ewing, C.M.; Sauvageot, J.; Isaacs, W.B.; De Marzo, A.M.; et al. A novel role of myosin VI in human prostate cancer. *Am. J. Pathol.* **2006**, *169*, 1843–1854. [CrossRef]
195. Palmer, K.J.; Hughes, H.; Stephens, D.J. Specificity of cytoplasmic dynein subunits in discrete membrane-trafficking steps. *Mol. Biol. Cell* **2009**, *20*, 2885–2899. [CrossRef]
196. Sandvig, K.; Llorente, A. Proteomic analysis of microvesicles released by the human prostate cancer cell line PC-3. *Mol. Cell. Proteomics* **2012**, *11*, M111.012914-1–M111.012914-11. [CrossRef]
197. Humphrey, P.A. Histopathology of Prostate Cancer. *Cold Spring Harb. Perspect. Med.* **2017**, *7*, a030411. [CrossRef]

198. Lazo, O.M.; Schiavo, G. Rab10 regulates the sorting of internalised TrkB for retrograde axonal transport. *eLife* **2023**, *12*, e81532. [CrossRef]
199. Shi, A.; Liu, O.; Koenig, S.; Banerjee, R.; Chen, C.C.H.; Eimer, S.; Grant, B.D. RAB-10-GTPase-mediated regulation of endosomal phosphatidylinositol-4,5-bisphosphate. *Proc. Natl. Acad. Sci. USA* **2012**, *109*, E2306–E2315. [CrossRef]
200. Yamamoto, H.; Koga, H.; Katoh, Y.; Takahashi, S.; Nakayama, K.; Shin, H.-W. Functional cross-talk between Rab14 and Rab4 through a dual effector, RUFY1/Rabip4. *Mol. Biol. Cell* **2010**, *21*, 2746–2755. [CrossRef]
201. Darshan, M.S.; Loftus, M.S.; Thadani-Mulero, M.; Levy, B.P.; Escuin, D.; Zhou, X.K.; Gjyrezi, A.; Chanel-Vos, C.; Shen, R.; Tagawa, S.T.; et al. Taxane-induced blockade to nuclear accumulation of the androgen receptor predicts clinical responses in metastatic prostate cancer. *Cancer Res.* **2011**, *71*, 6019–6029. [CrossRef] [PubMed]
202. Hung, S.-C.; Chang, L.-W.; Li, J.-R.; Wang, S.-S.; Yang, C.-K.; Chen, C.-S.; Lu, K.; Chen, C.-C.; Wang, S.-C.; Lin, C.-Y.; et al. Docetaxel rechallenge improves survival in patients with metastatic Castration-resistant Prostate Cancer: A retrospective study. *In Vivo* **2021**, *35*, 3509–3519. [CrossRef] [PubMed]
203. Rath, O.; Kozielski, F. Kinesins and cancer. *Nat. Rev. Cancer* **2012**, *12*, 527–539. [CrossRef] [PubMed]
204. Tan, M.H.; De, S.; Bebek, G.; Orloff, M.S.; Wesolowski, R.; Downs-Kelly, E.; Budd, G.T.; Stark, G.R.; Eng, C. Specific kinesin expression profiles associated with taxane resistance in basal-like breast cancer. *Breast Cancer Res. Treat.* **2012**, *131*, 849–858. [CrossRef]
205. De, S.; Cipriano, R.; Jackson, M.W.; Stark, G.R. Overexpression of kinesins mediates docetaxel resistance in breast cancer cells. *Cancer Res.* **2009**, *69*, 8035–8042. [CrossRef]
206. Gjyrezi, A.; Xie, F.; Voznesensky, O.; Khanna, P.; Calagua, C.; Bai, Y.; Kung, J.; Wu, J.; Corey, E.; Montgomery, B.; et al. Taxane resistance in prostate cancer is mediated by decreased drug-target engagement. *J. Clin. Investig.* **2020**, *130*, 3287–3298. [CrossRef]
207. Marín-Aguilera, M.; Codony-Servat, J.; Reig, Ò.; Lozano, J.J.; Fernández, P.L.; Pereira, M.V.; Jiménez, N.; Donovan, M.; Puig, P.; Mengual, L.; et al. Epithelial-to-mesenchymal transition mediates docetaxel resistance and high risk of relapse in prostate cancer. *Mol. Cancer Ther.* **2014**, *13*, 1270–1284. [CrossRef]
208. Marín-Aguilera, M.; Codony-Servat, J.; Kalko, S.G.; Fernández, P.L.; Bermudo, R.; Buxo, E.; Ribal, M.J.; Gascón, P.; Mellado, B. Identification of docetaxel resistance genes in castration-resistant prostate cancer. *Mol. Cancer Ther.* **2012**, *11*, 329–339. [CrossRef]
209. Nikitas, J.; Kishan, A.; Chang, A.; Duriseti, S.; Nichols, N.G.; Reiter, R.; Rettig, M.; Brisbane, W.; Steinberg, M.L.; Valle, L. Treatment intensification strategies for men undergoing definitive radiotherapy for high-risk prostate cancer. *World J. Urol.* **2024**, *42*, 165. [CrossRef]
210. Nosti, C.; Yu, M.; Jean-Baptiste, L.; Jaramillo, M.; Siretskiy, R. Radioresistance mechanisms in prostate cancer. In *Therapy Resistance in Prostate Cancer*; Elsevier: Amsterdam, The Netherlands, 2024; pp. 213–233.
211. Hendricks, A.G.; Lazarus, J.E.; Perlson, E.; Gardner, M.K.; Odde, D.J.; Goldman, Y.E.; Holzbaur, E.L. Dynein tethers and stabilizes dynamic microtubule plus ends. *Curr. Biol.* **2012**, *22*, 632–637. [CrossRef]
212. Chang, C.-C.; Chien, C.-C.; Yang, S.-H.; Chen, S.-H.; Huang, C.-J. Identification and clinical correlation of decreased expression of cytoplasmic dynein heavy chain 1 in patients with colorectal cancer. *Clin. Mol. Med.* **2008**, *1*, 6–10.
213. Swanton, C.; Tomlinson, I.; Downward, J. Chromosomal instability, colorectal cancer and taxane resistance. *Cell Cycle* **2006**, *5*, 818–823. [CrossRef] [PubMed]
214. Höing, S.; Yeh, T.Y.; Baumann, M.; Martinez, N.E.; Habenberger, P.; Kremer, L.; Drexler, H.C.; Küchler, P.; Reinhardt, P.; Choidas, A.; et al. Dynarrestin, a Novel Inhibitor of Cytoplasmic Dynein. *Cell Chem. Biol.* **2018**, *25*, 357–369.e6. [CrossRef] [PubMed]
215. Huszar, D.; Theoclitou, M.-E.; Skolnik, J.; Herbst, R. Kinesin motor proteins as targets for cancer therapy. *Cancer Metastasis Rev.* **2009**, *28*, 197–208. [CrossRef]
216. Priyanga, J.; Guha, G.; Bhakta-Guha, D. Microtubule motors in centrosome homeostasis: A target for cancer therapy? *Biochim. Biophys. Acta Rev. Cancer* **2021**, *1875*, 188524. [CrossRef]
217. Wang, J.-M.; Zhang, F.-H.; Liu, Z.-X.; Tang, Y.-J.; Li, J.-F.; Xie, L.-P. Cancer on motors: How kinesins drive prostate cancer progression? *Biochem. Pharmacol.* **2024**, *224*, 116229. [CrossRef]
218. Yu, H.; Pan, J.; Guo, Z.; Yang, C.; Mao, L. CART cell therapy for prostate cancer: Status and promise. *OncoTargets Ther.* **2019**, *12*, 391–395. [CrossRef]
219. Priceman, S.J.; Gerdts, E.A.; Tilakawardane, D.; Kennewick, K.T.; Murad, J.P.; Park, A.K.; Jeang, B.; Yamaguchi, Y.; Yang, X.; Urak, R.; et al. Co-stimulatory signaling determines tumor antigen sensitivity and persistence of CAR T cells targeting PSCA+ metastatic prostate cancer. *Oncoimmunology* **2018**, *7*, e1380764. [CrossRef]
220. Bhatia, V.; Kamat, N.V.; Pariva, T.E.; Wu, L.-T.; Tsao, A.; Sasaki, K.; Sun, H.; Javier, G.; Nutt, S.; Coleman, I.; et al. Targeting advanced prostate cancer with STEAP1 chimeric antigen receptor T cell and tumor-localized IL-12 immunotherapy. *Nat. Commun.* **2023**, *14*, 2041. [CrossRef]
221. Jiang, Y.; Wen, W.; Yang, F.; Han, D.; Zhang, W.; Qin, W. Prospect of Prostate Cancer Treatment: Armed CAR-T or Combination Therapy. *Cancers* **2022**, *14*, 967. [CrossRef]
222. Cheng, H.H.; Sokolova, A.O.; Schaeffer, E.M.; Small, E.J.; Higano, C.S. Germline and somatic mutations in prostate cancer for the clinician. *J. Natl. Compr. Cancer Netw.* **2019**, *17*, 515–521. [CrossRef] [PubMed]

223. Taylor, R.A.; Fraser, M.; Rebello, R.J.; Boutros, P.C.; Murphy, D.G.; Bristow, R.G.; Risbridger, G.P. The influence of BRCA2 mutation on localized prostate cancer. *Nat. Rev. Urol.* **2019**, *16*, 281–290. [CrossRef] [PubMed]
224. Hoang, H.T.; Schlager, M.A.; Carter, A.P.; Bullock, S.L. DYNC1H1 mutations associated with neurological diseases compromise processivity of dynein–dynactin–cargo adaptor complexes. *Proc. Natl. Acad. Sci. USA* **2017**, *114*, E1597–E1606. [CrossRef]
225. Blatner, N.R.; I Wilson, M.; Lei, C.; Hong, W.; Murray, D.; Williams, R.L.; Cho, W. The structural basis of novel endosome anchoring activity of KIF16B kinesin. *EMBO J.* **2007**, *26*, 3709–3719. [CrossRef]
226. Lindberg, J.; Mills, I.G.; Klevebring, D.; Liu, W.; Neiman, M.; Xu, J.; Wikström, P.; Wiklund, P.; Wiklund, F.; Egevad, L.; et al. The mitochondrial and autosomal mutation landscapes of prostate cancer. *Eur. Urol.* **2013**, *63*, 702–708. [CrossRef]
227. Doan, D.K.; Schmidt, K.T.; Chau, C.H.; Figg, W.D. Germline genetics of prostate cancer: Prevalence of risk variants and clinical implications for disease management. *Cancers* **2021**, *13*, 2154. [CrossRef]
228. Messina, C.; Cattrini, C.; Soldato, D.; Vallome, G.; Caffo, O.; Castro, E.; Olmos, D.; Boccardo, F.; Zanardi, E. BRCA mutations in prostate cancer: Prognostic and predictive implications. *J. Oncol.* **2020**, *2020*, 4986365. [CrossRef]
229. Cussenot, O.; Cancel-Tassin, G.; Rao, S.R.; Woodcock, D.J.; Lamb, A.D.; Mills, I.G.; Hamdy, F.C. Aligning germline and somatic mutations in prostate cancer. Are genetics changing practice? *BJU Int.* **2023**, *132*, 472–484. [CrossRef]
230. Burchardt, M.; Burchardt, T.; Shabsigh, A.; Ghafar, M.; Chen, M.; Anastasiadis, A.; de la Taille, A.; Kiss, A.; Buttyan, R. Reduction of wild type p53 function confers a hormone resistant phenotype on LNCaP prostate cancer cells. *Prostate* **2001**, *48*, 225–230. [CrossRef]
231. Maughan, B.L.; Guedes, L.B.; Boucher, K.; Rajoria, G.; Liu, Z.; Klimek, S.; Zoino, R.; Antonarakis, E.S.; Lotan, T.L. P53 status in the primary tumor predicts efficacy of subsequent abiraterone and enzalutamide in castration-resistant prostate cancer. *Prostate Cancer Prostatic Dis.* **2018**, *21*, 260–268. [CrossRef]
232. Morand, S.; Stanbery, L.; Walter, A.; Rocconi, R.P.; Nemunaitis, J. *BRCA1/2* mutation status impact on autophagy and immune response: Unheralded target. *JNCI Cancer Spectr.* **2020**, *4*, pkaa077. [CrossRef] [PubMed]
233. Lin, J.Z.; Wang, W.W.; Hu, T.T.; Zhu, G.Y.; Li, L.N.; Zhang, C.Y.; Xu, Z.; Yu, H.B.; Wu, H.F.; Zhu, J.G. FOXM1 contributes to docetaxel resistance in castration-resistant prostate cancer by inducing AMPK/mTOR-mediated autophagy. *Cancer Lett.* **2020**, *469*, 481–489. [CrossRef] [PubMed]
234. Nguyen, H.G.; Yang, J.C.; Kung, H.-J.; Shi, X.-B.; Tilki, D.; Lara, P.N.; White, R.W.D.; Gao, A.C.; Evans, C.P. Targeting autophagy overcomes Enzalutamide resistance in castration-resistant prostate cancer cells and improves therapeutic response in a xenograft model. *Oncogene* **2014**, *33*, 4521–4530. [CrossRef] [PubMed]
235. Giannakakou, P.; Sackett, D.L.; Ward, Y.; Webster, K.R.; Blagosklonny, M.V.; Fojo, T. P53 is associated with cellular microtubules and is transported to the nucleus by dynein. *Nat. Cell Biol.* **2000**, *2*, 709–717. [CrossRef]
236. Lakoduk, A.M.; Roudot, P.; Mettlen, M.; Grossman, H.M.; Schmid, S.L.; Chen, P.-H. Mutant p53 amplifies a dynamin-1/APPL1 endosome feedback loop that regulates recycling and migration. *J. Cell Biol.* **2019**, *218*, 1928–1942. [CrossRef]
237. Gan, Y.; Shi, C.; Inge, L.; Hibner, M.; Balducci, J.; Huang, Y. Differential roles of ERK and Aktpathways in regulation of EGFR-mediated signaling and motility in prostate cancer cells. *Oncogene* **2010**, *29*, 4947–4958. [CrossRef]
238. Morton, D.J.; Patel, D.; Joshi, J.; Hunt, A.; Knowell, A.E.; Chaudhary, J. ID4 regulates transcriptional activity of wild type and mutant p53 via K373 acetylation. *Oncotarget* **2017**, *8*, 2536–2549. [CrossRef]
239. Xu, S.; Weihua, Z. Loss of EGFR induced autophagy sensitizes hormone refractory prostate cancer cells to adriamycin: Loss of EGFR Sensitizes Cells to Adriamycin. *Prostate* **2011**, *71*, 1216–1224. [CrossRef]
240. Hashimoto, S.; Furukawa, S.; Hashimoto, A.; Tsutaho, A.; Fukao, A.; Sakamura, Y.; Parajuli, G.; Onodera, Y.; Otsuka, Y.; Handa, H.; et al. ARF6 and AMAP1 are major targets of KRAS and TP53 mutations to promote invasion, PD-L1 dynamics, and immune evasion of pancreatic cancer. *Proc. Natl. Acad. Sci. USA* **2019**, *116*, 17450–17459. [CrossRef]
241. Sanmamed, N.; Gómez-Rivas, J.; Buchser, D.; Montijano, M.; Gómez-Aparicio, M.A.; Duque-Santana, V.; Torres, L.; Zilli, T.; Ost, P.; Maldonado, A.; et al. Docetaxel provides oncological benefits in the era of new-generation androgen receptor inhibitors—Or is three a crowd? *Clin. Genitourin. Cancer* **2024**, *22*, 56–66. [CrossRef]
242. Fletcher, K.A.; Alkurashi, M.H.; Lindsay, A.J. Endosomal recycling inhibitors downregulate the androgen receptor and synergise with enzalutamide. *Investig. New Drugs* **2024**, *42*, 14–23. [CrossRef] [PubMed]
243. Christensen, J.R.; A Kendrick, A.; Truong, J.B.; Aguilar-Maldonado, A.; Adani, V.; Dzieciatkowska, M.; Reck-Peterson, S.L. Cytoplasmic dynein-1 cargo diversity is mediated by the combinatorial assembly of FTS-Hook-FHIP complexes. *eLife* **2021**, *10*, e74538. [CrossRef] [PubMed]

cancers

MDPI

Review

TSGA10 as a Model of a Thermal Metabolic Regulator: Implications for Cancer Biology

Ali Amini [1], Farzad Taghizadeh-Hesary [2], John Bracht [3,*] and Babak Behnam [3,4,*]

[1] Center for Data Science, American University, 4400 Massachusetts Avenue NW, Washington, DC 20016, USA
[2] Ear, Nose, Throat, and Head and Neck Research Center and Department, The Five Senses Health Institute, School of Medicine, Iran University of Medical Sciences, Tehran 14496-14535, Iran
[3] Biology Department, American University, 4400 Massachusetts Avenue NW, Washington, DC 20016, USA
[4] Avicenna Biotech Research, Germantown, MD 20854, USA
* Correspondence: jbracht@american.edu (J.B.); bbehnam@american.edu or babak.behnam@gmail.com (B.B.)

Simple Summary: We outline a potential mechanism by which Testis-Specific Gene 10 (TSGA10) helps mitochondria produce energy efficiently while reducing harmful byproducts like reactive oxygen species (ROS) and heat. By interacting with cytochrome c1 (CytC1) in mitochondrial Complex III, TSGA10 may stabilize electron transfer, minimizing leakage to oxygen and reducing ROS and heat production. Our model predicts that in cancer, TSGA10 acts as a double-edged sword: Low levels allow ROS buildup and force cancer cells to rely on less efficient energy methods (like glycolysis), promoting tumor growth. Conversely, high TSGA10 levels can suppress tumors by maintaining mitochondrial health and blocking cancer-friendly pathways. TSGA10 and Hypoxia-inducible factor 1-alpha (HIF-1α—a protein that helps cancer cells thrive in low oxygen) regulate each other, creating a balance that shapes tumor behavior. TSGA10's varying levels in tissues—high in the brain and testes (protecting against damage) and low in the liver (enabling cancer growth)—make it a promising target for therapies aimed at disrupting cancer's energy strategies.

Abstract: TSGA10, a multifunctional protein critical for mitochondrial coupling and metabolic regulation, plays a paradoxical role in cancer progression and carcinogenesis. Here, we outline a potential mechanism by which TSGA10 mediates metabolism in oncogenesis and thermal modulation. Initially identified in spermatogenesis, TSGA10 interacts with mitochondrial Complex III: it directly binds cytochrome c1 (CytC1). In our model, TSGA10 optimizes electron transport to minimize reactive oxygen species (ROS) and heat production while enhancing Adenosine Triphosphate (ATP) synthesis. In cancer, TSGA10's expression is context-dependent: Its downregulation in tumors like glioblastoma might disrupt mitochondrial coupling, promoting electron leakage, ROS accumulation, and genomic instability. This dysfunction would be predicted to contribute to a glycolytic shift, facilitating tumor survival under hypoxia. Conversely, TSGA10 overexpression in certain cancers suppresses HIF-1α, inhibiting glycolysis and metastasis. TSGA10 and HIF-1α engage in mutual counter-regulation—TSGA10 represses HIF-1α to sustain oxidative phosphorylation (OXPHOS), while HIF-1α suppression of TSGA10 under hypoxia or thermal stress amplifies glycolytic dependency. This interplay is pivotal in tumors adapting to microenvironmental stressors, such as cold-induced mitochondrial uncoupling, which mimics brown adipose tissue thermogenesis to reduce ROS and sustain proliferation. Tissue-specific TSGA10 expression further modulates cancer susceptibility: high levels in the testes and brain may protect against thermal and oxidative damage, whereas low expression in the liver permits HIF-1α-driven metabolic plasticity. Altogether, our model suggests that TSGA10 plays a central role in mitochondrial fidelity. We suggest that its crosstalk with oncogenic pathways position it as a metabolic rheostat, whose dysregulation

https://doi.org/10.3390/cancers17111756

fosters tumorigenesis through ROS-mediated mutagenesis, metabolic reprogramming, and microenvironmental remodeling. Targeting the hypothesized TSGA10-mediated mitochondrial coupling may offer therapeutic potential to disrupt cancer's adaptive energetics and restore metabolic homeostasis.

Keywords: TSGA10; mitochondrial coupling; Complex III (CytC1); carcinogenesis; HIF-1α; Warburg effect

1. Introduction

TSGA10, a gene encoding a 698-amino acid protein, was first identified in the testis and is involved in spermatogenesis, cytoskeletal organization, and metabolic regulation [1,2]. It undergoes alternative splicing, producing multiple transcript variants, and is expressed in tissues such as the brain, retina, and germ cell tumors [3,4]. Processed into N-terminal (27-kD) and C-terminal (55-kD) components, TSGA10 localizes to sperm and is expressed in other tissues, including the liver, brain, and kidney [3,5]. It interacts with Hypoxia-inducible factor 1-alpha (HIF-1α), modulating its nuclear localization and transcriptional activity, and inhibits cancer cell growth, motility, and invasion by downregulating HIF-1α target genes like Vascular Endothelial Growth Factor A (*VEGFA*), Matrix Metalloproteinase-2 (*MMP2*), and Matrix Metalloproteinase-9 (*MMP9*) [6,7]. TSGA10 also interacts with vimentin, a cytoskeletal protein involved in oxygen sensing [8].

TSGA10 is overexpressed in cancers such as melanoma, colon cancer, hepatocellular carcinoma, ovarian cancer, prostate cancer, and leukemia [9,10]. Paradoxically, its downregulation in some contexts is also linked to cancer progression, highlighting its dual role as both a tumor suppressor and potential oncogene [2]. Recent studies suggest that TSGA10 regulates mitochondrial organization and perinuclear translocation, potentially regulating metabolic activity [2,11], positioning it as a key modulator of cellular energy metabolism and stress responses.

Mitochondria, beyond their role in ATP production, are sophisticated regulators of stress responses, calcium signaling, and reactive oxygen species (ROS) production and signaling [12]. They act as molecular thermostats, modulating energy dissipation as heat through controlled coupling and uncoupling of chemiosmotic gradients, which are critical for tissue-specific metabolic demands and systemic temperature homeostasis.

Here, we propose that TSGA10, highly expressed in metabolically active cells, may optimize mitochondrial coupling efficiency by interacting with CytC1, enhancing electron transport through Complex III, promoting ATP synthesis, and minimizing electron leak, heat production, and ROS generation [2]. This mechanism may enable cells to adapt to fluctuating oxygen levels by switching between oxidative phosphorylation and glycolysis.

TSGA10's high expression in neurons suggests a critical role in protecting against thermal stress. Neurons are highly vulnerable to heat due to their complex structure, temperature-sensitive ion channels, and reliance on ATP-dependent processes, including synaptic function and membrane stability [13–15]. Thermal stress disrupts oxidative phosphorylation, increases ROS production, and compromises mitochondrial efficiency, leading to neuronal dysfunction and degeneration [16,17]. Neurons' limited regenerative capacity makes thermal-induced damage potentially permanent [18,19]. TSGA10 may enhance neuronal thermoregulation by optimizing mitochondrial coupling efficiency and reducing ROS levels, which is particularly vital in the brain, given its high metabolic demands and susceptibility to thermal stress. TSGA10's role at the intersection of mitochondrial regulation and environmental adaptation highlights its potential as a therapeutic target for neurodegenerative diseases and cancer.

2. Materials and Methods

The yeast strain AH109 was transformed with pGBKT7-TSGA10 and a rat testis library. Positive clones were identified using selective media, and TSGA10-prey gene expression was confirmed via PCR. CYTC1 emerged as a strong interactor, validated through sequencing and galactosidase activity assays. Additionally, yeast mating was performed with PGAD424-transformed Y187 yeast, and confirmed cDNA inserts were sequenced for accuracy.

To visualize the interaction, NIH3T3 cells were cultured and transfected with GFP-tagged TSGA10 using Lipofectamine (Invitrogen, Carlsbad, CA, USA) 3000. After 24–48 h, cells were stained with MitoTracker Red CMXRos Thermo Fisher Scientific, Cat. No. M7512, Waltham, MA, USA) to label mitochondria, and then fixed with 4% paraformaldehyde.

Using confocal microscopy, colocalization of GFP-TSGA10 (green) and mitochondria (red) appeared as yellow regions in merged images. Image quantification was performed using Zeiss LSM 510 i-UV laser scanning confocal microscope (Carl Zeiss AG, Oberkochen, Germany).

Mitochondrial fractions were isolated from NIH3T3 cells via differential centrifugation. The mitochondrial pellet was resuspended in RIPA buffer with protease inhibitors. Lysates were pre-cleared with protein A/G beads, and then incubated overnight with a mouse monoclonal anti-CytC1 antibody. Captured protein complexes were washed, eluted, and then analyzed via SDS-PAGE and Western blot. The membrane was probed with a rabbit anti-TSGA10 antibody to detect TSGA10 within the CytC1 immunoprecipitate and an anti-GFP antibody to identify GFP-TSGA10 fusion proteins and their proteolytic fragments (27 kDa N-terminus and 55 kDa C-terminus). Chemiluminescence was used for visualization.

To ensure specificity, a negative control using an irrelevant mouse IgG antibody was included in Co-IP, while untransfected or empty vector-transfected NIH3T3 cells served as controls for fluorescent imaging.

3. Mitochondrial Role in Heat Production and Uncoupling

3.1. Mitochondrial (De-)Coupling

In coupled mitochondria, the electron transport chain generates a proton gradient that drives ATP synthesis with high efficiency, while uncoupled mitochondria dissipate this gradient energy as heat through regulated proton leak mechanisms [20]. Decoupling of mitochondria is generally thought of as a way to slow down ATP production when ROS accumulate as a toxic byproduct [21]. However, decoupling is also a mechanism by which tissues can increase their operating temperature. This complex balance gives cells a control point for adjusting ATP, ROS, and temperature to meet the metabolic requirements of each tissue. Given that blood flow is also a thermal distribution system regulated by the central nervous system, the organism can use temperature both as input (to mediate metabolism) and as output (using mitochondrial uncoupling to alter local thermogenesis). In this way, an organism can raise or lower the thermal set-point of each tissue, embodying a shifting medley of metabolic states over time.

3.2. Mitochondrial Coupling Across Healthy Tissues

Mitochondrial metabolism exhibits a spectrum of coupling efficiency across tissues within the organism. At one end, spermatocytes in the testis prioritize tightly coupled oxidative phosphorylation (OXPHOS) to support their high energy needs during meiosis, while keeping temperatures low. Efficient ATP production via coupled mitochondria ensures genomic stability, but even this efficient process generates some ROS (which inescapably increases with OXPHOS activity), posing risks to DNA integrity. To mitigate this, spermatocytes employ robust antioxidant systems, crucial in the cooler testicular environment (2–8 °C below core body temperature), where lower thermal energy may reduce enzymatic repair efficiency [22].

3.3. Mitochondrial Decoupling in Brown Adipose Tissue

In stark contrast, brown adipose tissue (BAT) dissipates proton gradients to produce heat instead of ATP via the expression of uncoupling protein 1 (UCP1). The inverse relationship between coupling efficiency and thermal output highlights a metabolic trade-off: ATP synthesis versus heat. Cancer cells, facing microenvironmental stress, exploit metabolic plasticity. Under cold stress, they may adopt BAT-like uncoupling to reduce ROS and generate heat, sustaining proliferation. Concurrently, cold impairs OXPHOS enzymes like Complex III, promoting ROS accumulation and a glycolytic shift akin to the Warburg effect [23]. This dual adaptation—uncoupling for survival and glycolysis for ATP—reflects metabolic reprogramming under thermal duress [24]. Thus, tissue-specific mitochondrial strategies, from spermatocyte coupling to BAT uncoupling and cancer's hybrid approach, illustrate how metabolic pathways change across and within tissues to meet physiological and environmental demands.

3.4. Mitochondrial Uncoupling and Cancer

With emerging evidence linking metabolic reprogramming, heat production, and uncoupling to carcinogenesis, mitochondria play a complex role in cancer. Cancer cells often exhibit mitochondrial uncoupling, a process where proton leakage across the inner mitochondrial membrane reduces ATP synthesis efficiency, diverting energy into heat production (thermogenesis). This uncoupling, mediated by proteins like UCP1/UCP2, may support tumor survival by mitigating oxidative stress (ROS) or adapting to hypoxic microenvironments. Excessive ROS from dysfunctional electron transport chains can damage DNA and drive oncogenic mutations, while uncoupling-induced heat might alter the tumor microenvironment, promoting angiogenesis or metastasis. The Warburg effect (aerobic glycolysis) and mitochondrial uncoupling together reflect a metabolic shift that prioritizes rapid energy and biomass production over efficiency, enabling cancer progression. Thus, mitochondrial thermogenesis and uncoupling are not merely metabolic quirks but potential contributors to tumorigenesis and therapeutic targets.

4. TSGA10 and Metabolic Activity

TSGA10 exhibits context-dependent roles in metabolic regulation across postmitotic cells and cancer, with its expression levels aligning with distinct cellular energy demands. *TSGA10* expression across human tissues appears to show elevated expression in thermosensitive and energy-demanding tissues (e.g., testis and brain) and reduced levels in thermally resilient tissues (e.g., liver), reflecting a potential role in mitochondrial coupling and metabolic adaptation (Figure 1). Consistent with this observation, postmitotic cells like spermatocytes exhibit high TSGA10 expression, which also correlates with improved sperm quality [25]. This aligns with the high energy demands of spermatozoa, which rely on mitochondrial OXPHOS for motility. Similarly, in other postmitotic cells (e.g., neurons and nephrons), TSGA10's suppression of HIF-1α [6] may promote oxidative metabolism over glycolysis, ensuring efficient ATP production in these non-proliferative, long-lived cells.

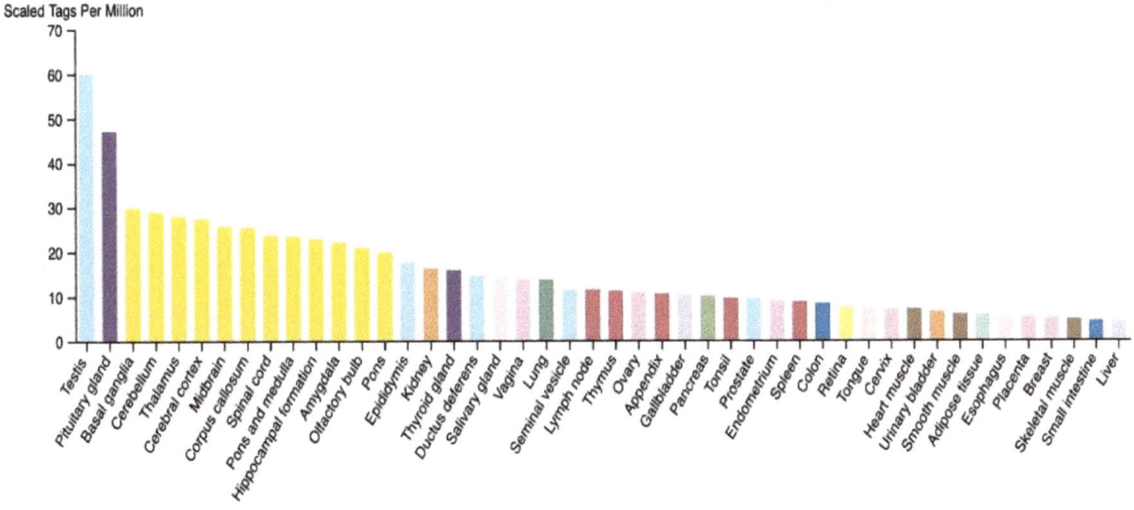

Figure 1. RNA expression level of TSGA10 in different human tissues (courtesy of the open-source The Human Atlas Protein [26]). Colors indicate tissue types, with male reproductive being light blue and neuronal being yellow.

In contrast, cancer cells exhibit a metabolic inconsistency. TSGA10 overexpression in breast cancer suppresses HIF-1α-mediated glycolysis and metastatic activity [27], effectively countering the Warburg effect (aerobic glycolysis). This suggests that high TSGA10 levels in tumors are associated with reduced glycolytic flux and attenuated malignancy. However, in hepatocellular carcinoma, TSGA10's link to drug resistance via lncRNAs [28] implies a more complex metabolic interplay, potentially involving stress-induced adaptations.

Overall, consistency across some cell types is observed. In postmitotic cells, the upregulated TSGA10 may support oxidative metabolism (e.g., in ciliary-riched spermatozoa and inferred neurons/nephrons), consistent with their reliance on OXPHOS. Also, in cancer cells, we predict that higher TSGA10 expression suppresses glycolysis (via HIF-1α inhibition), aligning with reduced tumor aggressiveness. Conversely, low TSGA10 may permit glycolytic dominance, fueling proliferation.

4.1. Transcriptional and Post-Transcriptional Regulation of TSGA10

The dynamic expression of TSGA10 across developmental stages and tissues underscores its context-dependent roles, which are tightly regulated by transcriptional and post-transcriptional mechanisms. Transcriptional regulation of TSGA10 involves nuclear transcription factor Y (NF-Y), which binds to the CCAAT box in its promoter to activate expression, as demonstrated in zebrafish and conserved in mammals [4,29]. This aligns with findings that NF-Y often regulates cell cycle-related genes, supporting TSGA10's potential role in mitosis and DNA repair via its conserved SMC domain. Epigenetic modifications further fine-tune TSGA10 expression. For instance, DNA hypomethylation at the TSGA10 promoter has been linked to its overexpression in cancers like transitional cell carcinoma [30], while hypermethylation in other contexts may suppress its expression, contributing to tissue-specific regulation. Histone modifiers, such as HDACs, may also modulate TSGA10 chromatin accessibility, though this warrants further investigation [31].

Post-transcriptional control mechanisms, particularly miRNA-mediated regulation, play pivotal roles in adapting TSGA10 levels to environmental stimuli. Hypoxia-induced miR-10b-3p directly targets TSGA10, promoting angiogenesis and metastasis in esophageal squamous cell carcinoma [32]. Similarly, exosomal miR-23a in nasopharyngeal carcinoma downregulates TSGA10 to enhance vascularization [25]. These findings highlight a conserved feedback loop where hypoxia-driven HIF-1α not only induces angiogenic miRNAs but also suppresses TSGA10, which itself inhibits HIF-1α nuclear localization [6]. This reciprocal regulation allows TSGA10 to act as a hypoxia-responsive brake on angiogenesis, balancing pro- and anti-tumorigenic signals.

Environmental stressors like hypoxia and hormonal signals further diversify TSGA10's regulatory landscape. In the testis, TSGA10's constitutive expression is likely maintained by androgen-responsive elements [25], whereas in cancers, oxidative stress and metabolic changes may alter its epigenetic or miRNA-mediated regulation. Such plasticity enables TSGA10 to adapt its roles—from supporting spermatogenesis in normal testis to suppressing angiogenesis in hypoxic tumors or promoting developmental patterning in embryos. Future studies should explore tissue-specific enhancers, lncRNAs, and additional post-translational modifications that may refine TSGA10's functional versatility in response to diverse stimuli.

4.2. *TSGA10 and Mitochondria*

TSGA10 is strongly associated with mitochondrial function, particularly in the context of spermatogenesis and male fertility. TSGA10 was shown to be essential for the proper arrangement of the mitochondrial sheath during spermatid differentiation [11]. TSGA10 has been shown to colocalize with mitochondria [11,33], shown in Figure 2, to bind to the mitochondrial protein NSUN2 [34].

TSGA10 is critical for sperm development and mitochondrial organization. Disruption of TSGA10 function led to defects in mitochondrial organization, resulting in male infertility. Similarly, the loss-of-function mutations in TSGA10 cause acephalic spermatozoa syndrome in humans, characterized by abnormal sperm head formation and mitochondrial sheath disorganization, further underscoring its role in mitochondrial integrity during spermatogenesis [35]. Additionally, a connection between TSGA10 and autophagy has been explored [36], suggesting that TSGA10 may regulate mitochondrial quality control through autophagy pathways, which is vital for maintaining mitochondrial function during sperm maturation. Collectively, these studies highlight TSGA10 as a key factor in mitochondrial organization and function, particularly in the context of male fertility, where its proper expression is crucial for the structural and functional integrity of sperm mitochondria.

TSGA10 is not merely associated with mitochondria; it directly interacts with cytochrome c1 (CytC1), a subunit of Complex III in the mitochondrial electron transport chain (ETC) [33]. The interaction between TSGA10 and CytC1 was discovered by one of us (B. Behnam) by a yeast two-hybrid (Y2H) screening assay and confirmed by both colocalization and co-immunoprecipitation [33]. This functional interaction suggests that TSGA10 may play a crucial role in mitochondrial energy metabolism or regulation. Figure 2 illustrates the comprehensive experimental validation of this interaction, demonstrating TSGA10's mitochondrial association through multiple complementary techniques. The combined evidence from yeast two-hybrid screening, co-immunoprecipitation studies, and immunofluorescence assays collectively establishes TSGA10 as a bona fide mitochondrial-interacting protein with specific affinity for CytC1. CytC1 directly transfers an electron from cytochrome b1 to cytochrome c1 [37], and we therefore propose that TSGA10 is a previously unknown metabolic regulator, leveraging its interaction with cytochrome c1 (CytC1) within Complex III of the electron transport chain to stabilize OXPHOS and modulate ROS production (Figure 3).

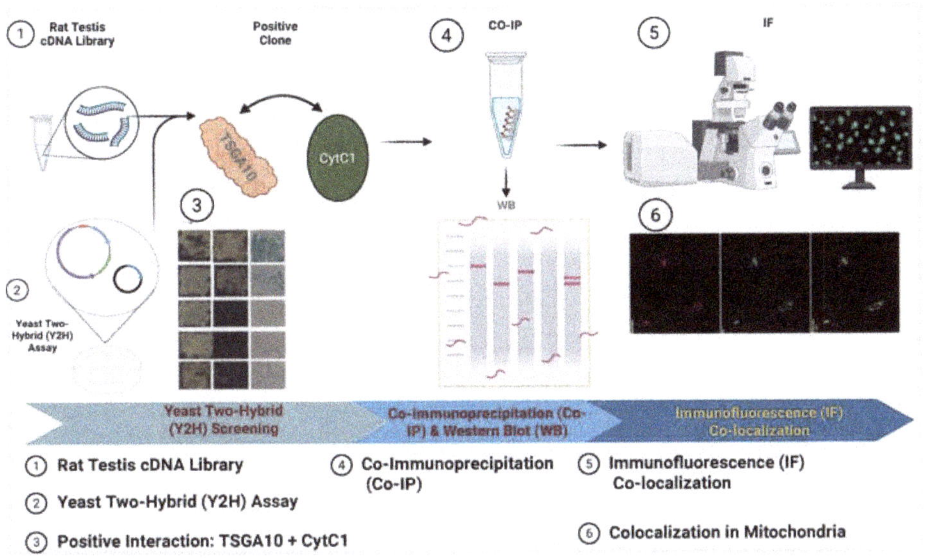

Figure 2. TSGA10 protein association with mitochondria in NIH3T3 cells. This schematic outlines the experimental workflow to demonstrate TSGA10's association with mitochondria, where TSGA10 interacts with CytC1. The process includes a yeast two-hybrid (Y2H) assay; TSGA10 is used as the bait protein to screen a testis cDNA library, leading to the identification of CytC1 as an interacting partner. Then, the interaction is validated in mammalian cells through co-immunoprecipitation (CO-IP) followed by Western blot (WB); CytC1 antibody is used for immunoprecipitation, and the presence of TSGA10 in the complex is confirmed using TSGA10 and GFP antibodies during WB. Finally, a immunofluorescence (IF) colocalization study isperformed to visualize the subcellular localization of TSGA10 and CytC1 in mitochondria. Together, these experiments confirm the physical interaction and mitochondrial co-localization of TSGA10 and CytC1.

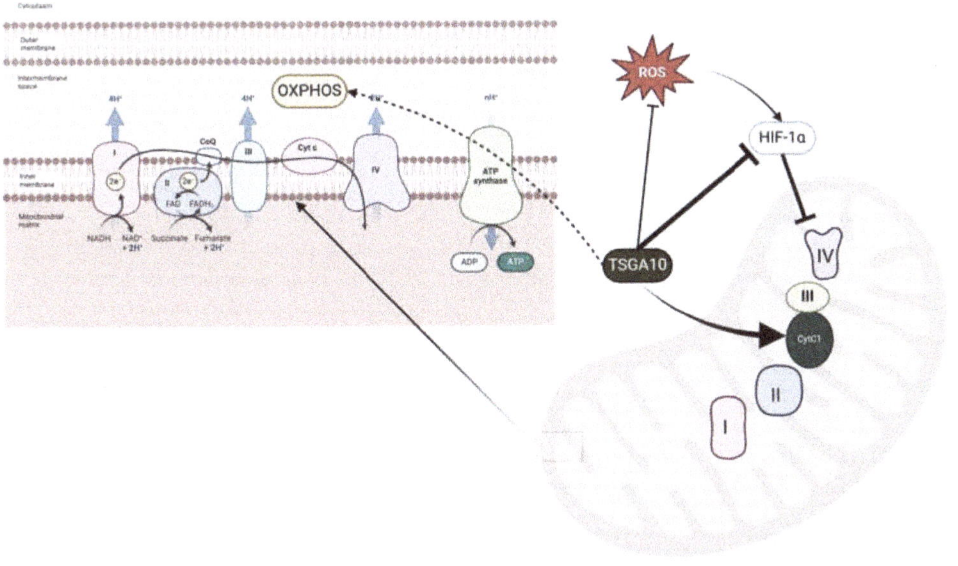

Figure 3. Proposed role of TSGA10 interaction with cytochrome c1 in mitochondria. The left panel illustrates the electron transport chain (ETC) in the mitochondrial inner membrane, where Complexes I–IV facilitate electron transfer and proton pumping to generate ATP through oxidative phosphorylation.

The right panel highlights the regulatory role of TSGA10 in mitochondrial function, where it interacts with cytochrome c1 (Cyt C1) of Complex III, potentially modulating electron transport efficiency. TSGA10 is also implicated in the suppression of HIF-1α, a key transcription factor involved in hypoxia adaptation, which can downregulate Complex IV activity. Additionally, TSGA10 may influence ROS production, linking its function to oxidative stress responses. The black arrows indicate proposed direct interactions, while dashed lines represent indirect or putative pathways.

4.3. *TSGA10 and Its Potential Role in Mitochondrial Coupling*

4.3.1. Mitochondrial Complex III and the Electron Transport Chain

As a subunit of Complex III in the mitochondrial ETC, CytC1 plays a critical role in transferring electrons from ubiquinol to cytochrome c. Complex III is an electrically driven proton pump in which subunit b1 mediates electron flow from ubiquinol along with proton uptake and release; CytC1 is involved in the terminal transfer of the electron to cytochrome c [38] (Figure 3).

The interaction between TSGA10 and CytC1 suggests that TSGA10 may have a role in modulating the function of Complex III in electron transfer to cytochrome c. Therefore, a plausible hypothesis for TSGA10's role in mitochondrial coupling can be proposed. We hypothesize that TSGA10 directly influences mitochondrial coupling by adjusting electron transfer efficiency, modulating the choice between the transfer of electrons to cytochrome c versus oxygen and ROS production. In essence, we propose that TSGA10 is an extra insulator ensuring the electron moves to cytochrome c instead of leaking to oxygen prematurely (i.e., before Complex IV, where it is supposed to combine with oxygen in a coordinated response, minimizing ROS production and generating water) (Figure 3). One could hypothesize that TSGA10 acts positively (leading to stronger coupling and lower ROS production) or negatively (leading to weaker coupling and increased ROS). Given that TSGA10 is expressed in energy-demanding cells, we propose that its role is to strengthen the coupling of electron flow within OXPHOS (i.e., promoting electron flow into cytochrome c and minimizing ROS production), thereby improving mitochondrial efficiency in ATP generation [2,39]. This regulatory role could be crucial for maintaining cellular energy homeostasis, particularly in tissues with high metabolic demands, such as the testis, where TSGA10 is predominantly expressed. However, the mechanistic details of how TSGA10 might do this remain to be investigated.

4.3.2. A Potential Thermal Role for TSGA10

There is an important thermal component to TSGA10's potential coupling function, which arises by minimizing the inefficiencies associated with electron leakage and ROS production (and as discussed earlier in Section 3, "Mitochondrial Role in Heat Production and Uncoupling"). When electron flux is high, excess electrons escape the ETC in a process known as electron leakage, contributing to ROS generation [21]. This premature donation of an electron to oxygen generates a superoxide anion, a potent ROS, but it also short-circuits the electron transport chain [40]. As discussed in Section 3 earlier, brown adipose tissue uses UCP1 to decouple the proton gradient, allowing protons to diffuse back across the inner mitochondrial membrane. Not only does this produce heat by non-shivering thermogenesis, but it also slows down an overactive ETC, thereby reducing toxic ROS production generated by electron leakage [41]. This protective effect is thermogenic: short-circuiting the ETC releases energy that would ordinarily go to making ATP; this energy produces heat instead [20].

A direct implication of the preceding discussion is that by binding Complex III and reducing electron leakage, TSGA10 could minimize both ROS and the need for uncoupling of the mitochondrial proton gradient. In turn, this would reduce the production of uncoupling-associated heat. TSGA10 thus would act as a cooling and coupling agent for the mitochondria. This mechanism could be particularly advantageous in scenarios where thermal regulation is essential, such as in protecting cells from thermal stress or in adapting to environments with limited energy resources.

4.3.3. A Potential Complex III Assembly Role for TSGA10

We have outlined a mechanism of direct regulation, but another possibility is that TSGA10 plays a role in the assembly of Complex III itself. Rip1, a Rieske Fe/S subunit of Complex III, is essential for electron transfer [42], and its proper assembly relies on the multi-protein mitochondrial contact site and cristae organizing system (MICOS) [43]. In particular, Mar26/Fmp10 coordinates Complex III assembly by linking Rieske Fe/S protein-containing intermediates to crista junctions for Rip1 maturation [43]. We hypothesize that TSGA10 functions similarly, helping to construct OXPHOS-competent Complex III.

4.3.4. Disruption of Mitochondrial Coupling in Health and Disease

Dysregulation of mitochondrial coupling is implicated in diseases such as cancer, neurodegeneration, and metabolic syndromes. If TSGA10 dysfunction disrupts coupling efficiency, it could contribute to these pathologies. Conversely, therapeutic targeting of the TSGA10-CytC1 axis might restore apoptosis and autophagy/mitophagy in cancer, and energy balance in mitochondrial disorders. For instance, enhancing TSGA10's activity could improve ATP synthesis in energy-deficient conditions, while inhibiting it might mitigate pathologies driven by excessive coupling.

Moreover, TSGA10's proposed role in optimizing mitochondrial electron transfer and coupling through Complex III may hold significant implications for cancer biology. By stabilizing electron flow to cytochrome c and minimizing leakage to oxygen, TSGA10 could suppress ROS generation and reduce heat production—a critical balance disrupted in carcinogenesis. Cancer cells often exhibit mitochondrial dysfunction, including elevated ROS, which drives genomic instability and oncogenic signaling. If TSGA10 expression is compromised, inefficient electron transfer at Complex III could exacerbate ROS leakage, promoting DNA damage and mutations that fuel tumorigenesis. Furthermore, TSGA10's potential role in Complex III assembly, akin to the mitochondrial contact site and cristae organizing system (MICOS)-dependent Rip1 maturation, suggests that its dysfunction might impair OXPHOS integrity, forcing cells to rely on glycolytic metabolism (the Warburg effect), a hallmark of cancer. Conversely, tumoral cells, especially the tumors in energy-demanding tissues (e.g., testicular cancers) might exploit TSGA10's coupling efficiency to sustain ATP production while evading ROS-induced apoptosis. The thermal component of TSGA10's function—reducing heat from electron leakage—could also influence the tumor microenvironment, where excess heat may promote angiogenesis or immune evasion. Thus, TSGA10 emerges as a potential metabolic gatekeeper whose dysregulation could link mitochondrial coupling defects, ROS-driven mutagenesis, and metabolic reprogramming in cancer progression, offering novel therapeutic targets to disrupt tumor energetics.

4.4. TSGA10 and Oxygen Sensing

Previous work has shown that TSGA10 also interacts with HIF-1α [6]. This positions TSGA10 as an oxysensor or member of the oxygen sensor complex protein, including HIF-1α, Complex III (bc1 complex), intermediate filament network, and TSGA10 itself, to modulate an appropriate response to stress and hypoxia. The mitochondrial Complex III regulates hypoxic activation of HIF-1α via enabling cells to sense hypoxia. Under normoxic conditions, HIF-1α is rapidly degraded by prolyl hydroxylases (PHDs), which require oxygen as a substrate. However, during hypoxia, Complex III of the electron transport chain generates ROS as a byproduct of incomplete oxygen reduction. These ROS inhibit PHD activity, stabilizing HIF-1α and allowing it to translocate to the nucleus. There, HIF-1α

dimerizes with HIF-1 β, activating the transcription of genes involved in angiogenesis, metabolism, and survival, thereby enabling cellular adaptation to hypoxia [44]. Given our hypothesis that TSGA10 directly affects electron transfer either to cytochrome c or oxygen (thereby generating the ROS involved in HIF-1α activation), it is positioned to be a direct player in the process of oxygen sensing by HIF-1α and Complex III.

4.5. HIF-1α in Thermoregulation

As described above, the interaction between HIF-1α, TSGA10, CytC1, and mitochondrial metabolism may play a role in dynamically adjusting energy production and thermoregulation in response to cellular and environmental stressors. While HIF-1's role in metabolism is well established, emerging evidence suggests that hypoxia-inducible factors can modulate thermogenic pathways. For instance, HIF-2α has been shown to function as a thermostat in beige adipocytes by regulating PKA activity and mitochondrial abundance in response to cold and re-warming [45]. These findings raise the intriguing possibility that HIF-1α may both respond to temperature and regulate mitochondrial heat production, acting as a molecular thermostat in conjunction with the TSGA10 feedback loop outlined above.

5. TSGA10 and HIF-1α Mutual Counter Repression in Thermoregulation

5.1. HIF-1α-Dependent Regulation of ROS Generation and Thermogenic Mechanisms in Hypoxic Conditions

HIF-1α has been demonstrated to play a crucial role in ROS generation, particularly under hypoxic conditions [46–48]. The landmark study by Chandel et al. [46] established that mitochondrial ROS are not only elevated during hypoxia but are essential for HIF-1α stabilization and activity, creating a feedback loop in which HIF-1α and ROS regulate each other. Multiple mechanisms underlying this relationship have been identified: HIF-1α reprograms mitochondrial metabolism under hypoxia, leading to electron leakage and enhanced ROS production at Complex III of the electron transport chain [47]; it transcriptionally induces pro-oxidant enzymes such as the NADPH oxidases (NOX family), which directly generate ROS [48]; and while HIF-1α downregulates mitochondrial respiration through mediators like pyruvate dehydrogenase kinase 1 (PDK1), this metabolic shift paradoxically causes transient mitochondrial dysfunction and ROS bursts during cellular adaptation. Guzy and Schumacker [47] specifically addressed this paradox of increased ROS generation during hypoxia, while Semenza's review [48] highlighted the significance of these HIF-1α-mediated mechanisms in cancer progression and potential therapeutic interventions.

Although HIF-1α does not directly increase heat generation as a primary function, it indirectly contributes to thermogenesis through multiple pathways under specific physiological and pathological conditions [49]. HIF-1α upregulates glycolytic enzymes and glucose transporters (e.g., GLUT1), favoring anaerobic metabolism, which, despite being less efficient for ATP production, generates more heat per mole of ATP than oxidative phosphorylation—particularly evident in tumors and inflammatory states [50]. In brown adipose tissue, HIF-1α may influence thermogenesis by modulating the expression of uncoupling protein 1 (UCP1), which dissipates the proton gradient as heat rather than producing ATP, though this regulation is context-dependent and influenced by factors such as PGC-1α and β-adrenergic signaling. Additionally, tumors with stabilized HIF-1α often exhibit elevated local temperatures due to increased glycolytic activity and altered mitochondrial function, contributing to metabolic heat production in the tumor microenvironment. Lee et al. [49] specifically discussed HIF-1α as part of the regulatory network in BAT thermogenesis under cold-induced or hypoxic conditions, while Wang et al. [50] demonstrated how HIF-1α-driven glycolysis increases metabolic activity, indirectly contributing to heat generation in tumors.

Because Complex III is a main producer of ROS in the ETC, TSGA10 sits at a three-way nexus in the flow of electrons, protons, and ROS. Given the regulatory interaction between TSGA10 and HIF-1α [6], we hypothesize that TSGA10 is able to coordinate environmental conditions (thermal and oxygen levels) and metabolic outputs (ATP, heat, and ROS).

HIF-1α is a key regulator of cellular adaptation to hypoxia, promoting glycolysis while suppressing OXPHOS [51]. The shift into glycolysis allows cells to generate ATP under low-oxygen conditions but reduces mitochondrial efficiency overall because less ATP is generated per input glucose molecule. But HIF-1α is not just activated by hypoxia: it responds to temperature also. One study on kidney tissue demonstrated that HIF-1α is significantly upregulated during cold ischemia, promoting cellular adaptations that enhance survival under low-temperature stress [52]. Another noted that HIF-1α mediates the response of skeletal muscle to cold stress [53]. These data suggest a thermal equilibrium, because TSGA10 regulates HIF-1α, suppressing its expression [2,6,7]. Furthermore, data support the counter-regulatory effect of HIF-1α acting as a suppressor of TSGA10 expression [32]. This mutual regulation makes for a tightly linked feedback loop, maintaining metabolic homeostasis and linking oxygen, ATP, and thermoregulatory roles (Figure 4).

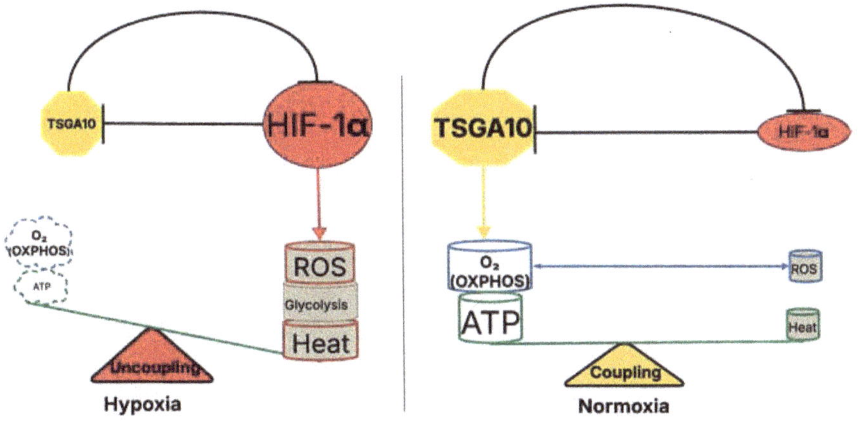

Figure 4. TSGA10-mediated regulation of mitochondrial coupling and metabolic balance in hypoxia and normoxia. The left panel represents the hypoxic state, where TSGA10 downregulation and increased HIF-1α activity are observed, promoting glycolysis and heat production while increasing ROS generation. This results in mitochondrial uncoupling, reducing ATP synthesis through OXPHOS. The right panel illustrates the normoxic condition, where TSGA10 enhances mitochondrial coupling, supporting efficient ATP production via OXPHOS while minimizing ROS accumulation and HIF-1α activity. The balance between mitochondrial coupling (normoxia) and uncoupling (hypoxia) is depicted as a metabolic scale, highlighting the role of TSGA10 in regulating cellular energy homeostasis.

In tightly-coupled tissues operating at cooler temperatures such as the testis, expression of TSGA10 tamps down HIF-1α, ensuring heat is not generated. However, if the temperature drops too low, HIF-1α may be activated, upregulating cellular temperature by downregulating TSGA10, thereby promoting uncoupling and heat generation. This interplay may position TSGA10 as a crucial mediator in the cellular thermoregulatory network, capable of fine-tuning mitochondrial metabolism in response to both hyperthermic and hypothermic stress. A similar logic could operate for ROS production. Under oxidative stress conditions, HIF-1α stabilization could lead to decreased expression of TSGA10, which in turn could decrease mitochondrial coupling efficiency, lowering OXPHOS and avoiding ROS toxicity. Overall, TSGA10's inhibition of HIF-1α concretely shifts metabolic priorities (glycolysis vs. OXPHOS), enhances mitochondrial efficiency (lowered ROS/heat), and protects thermosensitive tissues (e.g., brain) by stabilizing energy and redox balance. This interaction does not merely "influence" cellular responses to hypoxia and thermal stress; it mechanistically reprograms them, dynamically adapting to the distinct needs of different tissues throughout the body. Understanding this relationship may provide insights into how cells balance energy production, oxygen availability, and thermogenesis in both normal physiology and pathological conditions like cancer.

The mutual counter-regulation between TSGA10 and HIF-1α, which integrates thermoregulation with metabolic reprogramming, has profound implications for cancer. HIF-1α, a master regulator of hypoxia adaptation, drives the Warburg effect—promoting glycolysis and suppressing OXPHOS—to fuel tumor growth in low-oxygen microenvironments. However, TSGA10's suppression of HIF-1α introduces a critical metabolic checkpoint: by potentially stabilizing mitochondrial coupling at Complex III, TSGA10 might enhance OXPHOS efficiency, minimize ROS leakage, and reduce heat generation, thereby opposing HIF-1α's glycolytic shift. In cancers, dysregulation of this balance could drive malignancy. For instance, loss of TSGA10 expression—common in tumors like glioblastoma [2] —may unleash HIF-1α activity, locking cells into glycolysis, exacerbating ROS production (via mitochondrial uncoupling), and fostering genomic instability. Conversely, HIF-1α's suppression of TSGA10 under hypoxia or thermal stress (e.g., in poorly vascularized tumors) could amplify mitochondrial inefficiency, creating a pro-tumorigenic milieu of elevated ROS and heat. This ROS–heat axis may further remodel the tumor microenvironment, promoting angiogenesis or immune evasion. Notably, TSGA10's role in thermoregulation could also explain why certain cancers thrive in thermally unstable niches (e.g., testicular tumors), where disrupted TSGA10-HIF-1α feedback loops might permit metabolic plasticity. Thus, this bidirectional repression mechanism not only underscores how tumors hijack metabolic–thermogenic crosstalk to survive stress but also highlights TSGA10 as a potential therapeutic target to restore mitochondrial fidelity and disrupt cancer's adaptive energetics. In our model, the TSGA10 and HIF-1α feedback loop modulates cellular metabolism, favoring glycolysis, heat generation, and ROS production in hypoxia, while in normoxia the TSGA10-mediated tighter coupling of the ETC within mitochondria favors ATP synthesis, with lower heat and ROS production (Figure 4).

5.2. Role of Mitochondria Numbers and Blood Circulation in Thermal and Metabolic Stability

The vasculature's role in heat distribution complements mitochondrial activity, with blood flow patterns influencing both local and global temperature regulation. For example, compared to most tissues, BAT exhibits lower numbers of mitochondria per cell, but of higher quality as each mitochondrion is highly active, leveraging UCP1-mediated uncoupling to prioritize thermogenesis over ATP synthesis, with enhanced blood perfusion facilitating heat dissipation throughout the body [54,55].

Neurons, with their high energy demands, maintain densely packed mitochondria in synapses to sustain neurotransmission, balancing quantity and quality via mitophagy, while cerebral blood flow helps maintain optimal brain temperature [45,56].

Conversely, germ cells in testes prioritize mitochondrial quality over quantity, minimizing ROS to protect genomic integrity, with specialized testicular blood flow patterns contributing to the maintenance of lower scrotal temperatures [57,58]. Pathological states, such as cancer, disrupt this balance: tumors often amplify mitochondrial numbers but exhibit dysfunctional OXPHOS, favoring glycolysis, while simultaneously altering blood vessel architecture and flow patterns [59]. Thus, tissue-specific strategies in mitochondrial quantity and activity, coupled with blood-mediated thermal transport, show adaptability to physiological and environmental stressors [60].

5.3. TSGA10 as a Potential Mitochondrial Regulator in Cancer

Cancer cells exhibit metabolic plasticity, adapting their energy production depending on microenvironmental conditions such as hypoxia or nutrient stress. In some cases, tumors exploit mitochondrial uncoupling to minimize ROS accumulation, thereby preventing oxidative damage and supporting survival [61]. Given TSGA10's hypothesized role in promoting mitochondrial coupling, its downregulation in cancer may facilitate a shift toward glycolysis, favoring the Warburg effect, where cancer cells preferentially use glycolysis over OXPHOS, even in the presence of oxygen [2]. Indeed, it has been suggested that it is the uncoupling of mitochondrial OXPHOS that drives the shift toward aerobic glycolysis of cancer cells [62] (Figure 5).

This positions TSGA10 as a potential key mediator of either oncogenesis or tumor suppression, and may help explain the fact that it has been reported to have both properties [2]. Under cold stress, cancer cells may adopt BAT-like uncoupling to reduce ROS and generate heat, sustaining proliferation [61]. Concurrently, cold impairs OXPHOS enzymes like Complex III, promoting ROS accumulation and a glycolytic shift akin to the Warburg effect [63]. This dual adaptation—uncoupling OXPHOS to survive cold and shifting to glycolysis for ATP—reflects metabolic reprogramming under thermal duress, but also describes cancer metabolism: the Warburg effect [64].

It is reasonable to speculate that aberrant thermal conditions may be a driver of oncogenic metabolic states. The duality in TSGA10 expression and metabolic activity in postmitotic versus cancer cells highlights TSGA10 as a potential metabolic "switch" that reinforces the native metabolic state of the cell—OXPHOS in postmitotic cells and glycolysis suppression in cancer. Its role in stress responses (e.g., hypoxia and drug resistance) further underscores its adaptability to the cellular context [2].

TSGA10 expression levels reflect a cell's metabolic priorities—sustaining OXPHOS in postmitotic cells versus restraining glycolysis in cancer—making it a critical node in cellular energy homeostasis. Meanwhile, TSGA10's dual role in mitochondrial coupling and metabolic reprogramming positions it as a pivotal regulator in cancer biology. By promoting efficient electron transfer at Complex III and suppressing ROS leakage, TSGA10 potentially reinforces OXPHOS in energy-demanding postmitotic cells. However, its downregulation in tumors—a common feature observed in cancers like glioblastoma—may facilitate mitochondrial uncoupling, a metabolic strategy that minimizes ROS toxicity while driving the Warburg effect [62,65] (Figure 5). This uncoupling not only aligns with cancer cells' reliance on glycolysis for ATP production but also mirrors adaptations to thermal stress: under cold conditions, tumors may adopt BAT-like uncoupling to reduce ROS and generate heat, sustaining proliferation [61,63]. Here, TSGA10's loss could act as a metabolic "key", destabilizing OXPHOS fidelity to favor glycolysis, a hallmark of cancer metabolism. Conversely, in hypoxic or nutrient-deprived microenvironments, residual TSGA10 activity might paradoxically support tumor survival by balancing ROS and ATP production, explaining its context-dependent roles as both an oncogene and tumor suppressor. The interplay between thermal stress and TSGA10 dysfunction further underscores how aberrant environmental conditions (e.g., cold-induced OXPHOS impairment) could drive oncogenic metabolic states, akin to the Warburg effect. Ultimately, TSGA10's ability to enforce metabolic priorities—OXPHOS in healthy cells versus glycolysis in cancer—highlights its potential as a therapeutic target to disrupt tumor metabolic plasticity and restore mitochondrial fidelity in malignancies.

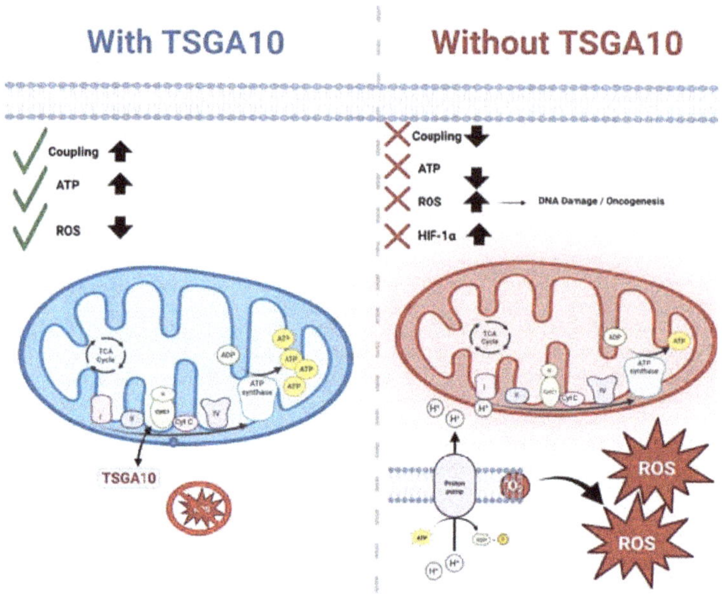

Figure 5. TSGA10's interaction with CytC1 and its downstream consequences: On the **left** (blue), TSGA10 binds to CytC1 in Complex III of the mitochondrial electron transport chain, enhancing electron transfer and promoting OXPHOS. This leads to improved mitochondrial coupling, increased ATP production (↑), and reduced reactive oxygen species (ROS) levels (↓), as indicated by green checkmarks. On the **right** (red), loss of TSGA10 disrupts mitochondrial coupling (×), decreases ATP output (↓), and increases ROS production (↑), which triggers HIF-1α stabilization, DNA damage, and oncogenic signaling due to mitochondrial dysfunction and metabolic reprogramming. TSGA10

5.4. TSGA10 Is Expressed in Postmitotic Energy-Demanding Cells

Body temperature, regulated by the hypothalamus, is maintained at approximately 37 °C, a condition under which TSGA10 exhibits minimal expression. However, mitochondria-rich postmitotic cells—such as those in the testis, retina, neurons, and nephrons—display significantly higher levels of ciliary–centrosomal proteins like TSGA10 [3,66]. These cells share common features, including a high metabolic rate, limited regenerative capacity due to their postmitotic nature, complex structure and function, temperature-sensitive ion channels, blood-barrier systems, and distinct Heat Shock Protein (HSP) expression. These metabolically active cells—especially spermatocytes—contain tightly coupled mitochondria to meet their substantial ATP demands for maintaining ion gradients, synaptic transmission (in neurons), and other energy-intensive processes. Their high coupling efficiency ensures that most energy from OXPHOS is directed toward ATP synthesis rather than heat production, minimizing proton leakage and optimizing the proton motive force for ATP generation.

In the testis, particularly in sperm cells, TSGA10 expression is exceptionally high, coinciding with a significantly lower tissue temperature (32–33 °C). While tissue temperature and TSGA10 expression are independently regulated, they operate synergistically, as a coordinated response to the high metabolic demands of sperm cells. Management of heat generation is a common theme in TSGA10-expressing cells. Excessive heat generation could lead to protein denaturation and impaired motility or function, so minimizing heat production through efficient coupling is vital for functional integrity.

In summary, some specialized postmitotic cells exhibit tightly coupled mitochondria to meet their high energy demands while minimizing thermal stress. This coupling reflects their unique functional roles and the critical need for efficient energy production, contrasting with other cell types that tolerate greater variability in energy and heat generation. The interplay between temperature, TSGA10 expression, and mitochondrial coupling underscores the sophisticated adaptations of these cells to their specific physiological environments.

6. Conclusions

The proposed role of TSGA10 in thermoregulation has several implications. In the testis, TSGA10's high expression aligns with its need for precise temperature regulation during spermatogenesis. Disruptions in TSGA10 expression or function could compromise mitochondrial coupling, leading to excessive heat production and impaired fertility. In the brain, TSGA10 may protect neurons from hyperthermic damage by maintaining mitochondrial efficiency. Given the brain's reliance on stable ion gradients and metabolic homeostasis, TSGA10 could play a crucial role in preventing heat-induced neuronal dysfunction. Meanwhile, the liver's low TSGA10 expression may reflect its inherent tolerance to higher physiological temperatures, underscoring the tissue-specific nature of TSGA10's thermoregulatory role. While the evidence supporting TSGA10's role in thermoregulation and cancer is compelling, several questions remain.

Does TSGA10 directly modulate cytochrome c reductase activity? Could TSGA10 act at the intermembrane space or inner boundary membrane to influence Complex III maturation? How does HIF-1α regulation of TSGA10 vary between thermosensitive and thermally robust tissues? Investigating this interplay could reveal novel insights into tissue-specific thermoregulation. Moreover, could TSGA10 modulation be leveraged to protect against hyperthermia-induced damage in conditions such as heatstroke and febrile illnesses? In conclusion, TSGA10 represents a promising candidate in the molecular regulation of cellular thermoregulation. By promoting mitochondrial coupling and minimizing uncoupling-related heat production, TSGA10 may act as a guardian of thermosensitive tissues. The interplay between TSGA10, CytC1, and HIF-1α highlights a complex regulatory network that balances energy efficiency and thermogenesis in response to physiological demands. Future research will illuminate whether this axis can be harnessed for therapeutic interventions, offering new avenues for protecting human cells from the dual threats of hyperthermia and metabolic stress.

TSGA10's dual association with Complex III (via CytC1) and Complex IV-linked HIF-1α pathways positions it as a critical orchestrator of mitochondrial fidelity, with profound implications for cancer. In our model, by stabilizing electron transfer at Complex III, TSGA10 can enhance mitochondrial coupling, ensuring efficient ATP synthesis while minimizing ROS leakage—a safeguard against genomic instability. As previously reported [2], TSGA10 expression exhibits two distinct phases—downregulation and upregulation—during cancer progression. TSGA10 overexpression may indicate a compensatory action of TSGA10 and its tumor suppressor functionality. However, in a phase of cancer progression where TSGA10 is downregulated (e.g., glioblastoma), perhaps under HIF-1α pressure, disrupted coupling at Complex III may promote electron leakage, elevating ROS to mutagenic levels that drive oncogenic mutations. Concurrently, TSGA10's mutual counter-regulation with HIF-1α hypoxia-activated promotion of glycolysis—creates a metabolic tug-of-war. In tumors, HIF-1α dominance suppresses TSGA10, locking cells into the Warburg effect [62], while TSGA10 deficiency might destabilize Complex III, exacerbating mitochondrial dysfunction and ROS-driven stress adaptation. This bidirectional axis may explain TSGA10's paradoxical roles in cancer: in thermosensitive tissues like the testis, its high expression could suppress HIF-1α to maintain OXPHOS and thermal stability, protecting against malignant transformation, whereas in thermally resilient tissues like the liver, low TSGA10 levels may permit HIF-1α-driven glycolytic reprogramming, facilitating tumorigenesis. The tissue specificity of TSGA10's function—critical in brain and testes but dispensable in liver—mirrors cancer vulnerabilities, where loss of TSGA10 in thermoregulatory tissues could permit heat-induced microenvironmental remodeling (e.g., angiogenesis and immune evasion). Thus, TSGA10 emerges as a metabolic–thermogenic rheostat, whose dysregulation at Complex III and crosstalk with HIF-1α may underpin the metabolic plasticity that fuels carcinogenesis, offering novel strategies to target mitochondrial vulnerabilities in tumors.

Author Contributions: Conceptualization, A.A., F.T.-H., J.B., and B.B.; methodology, A.A. and B.B.; software, A.A. and B.B.; validation, A.A. and B.B.; formal analysis, B.B.; investigation, A.A. and B.B.; resources, J.B. and B.B.; data curation, A.A. and B.B.; writing—original draft preparation, A.A., F.T.-H., J.B., and B.B.; writing—review and editing, A.A., J.B., and B.B.; visualization, A.A., J.B., and B.B.; supervision, J.B. and B.B.; project administration, J.B. and B.B.; funding acquisition, J.B. and B.B. All authors have read and agreed to the published version of the manuscript.

Funding: This work was supported by B.B.'s doctoral scholarship (Ministry of Health and Medical Education, Iran) and the NIH-NRSA T32 training grant, and by J.R.B's NIH grant 1R15GM146207.

Institutional Review Board Statement: Not applicable.

Informed Consent Statement: Not applicable.

Data Availability Statement: Data generated and analyzed during this study, including all raw and processed data from the transfection experiments, are available upon reasonable request from the corresponding author.

Acknowledgments: B.B. is deeply grateful to Jonathan Wolfe for his mentorship and support during his PhD at the Galton Laboratory, University College London (UCL), London, UK, which led to the successful execution of the preliminary yeast two-hybrid assays.

Conflicts of Interest: The authors declare that there are no potential commercial or financial conflicts of interest that could be perceived as influencing the research reported in this manuscript. The authors affiliated with MDPI confirm that their participation in the study was conducted in an academic and scientific capacity, independent of any commercial interests. All data analysis, interpretation, and conclusions were carried out without bias and in adherence to the ethical guidelines set forth by MDPI.

Abbreviations

The following abbreviations are used in this manuscript:

TSGA10	Testis-Specific Gene 10
HIF-1α	Hypoxia-Inducible Factor 1-Alpha
VEGFA	Vascular Endothelial Growth Factor A
MMP2	Matrix Metalloproteinase-2
MMP9	Matrix Metalloproteinase-9
OXPHOS	Oxidative Phosphorylation
ROS	Reactive Oxygen Species
ETC	Electron Transport Chain
CytC1	Cytochrome c1
UCP1	Uncoupling Protein 1
BAT	Brown Adipose Tissue
MICOS	Mitochondrial Contact Site and Cristae Organizing System
ATP	Adenosine Triphosphate
HSPs	Heat Shock Proteins
RISP	Rieske Iron–Sulfur Protein
COX4	Cytochrome c Oxidase Subunit 4

References

1. Modarressi, M.H.; Cameron, J.; Taylor, K.E.; Wolfe, J. Identification and characterisation of a novel gene, TSGA10, expressed in testis. *Gene* **2001**, *262*, 249–255. [CrossRef] [PubMed]
2. Taghizadeh-Hesary, F.; Ghadyani, M.; Kashanchi, F.; Behnam, B. Exploring TSGA10 Function: A Crosstalk or Controlling Mechanism in the Signaling Pathway of Carcinogenesis? *Cancers* **2024**, *16*, 3044. [CrossRef] [PubMed]
3. Behnam, B.; Modarressi, M.H.; Conti, V.; Taylor, K.E.; Puliti, A.; Wolfe, J. Expression of Tsga10 sperm tail protein in embryogenesis and neural development: From cilium to cell division. *Biochem. Biophys. Res. Commun.* **2006**, *344*, 1102–1110. [CrossRef]
4. Salehipour, P.; Nematzadeh, M.; Mobasheri, M.B.; Afsharpad, M.; Mansouri, K.; Modarressi, M.H. Identification of new TSGA10 transcript variants in human testis with conserved regulatory RNA elements in 5′untranslated region and distinct expression in breast cancer. *Biochim. Biophys. Acta (BBA) Gene Regul. Mech.* **2017**, *1860*, 973–982. [CrossRef]
5. Sha, Y.W.; Sha, Y.K.; Ji, Z.Y.; Mei, L.B.; Ding, L. TSGA10 is a novel candidate gene associated with acephalic spermatozoa. *Clin. Genet.* **2018**, *93*, 776–783. [CrossRef]
6. Hägele, S.; Behnam, B.; Borter, E.; Wolfe, J.; Paasch, U.; Lukashev, D.; Sitkovsky, M.; Wenger, R.H.; Katschinski, D.M. TSGA10 prevents nuclear localization of the hypoxia-inducible factor (HIF)-1α. *FEBS Lett.* **2006**, *580*, 3731–3738. [CrossRef] [PubMed]
7. Mansouri, K.; Mostafie, A.; Rezazadeh, D.; Shahlaei, M.; Modarressi, M.H. New function of TSGA10 gene in angiogenesis and tumor metastasis: A response to a challengeable paradox. *Hum. Mol. Genet.* **2016**, *25*, 233–244. [CrossRef]
8. Roghanian, A.; Jones, D.C.; Pattisapu, J.V.; Wolfe, J.; Young, N.T.; Behnam, B. Filament-associated TSGA10 protein is expressed in professional antigen presenting cells and interacts with vimentin. *Cell. Immunol.* **2010**, *265*, 120–126. [CrossRef]
9. Tanaka, R.; Ono, T.; Sato, S.; Nakada, T.; Koizumi, F.; Hasegawa, K.; Nakagawa, K.; Okumura, H.; Yamashita, T.; Ohtsuka, M.; et al. Over-expression of the testis-specific gene TSGA10 in cancers and its immunogenicity. *Microbiol. Immunol.* **2004**, *48*, 339–345. [CrossRef]
10. Mobasheri, M.B.; Modarressi, M.H.; Shabani, M.; Asgarian, H.; Sharifian, R.A.; Vossough, P.; Shokri, F. Expression of the testis-specific gene, TSGA10, in Iranian patients with acute lymphoblastic leukemia (ALL). *Leuk. Res.* **2006**, *30*, 883–889. [CrossRef]
11. Luo, G.; Hou, M.; Wang, B.; Liu, Z.; Liu, W.; Han, T.; Zhang, D.; Zhou, X.; Jia, W.; Tan, Y.; et al. Tsga10 is essential for arrangement of mitochondrial sheath and male fertility in mice. *Andrology* **2021**, *9*, 368–375. [CrossRef] [PubMed]
12. Brand, M.D.; Orr, A.L.; Perevoshchikova, I.V.; Quinlan, C.L. The role of mitochondrial function and cellular bioenergetics in ageing and disease. *Br. J. Dermatol.* **2013**, *169* (Suppl. 2), 1–8. [CrossRef]
13. Li, Y.; Li, X.M.; Wei, L.S.; Ye, J.F. Advancements in mitochondrial-targeted nanotherapeutics: Overcoming biological obstacles and optimizing drug delivery. *Front. Immunol.* **2024**, *15*, 1451989. [CrossRef] [PubMed]
14. Hille, B. *Ion Channels of Excitable Membranes*, 3rd ed.; Sinauer Associates Inc.: Sunderland, MA, USA, 2001.
15. Yeagle, P.L. *The Membranes of Cells*, 2nd ed.; Academic Press: Cambridge, MA, USA, 1993.
16. Alberts, B.; Johnson, A.; Lewis, J.; Raff, M.; Roberts, K.; Walter, P. *Molecular Biology of the Cell*, 6th ed.; Garland Science: New York, NY, USA, 2014.
17. Walker, B.R.; Moraes, C.T. Nuclear-Mitochondrial Interactions. *Biomolecules* **2022**, *12*, 427. [CrossRef] [PubMed]
18. Kandel, E.R.; Schwartz, J.H.; Jessell, T.M. *Principles of Neural Science*, 4th ed.; McGraw-Hill: New York, NY, USA, 2000.
19. Hertzler, J.I.; Bernard, A.R.; Rolls, M.M. Dendrite regeneration mediates functional recovery after complete dendrite removal. *Dev. Biol.* **2023**, *497*, 18–25. [CrossRef]
20. Jastroch, M.; Divakaruni, A.S.; Mookerjee, S.; Treberg, J.R.; Brand, M.D. Mitochondrial Proton and Electron Leaks. *Essays Biochem.* **2010**, *47*, 53–67. [CrossRef]
21. Sokolova, I. Mitochondrial adaptations to variable environments and their role in animals' stress tolerance. *Integr. Comp. Biol.* **2018**, *58*, 519–531. [CrossRef]
22. Hamilton, T.R.; Mendes, C.M.; de Castro, L.S.; de Assis, P.M.; Siqueira, A.F.; de Carvalho Delgado, J.; Goissis, M.D.; Muiño-Blanco, T.; Cebrián-Pérez, J.Á.; Nichi, M.; et al. Evaluation of Lasting Effects of Heat Stress on Sperm Profile and Oxidative Status of Ram Semen and Epididymal Sperm. *Oxidative Med. Cell. Longev.* **2016**, *2016*, 1687657. [CrossRef]
23. Park, A.; Kim, K.E.; Park, I.; Lee, S.H.; Park, K.Y.; Jung, M.; Li, X.; Sleiman, M.B.; Lee, S.J.; Kim, D.-S.; et al. Mitochondrial matrix protein LETMD1 maintains thermogenic capacity of brown adipose tissue in male mice. *Nat. Commun.* **2023**, *14*, 3746. [CrossRef]
24. Cannon, B.; Nedergaard, J. Brown adipose tissue: Function and physiological significance. *Physiol. Rev.* **2004**, *84*, 277–359. [CrossRef]
25. Mahani, S.T.; Behnam, B.; Abbassi, M.; Asgari, H.; Nazmara, Z.; Shirinbayan, P.; Joghataei, M.; Koruji, M. Tsga10 expression correlates with sperm profiles in the adult formalin-exposed mice. *Andrologia* **2016**, *48*, 1092–1099. [CrossRef] [PubMed]
26. The Human Protein Atlas Project. The Human Protein Atlas: The Open Access Resource for Human Proteins. 2024. Available online: https://www.proteinatlas.org (accessed on 29 April 2025).

27. Jahani, M.; Shahlaei, M.; Norooznezhad, F.; Miraghaee, S.S.; Hosseinzadeh, L.; Moasefi, N.; Khodarahmi, R.; Farokhi, A.; Mahnam, A.; Mansouri, K. TSGA10 Overexpression Decreases Metastatic and Metabolic Activity by Inhibiting HIF-1 in Breast Cancer Cells. *Arch. Med. Res.* **2020**, *51*, 41–53. [CrossRef] [PubMed]
28. Zhang, Z.; Chen, W.; Luo, C.; Zhang, W. Exploring a four-gene risk model based on doxorubicin resistance-associated lncRNAs in hepatocellular carcinoma. *Front. Pharmacol.* **2022**, *13*, 1015842. [CrossRef]
29. Asghari-Givehchi, S.; Hossein-Modarressi, M. Identification and expression analysis of zebrafish testis-specific gene 10 (tsga10). *Int. J. Dev. Biol.* **2019**, *63*, 623–629. [CrossRef] [PubMed]
30. Amoorahim, M.; Valipour, E.; Hoseinkhani, Z.; Mahnam, A.; Rezazadeh, D.; Ansari, M.; Shahlaei, M.; Gamizgy, Y.H.; Moradi, S.; Mansouri, K. TSGA10 overexpression inhibits angiogenesis of HUVECs: A HIF-2α biased perspective. *Microvasc. Res.* **2020**, *128*, 103952. [CrossRef]
31. Wei, W.; Gong, Y.; Guo, X.; Liu, M.; Zhou, Y.; Li, Z.; Zhou, L.; Wang, Z.; Gui, J. Gonadal transcriptomes reveal sex-biased expression genes associated with sex determination and differentiation in red-tail catfish (Hemibagrus wyckioides). *BMC Genom.* **2023**, *24*, 183. [CrossRef]
32. Zhang, Q.; Zhang, J.; Fu, Z.; Dong, L.; Tang, Y.; Xu, C.; Wang, H.; Zhang, T.; Wu, Y.; Dong, C.; et al. Hypoxia-induced microRNA-10b-3p promotes esophageal squamous cell carcinoma growth and metastasis by targeting TSGA10. *Aging* **2019**, *11*, 10374–10384. [CrossRef]
33. Behnam, B. Investigation of TSGA10 Gene Expression, Localization, and Protein Interaction in Human and Mouse Spermatogenesis. Ph.D. Thesis, University College London, London, UK, 2005.
34. Van Haute, L.; Lee, S.Y.; McCann, B.J.; Powell, C.A.; Bansal, D.; Vasiliauskaitė, L.; Garone, C.; Shin, S.; Kim, J.; Frye, M.; et al. NSUN2 introduces 5-methylcytosines in mammalian mitochondrial tRNAs. *Nucleic Acids Res.* **2019**, *47*, 8720–8733. [CrossRef]
35. Ye, Y.; Wei, X.; Sha, Y.; Li, N.; Yan, X.; Cheng, L.; Qiao, D.; Zhou, W.; Wu, R.; Liu, Q.; et al. Loss-of-function mutation in TSGA10 causes acephalic spermatozoa phenotype in human. *Mol. Genet. Genom. Med.* **2020**, *8*, e1284. [CrossRef]
36. Asgari, R.; Bakhtiari, M.; Rezazadeh, D.; Yarani, R.; Esmaeili, F.; Mansouri, K. TSGA10 as a Potential Key Factor in the Process of Spermatid Differentiation/Maturation: Deciphering Its Association with Autophagy Pathway. *Reprod. Sci.* **2021**, *28*, 3228–3240. [CrossRef]
37. Rich, P.R. The molecular machinery of Keilin's respiratory chain. *Biochem. Soc. Trans.* **2003**, *31*, 1095–1105. [CrossRef] [PubMed]
38. Esser, L.; Zhou, F.; Yu, C.A.; Xia, D. Crystal structure of bacterial cytochrome bc1 in complex with azoxystrobin reveals a conformational switch of the Rieske iron-sulfur protein subunit. *J. Biol. Chem.* **2019**, *294*, 12007–12019. [CrossRef] [PubMed]
39. Behnam, B.; Fazilaty, H.; Ghadyani, M.; Fadavi, P.; Taghizadeh-Hesary, F. Ciliated, Mitochondria-Rich Postmitotic Cells are Immune-privileged, and Mimic Immunosuppressive Microenvironment of Tumor-Initiating Stem Cells: From Molecular Anatomy to Molecular Pathway. *Front. Biosci. (Landmark Ed.)* **2023**, *28*, 261. [CrossRef]
40. Tabassum, N.; Kheya, I.S.; Asaduzzaman, S.A.I.; Maniha, S.M.; Fayz, A.H.; Zakaria, A.; Noor, R. A Review on the Possible Leakage of Electrons through the Electron Transport Chain within Mitochondria. *J. Biomed. Environ. Sci.* **2020**, *1*, 105–113. [CrossRef]
41. Demine, S.; Renard, P.; Arnould, T. Mitochondrial Uncoupling: A Key Controller of Biological Processes in Physiology and Diseases. *Cells* **2019**, *8*, 795. [CrossRef]
42. Conte, L.; Zara, V. The Rieske Iron-Sulfur Protein: Import and Assembly into the Cytochrome bc(1) Complex of Yeast Mitochondria. *Bioinorg. Chem. Appl.* **2011**, *2011*, 363941. [CrossRef]
43. Zerbes, R.M.; Colina-Tenorio, L.; Bohnert, M.; von der Malsburg, K.; Peikert, C.D.; Mehnert, C.S.; Perschil, I.; Klar, R.F.U.; de Boer, R.; Kram, A.; et al. Coordination of cytochrome bc1 complex assembly at MICOS. *EMBO Rep.* **2025**, *26*, 353–384. [CrossRef]
44. Klimova, T.; Chandel, N. Mitochondrial complex III regulates hypoxic activation of HIF. *Cell Death Differ.* **2008**, *15*, 660–666. [CrossRef] [PubMed]
45. Han, S.; Zhang, M.; Jeong, Y.Y.; Margolis, D.J.; Cai, Q. The role of mitophagy in the regulation of mitochondrial energetic status in neurons. *Autophagy* **2021**, *17*, 4182–4201. [CrossRef]
46. Chandel, N.S.; Maltepe, E.; Goldwasser, E.; Mathieu, C.E.; Simon, M.C.; Schumacker, P.T. Mitochondrial reactive oxygen species trigger hypoxia-induced transcription. *Proc. Natl. Acad. Sci. USA* **1998**, *95*, 11715–11720. [CrossRef]
47. Guzy, R.D.; Schumacker, P.T. Oxygen sensing by mitochondria at complex III: The paradox of increased reactive oxygen species during hypoxia. *Exp. Physiol.* **2006**, *91*, 807–819. [CrossRef] [PubMed]
48. Semenza, G.L. Hypoxia-inducible factors: Mediators of cancer progression and targets for cancer therapy. *Trends Pharmacol. Sci.* **2012**, *33*, 207–214. [CrossRef] [PubMed]
49. Lee, P.; Smith, S.; Linderman, J.; Courville, A.B.; Brychta, R.J.; Dieckmann, W.; Werner, C.D.; Chen, K.Y.; Celi, F.S. Temperature-acclimated brown adipose tissue modulates insulin sensitivity in humans. *Diabetes* **2014**, *63*, 3686–3698. [CrossRef]
50. Wang, T.; Xu, Y. Tumor hypoxia, HIF-1α, and microRNA-210: An emerging hypoxia pathway in cancer. *Cell Cycle* **2010**, *9*, 1346–1353. [CrossRef]

51. Kierans, S.J.; Taylor, C.T. Regulation of glycolysis by the hypoxia-inducible factor (HIF): Implications for cellular physiology. *J. Physiol.* **2021**, *599*, 23–37. [CrossRef]
52. Choi, D.; Jeong, J.; Chung, S.; Chang, Y.; Na, K.; Lee, K. Preservation of Hypoxia-Inducible Factor-1 Induced By ERK Phosphorylation Is Involved in Hypothermic Protection of Renal Ischemia-Reperfusion Injury. *Transplantation* **2014**, *98*, 354. Available online: https://journals.lww.com/transplantjournal/fulltext/2014/07151/preservation_of_hypoxia_inducible_factor_1_induced.1145.aspx (accessed on 17 May 2025). [CrossRef]
53. Chen, W.; Zhao, H.; Li, Y. Mitochondrial dynamics in health and disease: Mechanisms and potential targets. *Signal Transduct. Target. Ther.* **2023**, *8*, 333. [CrossRef]
54. Yamashita, S.I.; Kanki, T. Mitophagy Responds to the Environmental Temperature and Regulates Mitochondrial Mass in Adipose Tissues. *Adv. Exp. Med. Biol.* **2024**, *1461*, 229–243. [CrossRef] [PubMed]
55. Ikeda, K.; Yamada, T. Adipose tissue thermogenesis by calcium futile cycling. *J. Biochem.* **2022**, *172*, 197–203. [CrossRef]
56. Lee, J.H.; Rao, M.V.; Yang, D.S.; Stavrides, P.; Im, E.; Pensalfini, A.; Huo, C.; Sarkar, P.; Yoshimori, T.; Nixon, R.A. Transgenic expression of a ratiometric autophagy probe specifically in neurons enables the interrogation of brain autophagy in vivo. *Autophagy* **2019**, *15*, 543–557. [CrossRef]
57. Aitken, R.J.; Lewis, S.E.M. DNA damage in testicular germ cells and spermatozoa. When and how is it induced? How should we measure it? What does it mean? *Andrology* **2023**, *11*, 1545–1557. [CrossRef] [PubMed]
58. Moustakli, E.; Zikopoulos, A.; Skentou, C.; Bouba, I.; Tsirka, G.; Stavros, S.; Vrachnis, D.; Vrachnis, N.; Potiris, A.; Georgiou, I.; et al. Sperm Mitochondrial Content and Mitochondrial DNA to Nuclear DNA Ratio Are Associated with Body Mass Index and Progressive Motility. *Biomedicines* **2023**, *11*, 3014. [CrossRef] [PubMed]
59. Di Gregorio, J.; Petricca, S.; Iorio, R.; Toniato, E.; Flati, V. Mitochondrial and metabolic alterations in cancer cells. *Eur. J. Cell Biol.* **2022**, *101*, 151225. [CrossRef]
60. García-Aguilar, A.; Cuezva, J.M. A Review of the Inhibition of the Mitochondrial ATP Synthase by IF1 in vivo: Reprogramming Energy Metabolism and Inducing Mitohormesis. *Front. Physiol.* **2018**, *9*, 1322. [CrossRef] [PubMed]
61. Cheng, C.K.; Ding, H.; Jiang, M.; Yin, H.; Gollasch, M.; Huang, Y. Perivascular adipose tissue: Fine-tuner of vascular redox status and inflammation. *Redox Biol.* **2023**, *62*, 102683. [CrossRef]
62. Samudio, I.; Fiegl, M.; Andreeff, M. Mitochondrial uncoupling and the Warburg effect: Molecular basis for the reprogramming of cancer cell metabolism. *Cancer Res.* **2009**, *69*, 2163–2166. [CrossRef]
63. Liu, H.; Wang, S.; Wang, J.; Guo, X.; Song, Y.; Fu, K.; Gao, Z.; Liu, D.; He, W.; Yang, L.L. Energy metabolism in health and diseases. *Signal Transduct. Target. Ther.* **2025**, *10*, 69. [CrossRef]
64. Chen, X.S.; Li, L.Y.; Guan, Y.D.; Yang, J.-M.; Cheng, Y. Anticancer strategies based on the metabolic profile of tumor cells: Therapeutic targeting of the Warburg effect. *Acta Pharmacol. Sin.* **2016**, *37*, 1013–1019. [CrossRef]
65. Taghizadeh-Hesary, F. Is Chronic Ice Water Ingestion a Risk Factor for Gastric Cancer Development? An Evidence-Based Hypothesis Focusing on East Asian Populations. *Oncol. Ther.* **2024**, *12*, 629–646. [CrossRef]
66. Behnam, B.; Mobahat, M.; Fazilaty, H.; Wolfe, J.; Omran, H. TSGA10 is a Centrosomal Protein, Interacts with ODF2 and Localizes to Basal Body. *J. Cell Sci. Ther.* **2015**, *6*, 217. [CrossRef]

cancers

MDPI

Article

Apalutamide and Stereotactic Body Radiotherapy in Metastatic Hormone-Sensitive Prostate Cancer: Multicenter Real-World Study

Juan A. Encarnación [1,2,3,*], Virginia Morillo Macías [4], Isabel De la Fuente Muñoz [1], Violeta Derrac Soria [5], Luis Fernández Fornos [6], María Albert Antequera [7], Osamah Amr Rey [8], Vicente García Martínez [9], José L. Alonso-Romero [10] and Raquel García Gómez [11,†]

1 Department Radiation Oncology, Hospital Clínico Universitario Virgen de la Arrixaca, 30120 Murcia, Spain; isabeldelafuente123@gmail.com
2 Faculty of Medicine, University of Murcia, 30100 Murcia, Spain
3 Murcian Institute of Biosanitary Research, 30120 Murcia, Spain
4 Department Radiation Oncology, Consorcio Hospitalario Provincial, 12002 Castellón, Spain; vmorill@gmail.com
5 Department Radiation Oncology, Hospital Universitario La Ribera, 46600 Alzira, Spain
6 Department Radiation Oncology, Hospital Universitario San Juan, 03550 Alicante, Spain; lferfor@gmail.com
7 Department Radiation Oncology, Consorcio Hospital General Universitario, 46014 Valencia, Spain; maria2albert@gmail.com
8 Department Radiation Oncology, Hospital Clínico Universitario, 46010 Valencia, Spain; oamrey@gmail.com
9 Department Radiation Oncology, Hospital General Universitario Santa Lucía, 30202 Cartagena, Spain; vgm20@hotmail.com
10 Department of Medical Oncology, Hospital Clínico Universitario Virgen de la Arrixaca, 30120 Murcia, Spain; josel.alonso2@carm.es
11 Department Radiation Oncology, Hospital Universitario y Politécnico La Fe, 46026 Valencia, Spain
* Correspondence: juanantonio.encarnacion@um.es; Tel.: +34-630185347
† Working Group Coordinator.

Simple Summary: Metastatic prostate cancer is a serious condition with limited treatment options that offer long-term control. In recent years, new oral hormonal therapies have improved outcomes in patients with hormone-sensitive metastatic disease. Additionally, focused radiation techniques such as stereotactic body radiotherapy (SBRT) have been used to target individual cancer lesions. However, it is still unclear how these strategies work together in real-life clinical practice. This study evaluated the combination of a hormonal therapy called apalutamide with SBRT in patients from multiple centers across Spain. We observed high rates of tumor control and very low levels of side effects. Our results suggest that combining these treatments may delay the need for more aggressive therapies and help patients maintain a better quality of life. These findings support the use of this approach in selected patients with metastatic prostate cancer.

Abstract: Background: The management of metastatic hormone-sensitive prostate cancer (mHSPC) has evolved with the integration of androgen receptor signaling inhibitors (ARSIs) and metastasis-directed therapies (MDTs). Stereotactic body radiotherapy (SBRT) offers precise local control, yet real-world data on its combination with apalutamide remain limited. Methods: We conducted a multicenter retrospective cohort study including 134 patients with mHSPC treated with apalutamide and SBRT between February 2021 and December 2024. The primary endpoints were progression-free survival (PFS), local control (LC), and treatment safety. PSA kinetics and radiologic response were evaluated, and outcomes were analyzed according to PSA thresholds and treatment timing. Results: Most patients (93.3%) had low-volume disease; 97.1% presented with $\leq$5 metastases. At a median follow-up of 28 months, LC was 99.3% and 95.5% of patients were progression-free.

 https://doi.org/10.3390/cancers17132216

Complete radiological response was achieved in 87.5% of patients, and 68.4% attained ultralow PSA levels (≤0.02 ng/mL). Undetectable PSA and radiologic complete response were independently associated with improved PFS ($p = 0.010$ and $p = 0.011$, respectively). Treatment was well tolerated, with grade ≥3 toxicity occurring in only 2.2% of patients. Conclusions: The combination of apalutamide and SBRT in mHSPC is associated with high local and systemic disease control and minimal toxicity in a real-world setting. This approach may delay systemic treatment intensification and the onset of castration resistance. Prospective studies are warranted to confirm these findings.

Keywords: metastatic prostate cancer; apalutamide; stereotactic body radiotherapy; SBRT; androgen receptor signaling inhibitors; real-world study; PSA response; metastasis-directed therapy; oligometastatic disease

1. Introduction

Prostate cancer (PC) is one of the most common malignancies among men worldwide, representing a major public health burden due to its high incidence and the increasing life expectancy of the population [1,2]. In Europe, it accounts for approximately one in ten new cancer cases, and in Spain, although the incidence exceeds 12%, the majority of patients exhibit a high five-year survival rate, reflecting the efficacy of treatments in localized stages [3,4]. Approximately 4% of cases are diagnosed at the metastatic stage, either synchronously or metachronously, which is associated with a poorer prognosis [5,6].

In metastatic hormone-sensitive prostate cancer (mHSPC), the response to treatment is heterogeneous and depends on both clinical and biological factors [7]. Stratification based on tumor burden and the timing of metastasis onset has shown prognostic relevance [8,9]. These distinctions have contributed to the development of individualized therapeutic strategies aimed at optimizing clinical outcomes [10]. Historically, androgen deprivation therapy (ADT) monotherapy has been the standard treatment for these patients. However, progression to castration-resistant disease in a proportion of cases has driven the adoption of combination therapeutic approaches [11,12].

Over the past decade, the treatment landscape has evolved significantly with the introduction of androgen receptor signaling inhibitors (ARSIs), which have demonstrated improvements in overall survival (OS) and progression-free survival (PFS) when administered in combination [13–20].

Despite the advances observed in clinical trials, their implementation in real-world practice continues to present challenges. Factors such as comorbidities, advanced age, patient preferences, and regional disparities in access to therapies may influence both treatment selection and effectiveness. Moreover, the limited representation of certain subgroups in pivotal trials restricts the generalizability of the findings [15–19].

In this context, it is essential to complement clinical trial data with observational studies that reflect the effectiveness of treatments in routine clinical practice. This type of evidence is key to optimizing therapeutic sequencing, improving patient selection, and reducing the gap between clinical guidelines and real-world care. Specifically, in the management of mHSPC, stereotactic body radiation therapy (SBRT) has emerged as a relevant therapeutic option for patients with oligometastatic disease, allowing effective local control (LC) of metastases with an acceptable toxicity profile [21–23].

The objective of the present study is to evaluate the efficacy and safety of metastasis-directed therapy (MDT) using SBRT in mHSPC patients treated with apalutamide. Through a retrospective cohort analysis, we aim to assess the impact of SBRT as a complementary strategy in the management of metastatic disease, with the goal of delaying the initiation of a new systemic therapy (SST) line. This study aims to address this gap by evaluating the real-world impact of SBRT combined with apalutamide in a population with heterogeneous metastatic burden, thereby exploring an intermediate therapeutic approach between systemic intensification alone and MDT in isolation.

2. Materials and Methods

2.1. Study Design

Following approval by the ethics committee, a retrospective observational study was conducted, including mHSPC patients treated with apalutamide who also received MDT. Apalutamide was selected as the androgen receptor signaling inhibitor for this study because it is the most widely used agent in our clinical setting for patients with metastatic hormone-sensitive prostate cancer. Consequently, the cohort reflects real-world clinical practice in our region. All eligible patients from the participating centers who received both apalutamide and metastasis-directed SBRT for metastatic hormone-sensitive prostate cancer. The inclusion period was from February 2021 to December 2024, and data were collected at 8 centers nationwide.

2.2. Data Collection

Clinical data were extracted from electronic medical records, and follow-up was maintained until the end of the observation period or death, whichever occurred first. Collected variables were categorized into baseline characteristics and clinical outcomes following MDT.

Baseline data included age, prostate-specific antigen (PSA) level, type and location of metastases, date of apalutamide initiation, and time from apalutamide initiation to SBRT administration.

Treatment outcomes included LC, PFS, OS, toxicity (classified according to CTCAE v5.0), and PSA kinetics following treatment.

PSA values were collected retrospectively from electronic medical records. There was no standardized prospective protocol in place to define specific time points for PSA measurement. However, in the majority of cases, PSA levels were available at key clinical milestones: at the initiation of apalutamide, approximately one month after treatment initiation, prior to metastasis-directed radiotherapy, and during follow-up visits—typically every three months. These time points were selected for analysis because they represented the most consistently documented measurements across participating centers. Comparisons were made between groups based on disease volume, metastasis location, and number of lesions treated. Further stratification was conducted according to the time interval between apalutamide initiation and SBRT ($\leq$ or >3 months). LC was defined as the absence of radiological progression in treated lesions, assessed by conventional imaging or next-generation imaging (NGI), based on the initial diagnostic imaging modality (PET/CT, CT, bone scan). In patients with de novo oligometastatic disease, the primary prostate tumor was treated with radiotherapy in all cases. This uniform approach was intended to optimize local control and is consistent with current evidence supporting the treatment of the primary tumor in the oligometastatic setting.

Toxicity data were collected retrospectively through a systematic review of electronic medical records at all participating centers. The evaluation was based on physician-reported adverse events documented during follow-up visits, and grading was assigned retrospectively according to CTCAE v5.0 criteria. Toxicity reporting was performed uniformly across centers following a standardized data collection template designed for the study.

2.3. Statistical Analysis

A descriptive analysis of baseline characteristics, progression patterns, and clinical events post-treatment was performed. Categorical variables were expressed as absolute frequencies and percentages, while continuous variables were summarized using medians and interquartile ranges.

PFS, LC, and OS were estimated using the Kaplan–Meier method and compared using the log-rank test. Cox regression models were employed to explore associations between clinical factors (metastatic burden, location, time to progression, PSA levels) and outcomes of interest. A p-value of <0.05 was considered statistically significant. Data analysis was conducted using SPSS for Windows, version 25.0 (IBM Corp. Armonk, NY, USA).

3. Results

3.1. Baseline Characteristics

The study cohort included 134 patients with mHSPC, the majority of whom were diagnosed using next-generation imaging (NGI) techniques (83.8%). Most patients presented with low-volume metastatic disease (93.3%). A detailed description of baseline characteristics is provided in Table 1.

Table 1. Baseline characteristics of patients.

	***n*: 134**
Median age at the start of treatment	73 years (56–87)
Type of patient (proportion, *n*)	
Synchronous debut	24.6% (33)
Metachronous biochemical recurrence	75.4% (101)
Diagnostic PSA (median)	8.38 ng/mL (0.25–158)
Metastasis location (proportion, *n*)	
Extrapelvic nodal (M1a)	26.8% (36)
Bone (M1b)	68.7% (92)
Visceral (M1c)	4.5% (6)

3.2. Lesion Distribution and Characteristics of Metastatic Disease

A total of 97.1% of patients presented with fewer than five lesions at diagnosis, although not all lesions were necessarily treated. Not all metastatic lesions were treated because some patients presented with multiple metastases, and the approach to SBRT varied across participating centers. In certain institutions, the clinical decision was made to target only selected lesions based on factors such as lesion size, location, symptomatology, and institutional protocols. Bone lesions accounted for 62.7% of SBRT treatments, extrapelvic nodal metastases for 29.1%, and visceral metastases for 3%. A single lesion was treated in 66.2% of patients, two lesions in 22.8%, three lesions in 8.8%, and four lesions in 2.2%.

3.3. Fractionation Scheme

The distribution of dose per fraction was analyzed, with the most common regimen being 10 Gy in 3 fractions (30.1%), followed by 9 Gy in 3 fractions (16.9%) and 7.5 Gy in 6 fractions (12.5%). All fractionation schemes used delivered a Biologically Effective Dose (BED) greater than 100 Gy.

3.4. PSA Levels

A PSA reduction of more than 90% from baseline during the first follow-up after ARSI initiation was achieved in 21.3% of patients. The median PSA level before SBRT was 0.96 ng/mL (range: 0.01–140 ng/mL), and a median PSA of 0.06 ng/mL was observed following metastasis-directed therapy.

In 12.5% of patients, no PSA response was observed after ARSI initiation; however, among these, 47% achieved a >90% PSA response following the addition of MDT.

Patients were stratified based on whether they achieved ultralow PSA levels, defined as UL1 (0.02–0.2 ng/mL) and UL2 ($\leq$0.02 ng/mL), in an effort to closely monitor those who did not reach UL2 values. At 28 months of follow-up, 68.4% of patients had reached UL2 levels and 25.7% UL1.

3.5. PFS, LC, and OS

The median follow-up was 28 months. Following combination treatment, 87.5% of patients achieved complete response (CR), 10.3% partial response (PR), and 2.2% stable disease (SD). A radiological response was observed in 99.3% of patients at the early assessment (3 months) after MDT. Radiological response was assessed using the same imaging modality initially employed to detect the metastatic lesions in the majority of patients, most commonly PET-CT or bone scintigraphy. Response was evaluated based on radiological criteria, defined as complete disappearance of the treated lesion(s) for complete response, a reduction in lesion size or metabolic activity for partial response, and no significant change for stable disease.

In PFS analysis, 95.5% of patients remained progression-free at the time of data cutoff. LC was 99.3%, with only one patient showing persistent disease after MDT.

Treatment response was identified as an independent prognostic factor for disease progression, with a 13-fold increased risk in patients who did not achieve CR (HR 13.144, 95% CI: 1.804–95.781; $p = 0.011$).

PFS was significantly longer in patients who achieved undetectable PSA levels ($\leq$0.02 ng/mL), reaching 44.68 months (95% CI: 44.02–45.29), compared to 42.51 months (95% CI: 39.57–46.20) in those with PSA > 0.02 ng/mL ($p = 0.010$).

Achieving a UL2 PSA level was associated with improved OS (HR 3.095, 95% CI: 0.868–11.039; $p = 0.082$), showing a trend without reaching statistical significance. In multivariate analysis, UL2 PSA levels were identified as an independent prognostic factor for disease progression (HR 9.949, 95% CI: 1.158–85.469; $p = 0.036$).

The Kaplan–Meier curves illustrating disease-free survival (DFS) and OS are shown in Figures 1–3.

3.6. Acute and Chronic Toxicity After MDT

Only 2.2% of patients experienced grade 3 (G3) acute toxicity, while G1 and G2 toxicity were observed in 22.1% and 8.1% of cases, respectively. Chronic toxicity was reported as G1 in 14.2% and G2 in 2.2%, with G3 events (2.2%) predominantly consisting of asthenia (1.5%) and pain (0.7%).

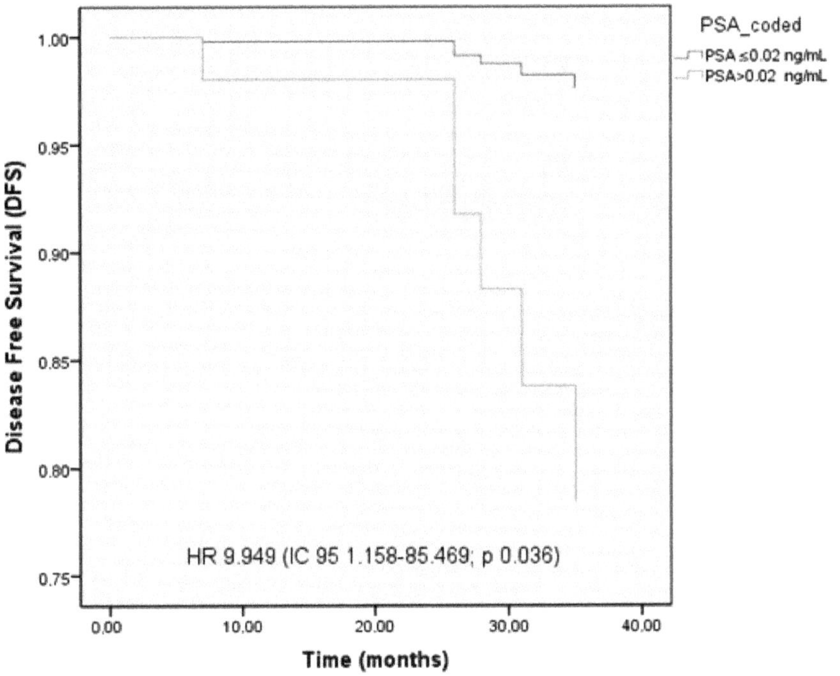

Figure 1. Kaplan–Meier curve of disease-free survival (DFS) stratified by PSA levels ≤ 0.02 ng/mL and >0.02 ng/mL.

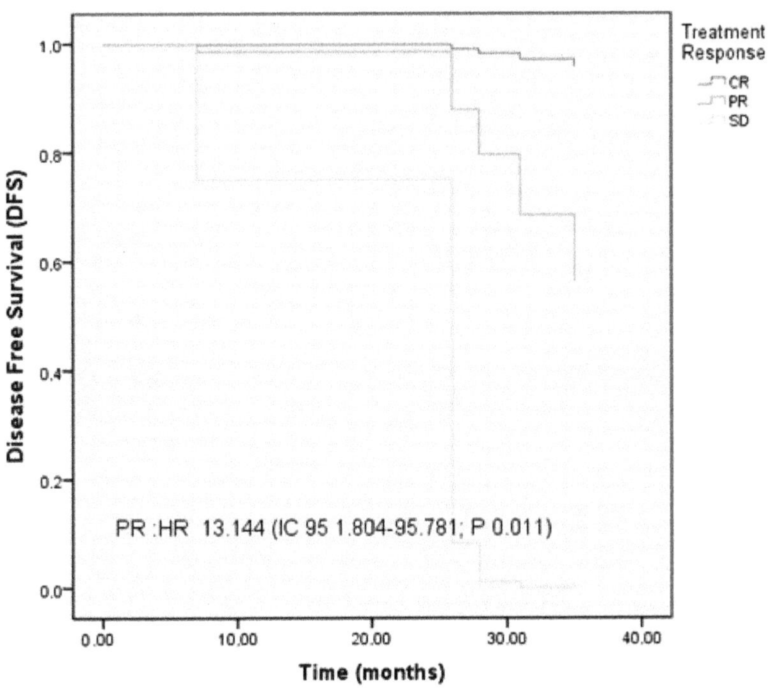

Figure 2. Kaplan–Meier curves of disease-free survival (DFS) according to treatment response.

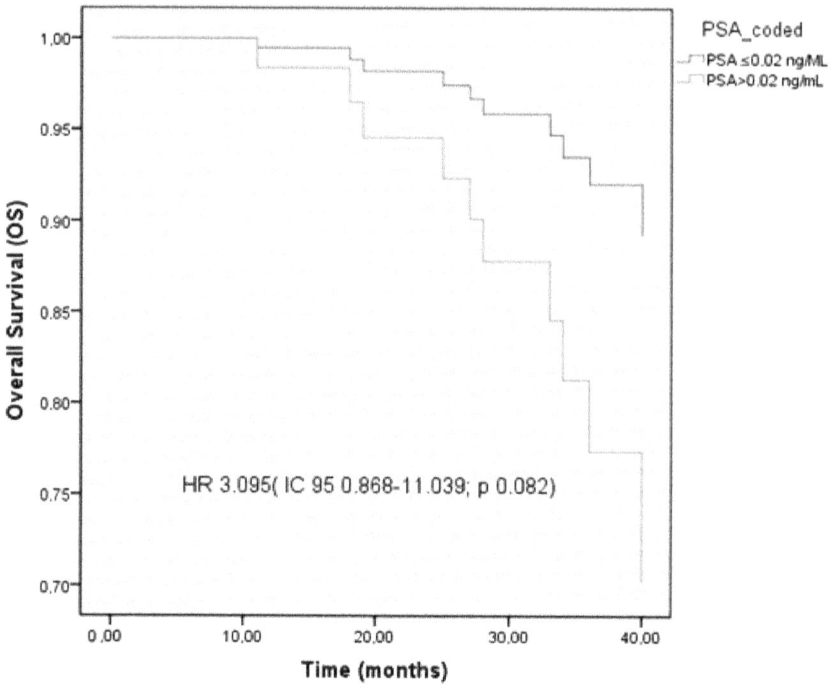

Figure 3. Kaplan–Meier curve of overall survival (OS) stratified by PSA levels ≤ 0.02 ng/mL and >0.02 ng/mL.

4. Discussion

In this multicenter retrospective study, we aimed to evaluate the synergistic effect of SBRT combined with ARSI in patients with mHSPC, without the emergence of significant toxicity when both therapies are administered together.

In a clinical context where optimizing treatment sequencing is crucial to prolong systemic therapy efficacy, our findings suggest that metastasis-directed therapy (MDT) via SBRT may provide meaningful clinical benefits. These include delaying the initiation of further systemic therapies and potentially postponing the onset of castration resistance. This approach aligns with the findings of the SABR-COMET trial [24], which demonstrated that SBRT in oligometastatic patients not only improves local control but is also associated with extended overall survival, supporting its role as an effective component of comprehensive metastatic disease management.

A key finding in our cohort was the high disease control rate. At three months post-treatment, 87.5% of patients achieved CR, a proportion that remained stable throughout follow-up. This suggests a substantial systemic and local impact of the combined therapeutic strategy. Moreover, the marked PSA decline following SBRT highlights the potential of MDT to intensify treatment in high-risk patients by reducing tumor clone burden.

Subgroup analysis identified both early clinical response and post-treatment PSA levels as independent prognostic factors. Patients achieving CR demonstrated significantly longer disease-free survival (DFS) compared to those with partial response or stable disease, emphasizing the clinical utility of early response as an efficacy marker. Additionally, PSA levels after SBRT were significantly correlated with outcomes; a PSA ≤ 0.02 ng/mL was associated with improved DFS and OS. These results support PSA not only as a marker of therapeutic response but also as a prognostic indicator following MDT, consistent with prior studies identifying PSA < 0.02 ng/mL as predictive of disease evolution [25–27].

Our study demonstrated excellent local control, reinforcing SBRT as a precise and effective modality for targeting progressive metastatic lesions. In the EXTEND study [28], combining systemic therapy with MDT in oligometastatic mHSPC improved radiographic PFS (rPFS) compared to systemic therapy alone, with benefits in PFS and delayed systemic treatment intensification—without a significant increase in toxicity. Although our event numbers are limited, our results are encouraging and support the use of SBRT as a strategy to prolong the duration of systemic response.

It is also noteworthy that most of the patients in our cohort had low-volume disease and predominantly bone metastases, which may have contributed positively to both local control and reduced toxicity. These findings are consistent with the literature, which identifies patients with low-volume disease as those most likely to benefit from local intensification strategies without altering systemic therapy [29–31], as exemplified by the PERSIAN trial [32] evaluating SBRT in oligometastatic HSPC.

In terms of safety, the observed toxicity profile was favorable, with most adverse events being mild to moderate, indicating good tolerability of the treatments in this patient population. This toxicity profile is similar to that reported in SBRT-treated patients with castration-resistant prostate cancer (CRPC) [33].

The combination of SBRT and hormonal therapy represents an emerging therapeutic strategy in mHSPC. The synergy lies in SBRT's ability to eradicate visible metastases while hormonal therapy targets undetectable micrometastatic disease, optimizing overall disease control. The RADIOSA trial [34] was the first randomized study in oligometastatic mHSPC to demonstrate a significant improvement in PFS with the addition of short-term ADT to SBRT compared to SBRT alone. However, it also suggests that carefully selected patients may benefit from SBRT monotherapy, highlighting the importance of individualized treatment selection. This approach could allow for initial SBRT alone, postponing or reducing the duration of systemic therapy and avoiding its cumulative toxicity.

Moreover, the potential immunomodulatory effect of SBRT, through mechanisms such as the abscopal effect and increased antigen presentation, could complement systemic treatments by enhancing anti-tumor responses beyond the irradiated sites. These biological interactions may partially explain the sustained disease control observed in a substantial proportion of our cohort. The favorable safety profile of SBRT further reinforces its role as a viable option, particularly in patients with limited disease burden and long disease-free intervals. Taken together, these findings support the rationale for integrating SBRT into the therapeutic algorithm of mHSPC, not only as a palliative measure but as a disease-modifying strategy within a multimodal approach.

The Wolverine meta-analysis [35], which pooled data from multiple randomized trials, supports the benefit of MDT across several survival-related outcomes in oligometastatic mHSPC. This combined analysis showed improved PFS with MDT compared to no MDT, regardless of whether patients presented de novo, the staging modalities used, castration sensitivity status, or the inclusion of ADT.

The main limitations of our study include its retrospective design and lack of a control group, limiting the ability to establish direct causality. Ongoing studies, such as the phase II Trial of Stereotactic Body Radiation Therapy and Androgen Deprivation for Oligometastases in Prostate Cancer (SBRT-SG 05) [36], aim to address these questions. Additionally, the limited number of progression events in our cohort restricts the statistical power for more robust multivariate analyses. Nevertheless, the multicenter design and inclusion of a cohort representative of real-world clinical practice lend strength and practical applicability to our findings.

Although our findings suggest favorable oncologic outcomes, the retrospective nature of the study and absence of a control group limit the ability to draw causal conclusions, particularly regarding survival benefit. Attempts to construct a matched or historical control cohort were not feasible due to variability in treatment approaches and incomplete data in non-SBRT populations. Therefore, our results should be interpreted as hypothesis-generating and reflective of real-world practice.

Nevertheless, the multicenter design and inclusion of a cohort representative of real-world clinical practice lend strength and practical applicability to our findings. The consistency of clinical benefit observed across participating centers reinforces the external validity of the results and supports the feasibility of incorporating metastasis-directed therapy into routine management of mHSPC.

5. Conclusions

SBRT appears to be an effective and well-tolerated therapeutic option in patients with mHSPC treated with apalutamide. Our results reinforce the synergy between both treatments, showing limited toxicity while potentially delaying the onset of castration resistance and the initiation of new systemic therapies. This may positively impact both disease control and patient quality of life. These findings support the value of MDT as a complementary tool in the personalized management of mHSPC and underscore the need for prospective studies to validate these outcomes in larger cohorts.

Author Contributions: Conceptualization: J.A.E., V.M.M. and R.G.G.; Methodology: J.A.E., V.M.M. and R.G.G.; Formal analysis: J.A.E., V.M.M. and R.G.G.; Investigation: all authors; Resources: J.A.E., V.M.M. and R.G.G.; Data curation: all authors; Writing—original draft preparation: J.A.E. and V.M.M.; Writing—review and editing: all authors; Visualization: J.A.E., R.G.G. and V.M.M.; Supervision: V.M.M. and R.G.G.; Project administration: J.A.E. All authors have read and agreed to the published version of the manuscript.

Funding: This research received no external funding.

Institutional Review Board Statement: The study was conducted in accordance with the Declaration of Helsinki and approved by the Ethics Committee of the Hospital Clínico Universitario Virgen de la Arrixaca (protocol code 2022-9-8-HCUVA, approved on 28 November 2022).

Informed Consent Statement: Informed consent was waived by the Ethics Committee due to the retrospective and observational nature of the study, in accordance with current legislation and institutional guidelines.

Data Availability Statement: The data presented in this study are available on reasonable request from the corresponding author. The data are not publicly available due to privacy and ethical restrictions.

Conflicts of Interest: The authors declare no conflict of interest.

Abbreviations

ADT	Androgen Deprivation Therapy
ARSI	Androgen Receptor Signaling Inhibitors
BED	Biologically Effective Dose
BS	Bone Scan
CR	Complete Response
CT	Computed Tomography
CTCAE	Common Terminology Criteria for Adverse Events
Fig	Figure
G	Grade

LC	Local Control
mHSPC	Metastatic Hormone-Sensitive Prostate Cancer
MDT	Metastasis-Directed Therapy
OS	Overall Survival
PET/CT	Positron Emission Tomography/Computed Tomography
PC	Prostate Cancer
PR	Partial Response
PSA	Prostate Specific Antigen
PFS	Progression-Free Survival
rPFS	Radiographic Progression–Free Survival
SBRT	Stereotactic Body Radiation Therapy
SD	Stable Disease
SST	Subsequent Systemic Therapy
UL1 PSA	Ultralow 1 Prostate Specific Antigen
UL2 PSA	Ultralow 2 Prostate Specific Antigen

References

1. Sung, H.; Ferlay, J.; Siegel, R.L.; Laversanne, M.; Soerjomataram, I.; Jemal, A.; Bray, F. Global Cancer Statistics 2020: GLOBOCAN Estimates of Incidence and Mortality Worldwide for 36 Cancers in 185 Countries. *CA A Cancer J. Clin.* **2021**, *71*, 209–249. [CrossRef] [PubMed]
2. Global Burden of Disease Cancer Collaboration; Fitzmaurice, C.; Abate, D.; Abbasi, N.; Abbastabar, H.; Abd-Allah, F.; Abdel-Rahman, O.; Abdelalim, A.; Abdoli, A.; Abdollahpour, I.; et al. Global, regional, and national cancer incidence, mortality, years of life lost, years lived with disability, and disability-Adjusted life-years for 29 cancer groups, 1990 to 2017: A systematic analysis for the global burden of disease study. *JAMA Oncol.* **2019**, *5*, 1749–1768. [PubMed]
3. Butler, S.S.; Muralidhar, V.; Zhao, S.G.; Sanford, N.N.; Franco, I.; Fullerton, Z.H.; Chavez, J.; D'Amico, A.V.; Feng, F.Y.; Rebbeck, T.R.; et al. Prostate cancer incidence across stage, NCCN risk groups, and age before and after USPSTF Grade D recommendations against prostate-specific antigen screening in 2012. *Cancer* **2020**, *126*, 717–724. [CrossRef] [PubMed]
4. Jemal, A.; Fedewa, S.A.; Ma, J.; Siegel, R.; Lin, C.C.; Brawley, O.; Ward, E.M. Prostate Cancer Incidence and PSA Testing Patterns in Relation to USPSTF Screening Recommendations. *JAMA* **2015**, *314*, 2054–2061. [CrossRef]
5. Jemal, A.; Culp, M.B.; Ma, J.; Islami, F.; A Fedewa, S. Prostate Cancer Incidence 5 Years After US Preventive Services Task Force Recommendations Against Screening. *JNCI J. Natl. Cancer Inst.* **2021**, *113*, 64–71. [CrossRef]
6. Desai, M.M.; Cacciamani, G.E.; Gill, K.; Zhang, J.; Liu, L.; Abreu, A.; Gill, I.S. Trends in Incidence of Metastatic Prostate Cancer in the US. *JAMA Netw. Open* **2022**, *5*, e222246. [CrossRef]
7. Mottet, N.; van den Bergh, R.C.N.; Briers, E.; Van den Broeck, T.; Cumberbatch, M.G.; De Santis, M.; Fanti, S.; Fossati, N.; Gandaglia, G.; Gillessen, S.; et al. EAU-EANM-ESTRO-ESUR-SIOG Guidelines on Prostate Cancer-2020 Update. Part 1: Screening, Diagnosis, and Local Treatment with Curative Intent. *Eur. Urol.* **2021**, *79*, 243–262. [CrossRef]
8. Siegel, R.L.; Miller, K.D.; Fuchs, H.E.; Jemal, A. Cancer statistics, 2022. *CA Cancer J. Clin.* **2022**, *72*, 7–33. [CrossRef]
9. Eastham, J.A.; Auffenberg, G.B.; Barocas, D.A.; Chou, R.; Crispino, T.; Davis, J.W.; Eggener, S.; Horwitz, E.M.; Kane, C.J.; Kirkby, E.; et al. Clinically Localized Prostate Cancer: AUA/ASTRO Guideline, Part I: Introduction, Risk Assessment, Staging, and Risk-Based Management. *J. Urol.* **2022**, *208*, 10–18. [CrossRef]
10. National Comprehensive Cancer Network (NCCN). NCCN Clinical Practice Guidelines in Oncology. Available online: https://www.nccn.org/ (accessed on 5 September 2024).
11. Cornford, P.; Bellmunt, J.; Bolla, M.; Briers, E.; De Santis, M.; Gross, T.; Henry, A.M.; Joniau, S.; Lam, T.B.; Mason, M.D.; et al. EAU-ESTRO-SIOG Guidelines on Prostate Cancer. Part II: Treatment of Relapsing, Metastatic, and Castration-Resistant Prostate Cancer. *Eur. Urol.* **2017**, *71*, 630–642. Available online: https://pubmed.ncbi.nlm.nih.gov/27591931/ (accessed on 28 October 2024). [CrossRef]
12. Dabkara, D.; Mondal, D.; Ghosh, J.; Biswas, B.; Ganguly, S. How I treat Metastatic Hormone-Sensitive Prostate Cancer? *Indian J. Med. Paediatr. Oncol.* **2021**, *42*, 100–107. [CrossRef]
13. Chi, K.N.; Chowdhury, S.; Bjartell, A.; Chung, B.H.; Gomes, A.J.P.d.S.; Given, R.; Juárez, A.; Merseburger, A.S.; Özgüroğlu, M.; Uemura, H.; et al. Apalutamide in Patients with Metastatic Castration-Sensitive Prostate Cancer: Final Survival Analysis of the Randomized, Double-Blind, Phase III TITAN Study. *J. Clin. Oncol.* **2021**, *39*, 2294–2303. [CrossRef] [PubMed]
14. Chi, K.N.; Agarwal, N.; Bjartell, A.; Chung, B.H.; Gomes, A.J.P.D.S.; Given, R.; Soto, A.J.; Merseburger, A.S.; Özgüroglu, M.; Uemura, H.; et al. Apalutamide for Metastatic, Castration-Sensitive Prostate Cancer. *N. Engl. J. Med.* **2019**, *381*, 13–24. [CrossRef] [PubMed]

15. Fizazi, K.; Tran, N.; Fein, L.; Matsubara, N.; Rodriguez-Antolin, A.; Alekseev, B.Y.; Özgüroğlu, M.; Ye, D.; Feyerabend, S.; Protheroe, A.; et al. Abiraterone acetate plus prednisone in patients with newly diagnosed high-risk metastatic castration-sensitive prostate cancer (LATITUDE): Final overall survival analysis of a randomised, double-blind, phase 3 trial. *Lancet Oncol.* **2019**, *20*, 686–700. [CrossRef]
16. Fizazi, K.; Foulon, S.; Carles, J.; Roubaud, G.; McDermott, R.; Fléchon, A.; Tombal, B.; Supiot, S.; Berthold, D.; Ronchin, P.; et al. Abiraterone plus prednisone added to androgen deprivation therapy and docetaxel in de novo metastatic castration-sensitive prostate cancer (PEACE-1): A multicentre, open-label, randomised, phase 3 study with a 2 × 2 factorial design. *Lancet* **2022**, *399*, 1695–1707. [CrossRef]
17. Caro Teller, J.M.; Cortijo Cascajares, S.; Escribano Valenciano, I.; Serrano Garrote, O.; Ferrari Piquero, J.M. Uso, efectividad y seguridad de abiraterona en cáncer de próstata. *Farm. Hosp.* **2014**, *38*, 118–122.
18. Davis, I.D.; Martin, A.J.; Stockler, M.R.; Begbie, S.; Chi, K.N.; Chowdhury, S.; Coskinas, X.; Frydenberg, M.; Hague, W.E.; Horvath, L.G.; et al. Enzalutamide with Standard First-Line Therapy in Metastatic Prostate Cancer. *N. Engl. J. Med.* **2019**, *381*, 121–131. [CrossRef]
19. Armstrong, A.J.; Szmulewitz, R.Z.; Petrylak, D.P.; Holzbeierlein, J.; Villers, A.; Azad, A.; Alcaraz, A.; Alekseev, B.; Iguchi, T.; Shore, N.D.; et al. ARCHES: A Randomized, Phase III Study of Androgen Deprivation Therapy with Enzalutamide or Placebo in Men with Metastatic Hormone-Sensitive Prostate Cancer. *J. Clin. Oncol.* **2019**, *37*, 2974–2986. [CrossRef]
20. Smith, M.R.; Hussain, M.; Saad, F.; Fizazi, K.; Sternberg, C.N.; Crawford, E.D.; Kopyltsov, E.; Park, C.H.; Alekseev, B.; Montesa-Pino, Á.; et al. Darolutamide and Survival in Metastatic, Hormone-Sensitive Prostate Cancer. *N. Engl. J. Med.* **2022**, *386*, 1132–1142. [CrossRef]
21. Glicksman, R.M.; Metser, U.; Vines, D.; Valliant, J.; Liu, Z.; Chung, P.W.; Bristow, R.G.; Finelli, A.; Hamilton, R.; Fleshner, N.E.; et al. Curative-intent Metastasis-directed Therapies for Molecularly defined Oligorecurrent Prostate Cancer: A Prospective Phase II Trial Testing the Oligometastasis Hypothesis. *Eur. Urol.* **2021**, *80*, 374–382. [CrossRef]
22. Supiot, S.; Vaugier, L.; Pasquier, D.; Buthaud, X.; Magné, N.; Peiffert, D.; Sargos, P.; Crehange, G.; Pommier, P.; Loos, G.; et al. OLIGOPELVIS GETUG P07, a Multicenter Phase II Trial of Combined High-dose Salvage Radiotherapy and Hormone Therapy in Oligorecurrent Pelvic Node Relapses in Prostate Cancer. *Eur. Urol.* **2021**, *80*, 405–414. [CrossRef] [PubMed]
23. Phillips, R.; Shi, W.Y.; Deek, M.; Radwan, N.; Lim, S.J.; Antonarakis, E.S.; Rowe, S.P.; Ross, A.E.; Gorin, M.A.; Deville, C.; et al. Outcomes of Observation vs Stereotactic Ablative Radiation for Oligometastatic Prostate Cancer: The ORIOLE Phase 2 Randomized Clinical Trial. *JAMA Oncol.* **2020**, *6*, 650–659. [CrossRef] [PubMed]
24. Palma, D.A.; Olson, R.; Harrow, S.; Gaede, S.; Louie, A.V.; Haasbeek, C.; Mulroy, L.; Lock, M.; Rodrigues, G.B.; Yaremko, B.P.; et al. Stereotactic Ablative Radiotherapy for the Comprehensive Treatment of Oligometastatic Cancers: Long-Term Results of the SABR-COMET Phase II Randomized Trial. *J. Clin. Oncol.* **2020**, *38*, 2830–2838. [CrossRef]
25. ESMO 2023: Effect of Rapid Ultra-Low Prostate-Specific Antigen Decline (UL PSA) in TITAN Patients with Metastatic Castration-Sensitive Prostate Cancer (mCSPC) Who Received Apalutamide Plus Androgen Deprivation Therapy [Internet]. Available online: https://www.urotoday.com/conference-highlights/esmo-2023/esmo-2023-prostate-cancer/147490-esmo-2023-effect-of-rapid-ultra-low-prostate-specific-antigen-decline-ul-psa-in-titan-patients-with-metastatic-castration-sensitive-prostate-cancer-mcspc-who-received-apalutamide-plus-androgen-deprivation-therapy.html (accessed on 28 October 2024).
26. López-Abad, A.; Backhaus, M.R.; Gómez, G.S.; Avellaneda, E.C.; Alarcón, C.M.; Cubillana, P.L.; Giménez, P.Y.; Rodríguez, P.d.P.; Fita, M.J.J.; Durán, M.C.; et al. Real-world prostate-specific antigen reduction and survival outcomes of metastatic hormone-sensitive prostate cancer patients treated with apalutamide: An observational, retrospective, and multicentre study. *Prostate Int.* **2023**, *12*, 20–26. [CrossRef]
27. Encarnación Navarro, J.A.; Morillo Macías, V.; Borrás Calbo, M.; De la Fuente Muñoz, I.; Lozano Martínez, A.; García Martínez, V.; Fernández Fornos, L.; Guijarro Roche, M.; Amr Rey, O.; García Gómez, R. Multicenter Real-World Study: 432 Patients with Apalutamide in Metastatic Hormone-Sensitive Prostate Cancer. *Curr. Oncol.* **2025**, *32*, 119. [CrossRef]
28. Tang, C.; Sherry, A.D.; Haymaker, C.; Bathala, T.; Liu, S.; Fellman, B.; Cohen, L.; Aparicio, A.; Zurita, A.J.; Reuben, A.; et al. Addition of Metastasis-Directed Therapy to Intermittent Hormone Therapy for Oligometastatic Prostate Cancer: The EXTEND Phase 2 Randomized Clinical Trial. *JAMA Oncol.* **2023**, *9*, 825–834. [CrossRef]
29. Sathianathen, N.J.; Koschel, S.; Thangasamy, I.A.; Teh, J.; Alghazo, O.; Butcher, G.; Howard, H.; Kapoor, J.; Lawrentschuk, N.; Siva, S.; et al. Indirect Comparisons of Efficacy between Combination Approaches in Metastatic Hormone-sensitive Prostate Cancer: A Systematic Review and Network Meta-analysis. *Eur. Urol.* **2020**, *77*, 365–372. [CrossRef]
30. Marchioni, M.; Di Nicola, M.; Primiceri, G.; Novara, G.; Castellan, P.; Paul, A.K.; Veccia, A.; Autorino, R.; Cindolo, L.; Schips, L. New Antiandrogen Compounds Compared to Docetaxel for Metastatic Hormone Sensitive Prostate Cancer: Results from a Network Meta-Analysis. *J. Urol.* **2020**, *203*, 751–758. [CrossRef]
31. Wang, L.; Paller, C.J.; Hong, H.; De Felice, A.; Alexander, G.C.; Brawley, O. Comparison of Systemic Treatments for Metastatic Castration-Sensitive Prostate Cancer: A Systematic Review and Network Meta-analysis. *JAMA Oncol.* **2021**, *7*, 412–420. [CrossRef]

32. Francolini, G.; Porreca, A.; Facchini, G.; Santini, D.; Bruni, A.; Simoni, N.; Trovò, M.; Osti, M.F.; Fornarini, G.; Sisani, M.; et al. PERSIAN trial (NCT05717660): An ongoing randomized trial testing androgen deprivation therapy, apalutamide and stereotactic body radiotherapy. An alternative "triplet" for oligometastatic hormone sensitive prostate cancer patients. *Med. Oncol.* **2023**, *41*, 39. [CrossRef]
33. Triggiani, L.; Mazzola, R.; Magrini, S.M.; Ingrosso, G.; Borghetti, P.; Trippa, F.; Lancia, A.; Detti, B.; Francolini, G.; Matrone, F.; et al. Metastasis-directed stereotactic radiotherapy for oligoprogressive castration-resistant prostate cancer: A multicenter study. *World J. Urol.* **2019**, *37*, 2631–2637. [CrossRef] [PubMed]
34. Marvaso, G.; Corrao, G.; Zaffaroni, M.; Vincini, M.G.; Lorubbio, C.; Gandini, S.; Fodor, C.; Netti, S.; Zerini, D.; Luzzago, S.; et al. ADT with SBRT versus SBRT alone for hormone-sensitive oligorecurrent prostate cancer (RADIOSA): A randomised, open-label, phase 2 clinical trial. *Lancet Oncol.* **2025**, *26*, 300–311. [CrossRef] [PubMed]
35. Tang, C.; Sherry, A.D.; Hwang, H.; Francolini, G.; Livi, L.; Tran, P.T.; Corn, P.G.; Aparicio, A.; Simontacchi, G.; Kiess, A.P.; et al. World-wide oligometastatic prostate cancer (omPC) meta-analysis leveraging individual patient data (IPD) from randomized trials (WOLVERINE): An analysis from the X-MET collaboration. *J. Clin. Oncol.* **2025**, *43* (Suppl. S5), 15. [CrossRef]
36. Conde-Moreno, A.J.; López-Campos, F.; Hervás, A.; Morillo, V.; Méndez, A.; Puertas, M.D.M.; Valero-Albarrán, J.; Gómez Iturriaga, A.; Rico, M.; Vázquez, M.L.; et al. A Phase II Trial of Stereotactic Body Radiation Therapy and Androgen Deprivation for Oligometastases in Prostate Cancer (SBRT-SG 05). *Pract. Radiat. Oncol.* **2024**, *14*, e344–e352. [CrossRef]

MDPI AG
Grosspeteranlage 5
4052 Basel
Switzerland
Tel.: +41 61 683 77 34

Cancers Editorial Office
E-mail: cancers@mdpi.com
www.mdpi.com/journal/cancers

www.ingramcontent.com/pod-product-compliance
Lightning Source LLC
Chambersburg PA
CBHW040752250926
51850CB00007BB/1479
* 9 7 8 3 7 2 5 8 6 7 3 0 1 *